Mastering ArcGIS Enterprise Administration

Install, configure, and manage ArcGIS Enterprise to publish, optimize, and secure GIS services

Chad Cooper

BIRMINGHAM - MUMBAI

Mastering ArcGIS Enterprise Administration

First published: October 2017

Production reference: 1241017

Published by Packt Publishing Ltd.
Livery Place
35 Livery Street
Birmingham
B3 2PB, UK.
ISBN 978-1-78829-749-3

www.packtpub.com

Credits

Author
Chad Cooper

Reviewers
Daniel Huber
Zebadiah K. Steeby

Commissioning Editor
Aaron Lazar

Acquisition Editor
Karan Sadawana

Content Development Editor
Akshada Iyer

Technical Editor
Tiksha Sarang

Copy Editor
Zainab Bootwala

Project Coordinator
Prajakta Naik

Proofreader
Safis Editing

Indexer
Francy Puthiry

Graphics
Abhinash Sahu

Production Coordinator
Shraddha Falebhai

About the Author

Chad Cooper has worked in the geospatial industry over the last 15 years as a technician, analyst, and developer, pertaining to state and local government, oil and gas, and academia. For the last 3 years, he has worked as a solutions engineer, consulting on the State and Local Government team with Geographic Information Services, Inc. At work, he couldn't be happier spending the day writing Python and helping clients get the most out of their data through the use of the Esri platform. At home, he enjoys hanging out with his gorgeous wife of 12 years and their 3 wonderful children. They enjoy hiking, fishing, and doing nothing on a nice beach. Chad has a bachelor's degree from the Mississippi State University and a master's degree from the University of Arkansas, both in geology.

Writing a book has been on my bucket list for quite a few years now. I've published articles in Python Magazine and Esri's ArcUser, but when the opportunity to write this book came along, I knew I had to take it. This book was written in my office, at airports, on airplanes, in hotel rooms, at conference centers, at coffee shops, in cars, at the beach, in a cabin, and on the couch while watching Care Bears with my amazing, beautiful daughter. Needless to say, I was always crunched for time. Without the support and help of my wonderful wife and kids, writing this book never would have been possible. Thank you so very much for helping me accomplish this goal. My employer, Geographic Information Services Inc. (GISinc), also played a vital role in this publication by providing necessary resources and being understanding on the days after 2 A.M. writing sessions the night before. I had an amazing crew of technical reviewers and am indebted to them, especially Daniel Huber and Zebadiah K. Steeby, both colleagues at GISinc, who have provided guidance, support, and continuing education over the years. Finally, the editorial staff at Packt Publishing was great to work with and provided support and guidance throughout the entire writing process.

About the Reviewers

Daniel Huber has been working in the GIS industry for 20 years--primarily in the DoD and Federal business space, supporting Facility Mapping, Command and Control Systems, and Resource Management. He has held the role of a GIS analyst, developer, and system architect and has worked at all levels within organizations, ranging from field level to headquarters. He currently supports his company's federal team as an enterprise architect, designing and implementing end-to-end enterprise GIS solutions and providing technical leadership across the company.

Dan has also been a bomb disposal technician and communications specialist in the US Air Force and currently experiments with home automation and electronics solutions when not supporting his community as an amateur radio operator.

Zebadiah K. Steeby has over 10 years of experience with designing and implementing GIS solutions. His career has consisted of working in a variety of roles ranging from that of an analyst to a database administrator. He has worked on both government and commercial solutions in a wide range of technologies. As a solutions engineer, his current responsibilities include assessing customers' existing GIS/IT environments and recommending areas of improvements in application technology, system performance, and software migration plans. He also implements the enterprise GIS system architecture, including the installation and configuration of software and deploying and configuring custom applications.

Customer Feedback

Thanks for purchasing this Packt book. At Packt, quality is at the heart of our editorial process. To help us improve, please leave us an honest review on this book's Amazon page at `https://www.amazon.com/dp/1788297490`.

If you'd like to join our team of regular reviewers, you can e-mail us at `customerreviews@packtpub.com`. We award our regular reviewers with free eBooks and videos in exchange for their valuable feedback. Help us be relentless in improving our products!

Table of Contents

Preface

When ArcGIS Enterprise 10.5 was released in December of 2016, it brought with it substantial changes to the Esri web GIS ecosystem. With that release, ArcGIS Server, Portal for ArcGIS, ArcGIS Data Store, and the ArcGIS Web Adaptor became the four main components of an ArcGIS Enterprise deployment. ArcGIS Enterprise 10.5 is a complete web GIS in your own infrastructure, whether that be on-premises, in the cloud, or a combination of the two.

This book will teach you how to properly install and configure all components of ArcGIS Enterprise, including setting up and maintaining an enterprise geodatabase on SQL Server. After all software components are ready, we will cover publishing content to ArcGIS Server and Portal for ArcGIS. Administration of the many pieces of ArcGIS Enterprise is a key concept that is central to the purpose of this book; we will cover the many ways we can administer, configure, and maintain each piece of the ArcGIS Enterprise platform. No GIS book would be complete without covering Python, and we will cover several ways to use Python along with Esri libraries to get creative and script out repetitive tasks as well as quick ad hoc jobs. Security is a paramount concern in any enterprise system, and we will discuss ways to keep your system safe and secure. Finally, we will wrap up coverage of standards and best practices along with ways to use those to help you efficiently and successfully troubleshoot errors and issues when they arise in your environment.

What this book covers

`Chapter 1`, *ArcGIS Enterprise Introduction and Installation*, introduces ArcGIS Enterprise and covers the installation and configuration of all aspects of ArcGIS Server, Portal for ArcGIS, ArcGIS Data Store, and ArcGIS Web Adaptor. Once you are done with this chapter, you will have a fully functioning instance of the ArcGIS Enterprise core software.

`Chapter 2`, *Enterprise Geodatabase Administration*, walks through the creation and configuration of an enterprise geodatabase on Microsoft SQL Server. You will learn how to connect to the geodatabase, load data, create users and roles, set privileges, and configure and perform geodatabase maintenance. Publishing to the ArcGIS Data Store is also discussed along with server-object extensions and server-object interceptors.

`Chapter 3`, *Publishing Content*, covers the different types of services available in ArcGIS Server and how to publish, configure, and manage those services.

`Chapter 4`, *ArcGIS Server Administration*, is a very important chapter as it introduces ways to access ArcGIS Server and carry out administrative tasks crucial to a smooth-running environment. We will discuss ArcGIS Server logs, accounts, and how to use the ArcGIS Server REST Administrator Directory efficiently to complete tasks.

`Chapter 5`, *Portal for ArcGIS Administration*, is another crucial chapter that shows how to access administrative functions of Portal for ArcGIS to customize the look and feel of your portal, how to manage content, and how to administer various pieces of your portal through the Portal REST Administrative Directory. Backing up and restoring your portal is discussed along with useful tools to manage Portal items.

`Chapter 6`, *Security*, is a chapter to pay close attention to as security always needs to be on your mind. We will discuss passwords, methods to keep ArcGIS Server and Portal for ArcGIS secure, and the details and benefits of federation.

`Chapter 7`, *Scripting Administrative Tasks*, is the first of our hands-on chapters. We will use Python to load data into your geodatabase, perform an inventory of your ArcGIS Server services, bulk publish services, and script the replication of one ArcGIS Server environment into another.

`Chapter 8`, *The ArcGIS Python API*, our second hands-on chapter, introduces the new and exciting ArcGIS API for Python, which allows Pythonic access to your entire web GIS. We will discuss the installation of the API and how to easily use it to work with services, Portal items and users, and even features in a feature layer.

`Chapter 9`, *ArcGIS Enterprise Standards and Best Practices*, discusses measures you can take to enforce integrity in your environment and applications using standards and best practices. Security, data, storage, and scripting, among other items, can all benefit from standards and best practices.

`Chapter 10`, *Troubleshooting ArcGIS Enterprise Issues and Errors*, brings this book to an end by bringing together many things you learned in previous chapters to help you track down issues, determine their causes, and come up with resolutions quickly and efficiently.

What you need for this book

Mastering ArcGIS Enterprise Administration is written for ArcGIS Enterprise 10.5.1, but version 10.5 can be used as well. You will need access to at least one Windows server with at least Windows Server 2008 as the operating system, access to ArcGIS Enterprise 10.5.1 installation files, and licensing for ArcGIS Enterprise. You will need Microsoft SQL Server 2012 SP3, 2014, or 2016 (Microsoft offers 180-day trial licenses for SQL Server) and ArcGIS Desktop 10.5.1 or ArcGIS Pro. For Python coding, you will need Python 2.7.x that installs with ArcGIS Desktop and Python 3.x that either comes with ArcGIS Pro or can be installed separately. A Python IDE is optional but recommended.

Who this book is for

This book is geared toward senior GIS analysts, GIS managers, GIS administrators, DBAs, GIS architects, and GIS engineers who need to install, configure, and administer ArcGIS Enterprise 10.5.1. Anyone wishing to become more comfortable working with the many administrative interfaces of ArcGIS Enterprise will benefit from this book.

Conventions

In this book, you will find a number of text styles that distinguish between different kinds of information. Here are some examples of these styles and an explanation of their meaning. Code words in text, database table names, folder names, filenames, file extensions, pathnames, dummy URLs, user input, and Twitter handles are shown as follows: "After we have our map, we can use `add_layer` to add our `stations_item` to the preceding `map1`."

A block of code is set as follows:

```
for fc in fcs:
  fields = arcpy.ListFields(os.path.join(input_ds, fc))
  field_names = [field.name for field in fields]
  for k, v in field_domains.iteritems():
  if k in field_names:
    arcpy.AssignDomainToField_management(fc, k, v)
```

Any command-line input or output is written as follows:

```
npm install -g json
```

New terms and **important words** are shown in bold. Words that you see on the screen, for example, in menus or dialog boxes, appear in the text like this: "Double-click on **Add ArcGIS Server**."

Warnings or important notes appear like this.

Tips and tricks appear like this.

Reader feedback

Feedback from our readers is always welcome. Let us know what you think about this book-what you liked or disliked. Reader feedback is important for us as it helps us develop titles that you will really get the most out of. To send us general feedback, simply e-mail `feedback@packtpub.com`, and mention the book's title in the subject of your message. If there is a topic that you have expertise in and you are interested in either writing or contributing to a book, see our author guide at `www.packtpub.com/authors`.

Customer support

Now that you are the proud owner of a Packt book, we have a number of things to help you to get the most from your purchase.

Downloading the example code

You can download the example code files for this book from your account at `http://www.packtpub.com`. If you purchased this book elsewhere, you can visit `http://www.packtpub.com/support` and register to have the files e-mailed directly to you. You can download the code files by following these steps:

1. Log in or register to our website using your e-mail address and password.
2. Hover the mouse pointer on the **SUPPORT** tab at the top.
3. Click on **Code Downloads & Errata**.
4. Enter the name of the book in the **Search** box.
5. Select the book for which you're looking to download the code files.
6. Choose from the drop-down menu where you purchased this book from.
7. Click on **Code Download**.

Once the file is downloaded, please make sure that you unzip or extract the folder using the latest version of:

- WinRAR / 7-Zip for Windows
- Zipeg / iZip / UnRarX for Mac
- 7-Zip / PeaZip for Linux

The code bundle for the book is also hosted on GitHub at `https://github.com/PacktPublishing/Mastering-ArcGIS-Enterprise-Administration`. We also have other code bundles from our rich catalog of books and videos available at `https://github.com/PacktPublishing/`. Check them out!

Downloading the color images of this book

We also provide you with a PDF file that has color images of the screenshots/diagrams used in this book. The color images will help you better understand the changes in the output. You can download this file from `https://www.packtpub.com/sites/default/files/downloads/MasteringArcGISEnterpriseAdministration_ColorImages.pdf`.

Errata

Although we have taken every care to ensure the accuracy of our content, mistakes do happen. If you find a mistake in one of our books-maybe a mistake in the text or the code-we would be grateful if you could report this to us. By doing so, you can save other readers from frustration and help us improve subsequent versions of this book. If you find any errata, please report them by visiting `http://www.packtpub.com/submit-errata`, selecting your book, clicking on the **Errata Submission Form** link, and entering the details of your errata. Once your errata are verified, your submission will be accepted and the errata will be uploaded to our website or added to any list of existing errata under the Errata section of that title. To view the previously submitted errata, go to `https://www.packtpub.com/books/content/support` and enter the name of the book in the search field. The required information will appear under the **Errata** section.

Piracy

Piracy of copyrighted material on the Internet is an ongoing problem across all media. At Packt, we take the protection of our copyright and licenses very seriously. If you come across any illegal copies of our works in any form on the Internet, please provide us with the location address or website name immediately so that we can pursue a remedy. Please contact us at `copyright@packtpub.com` with a link to the suspected pirated material. We appreciate your help in protecting our authors and our ability to bring you valuable content.

Questions

If you have a problem with any aspect of this book, you can contact us at `questions@packtpub.com`, and we will do our best to address the problem.

1

ArcGIS Enterprise Introduction and Installation

Since the release of ArcGIS 9 in 2004, ArcGIS Server has continued to grow and evolve. This evolution is ongoing and evident in the latest release of the ArcGIS platform, ArcGIS 10.5, released in December 2016. With the release of any new software version, comes changes in system requirements, licensing, and functionality. The 10.5 release of ArcGIS 10.5, now known as ArcGIS Enterprise, brought a substantial number of changes to administrators and users of this vastly popular and pervasive geographic information systems software package. At the time of this writing, ArcGIS Enterprise is at version 10.5.1, a quality improvement release set loose in the wild in the summer of 2017. This book will focus on ArcGIS Enterprise version 10.5.1. We will refer to both 10.5 and 10.5.1 versions, as many of the newest features were released at 10.5.

To fully understand how to install ArcGIS Enterprise, it is first important to know the structure of ArcGIS Enterprise, what it is and isn't, its different components, and, new to ArcGIS Enterprise at 10.5, server roles. This chapter will help you do just that; you will learn what ArcGIS Enterprise 10.5.1 is, how it differs from previous versions of ArcGIS, and how to install and initially configure the key components of ArcGIS Enterprise.

By the end of this chapter, you will be comfortable with the structure of ArcGIS Enterprise and capable of confidently installing and configuring it in your own environment.

In this chapter, we will cover the following topics:

- What is ArcGIS Enterprise and how is it different from previous versions of ArcGIS?
- What are the four components of ArcGIS Enterprise and how do they work together?
- What are server roles and how do they function?
- Installation and configuration of the following:
    - ArcGIS Server
    - Portal for ArcGIS
    - ArcGIS Web Adaptors for both ArcGIS Server and Portal for ArcGIS
    - ArcGIS Data Store

Introduction to ArcGIS Enterprise 10.5.1

ArcGIS Enterprise 10.5.1 is the latest version of the ArcGIS Server product line from Esri. Released in summer 2017, ArcGIS Enterprise represents a substantial shift in how ArcGIS Server and its components are structured, licensed, and deployed.

How ArcGIS Enterprise 10.5.1 is different

ArcGIS Enterprise 10.5.1 is a complete web GIS in your own infrastructure, whether on-premises, in the cloud, or a combination of the two. At 10.5.1, ArcGIS for Server now becomes ArcGIS Enterprise, consisting of the following four major components:

- ArcGIS Server
- Portal for ArcGIS
- ArcGIS Data Store
- ArcGIS Web Adaptor

The underlying technologies behind these components remain the same as in previous versions, with enhancements.

Also new at ArcGIS Enterprise 10.5 were licensing roles. Prior to 10.5, ArcGIS Server was licensed with varying levels and editions. Roles at 10.5 offer differing capabilities and types of services that can be published.

Components of ArcGIS Enterprise 10.5.1

The ArcGIS Enterprise product line consists of four software components that are designed to work together. These are as follows:

- **ArcGIS Server**: These are the core web services component to share maps authored in ArcGIS Desktop and ArcGIS Pro and perform geospatial analysis over the internet.
- **Portal for ArcGIS**: This allows users in your organization to share data, maps, and other geospatial content through application authoring (including Web AppBuilder) and hosting capabilities. Through federation with ArcGIS Server, Portal becomes the identity store for ArcGIS Enterprise, allowing for a single management point for access and authorization. Think of Portal for ArcGIS as an on-premises version of ArcGIS Online.
- **ArcGIS Data Store**: This is an application that will locally store your Portal's feature layer data, caches, and big data.
- **ArcGIS Web Adaptor**: This allows you to expose your ArcGIS Server through your organization's standard website and port, letting you easily share your map services over the internet. When paired with IIS and Active Directory, the Web Adaptor provides a smooth method for authentication and access using **Integrated Windows Authentication** (**IWA**).

A base ArcGIS Enterprise deployment consists of ArcGIS Server, Portal for ArcGIS, ArcGIS Data Store, and the Web Adaptor.

Server roles and extensions

New to ArcGIS Enterprise 10.5 was the concept of roles. Roles provide added functionality to ArcGIS Enterprise as deployed in your own infrastructure. Need to serve out and analyze imagery, rasters, or remotely sensed data? ArcGIS Image Server, formerly known as the Image Server Extension, allows you to serve massive imagery collections on the fly. At ArcGIS Enterprise 10.5.1, there are five licensing roles:

- GIS Server
- Image Server
- GeoEvent Server
- GeoAnalytics Server
- Business Analyst Server

Each server role requires its own instance of ArcGIS Server and a dedicated hardware resource; it is no longer recommended to deploy multiple roles to a single server for performance concerns. Many of these roles can also be deployed as distributed servers, allowing for the spreading out of processing across multiple servers.

GIS Server

The GIS Server role is core ArcGIS Server; it is the role that provided many of the ArcGIS Server capabilities prior to ArcGIS Enterprise 10.5. ArcGIS GIS Server is still offered in three editions, with each successive edition offering additional functionality:

- **Basic**: This manages your geodatabase and public feature services (without the ability to edit); it cannot be deployed with Portal for ArcGIS.
- **Standard**: This is everything in Basic, plus the ability to edit feature services and publish geoprocessing services from any tool included in ArcGIS Desktop Standard or ArcGIS Pro Standard; it can be implemented with Portal for ArcGIS.
- **Advanced**: This is everything in Standard, plus the ability to publish geoprocessing services from any tool included in ArcGIS Desktop Advanced or ArcGIS Pro Advanced. It also includes additional geostatistical and Spatial Analyst tools, and it can be implemented with Portal for ArcGIS.

Image Server

With ArcGIS Image Server, formerly known as Image Server Extension, large collections of satellite imagery, aerial photos, and rasters can be served dynamically on the fly. Image Server can also run raster processing models allowing distributed analysis of imagery and rasters.

GeoEvent Server

GeoEvent Server, known as the GeoEvent Extension prior to 10.5, enables the integration of real-time data into your enterprise GIS from a variety of sources and sensors. With GeoEvent Server, you can stream event data to client applications, view feature statuses with the Operations Dashboard for ArcGIS, filter geoevents, and detect and analyze the spatial proximity of events with geofences. With GeoEvent Server, real-time data can be published to a spatiotemporal big data store.

GeoAnalytics Server

With ArcGIS GeoAnalytics Server, new at 10.5, big data analysis can be distributed across multiple ArcGIS Server machines, allowing users to perform analyses more quickly on even larger amounts of data than before.

Business Analyst Server

ArcGIS Business Analyst Server, when used with ArcGIS Enterprise, enables your organization to host business analyst-based capabilities such as site analytics and custom reporting. Business Analyst Server also allows you to host the Esri GeoEnrichment service on-premise and behind your firewall.

Licensing

As in previous versions of ArcGIS Server, Enterprise is broken down by editions and levels.

ArcGIS Enterprise editions

As discussed earlier in this chapter, ArcGIS GIS Server is offered in three editions, with each successive edition offering additional functionality--Basic, Standard, and Advanced. Let's examine these editions a bit closer.

Basic edition

ArcGIS GIS Server Basic edition includes geodatabase management and the ability to publish read-only feature services. Also included are the geodata service and geometry service. Web editing is not available and this edition cannot be federated with Portal for ArcGIS. No ArcGIS Server extensions are available for purchase and implementation at the Basic edition.

Standard edition

The Standard edition of ArcGIS GIS Server adds all GIS web service types (cached map and image, dynamic map, feature, geocoding, geoprocessing, image from a single raster, print, and schematic) offered by the ArcGIS GIS Server. Geoprocessing services can utilize any tool included with ArcGIS Desktop Standard. The Standard edition can be deployed with Portal for ArcGIS, allowing hosted layer types such as feature layers, scene layers, and tile layers. Most ArcGIS Server extensions are available for purchase and implementation at the Standard edition.

Advanced edition

The Advanced edition includes everything at the Standard edition plus the ability to publish geoprocessing models and scripts utilizing any tool included in ArcGIS Desktop Advanced. The ArcGIS Network Analyst for Server extension is included, and all Server Extensions are available for purchase and implementation. Portal for ArcGIS can be implemented with the Advanced edition.

Levels of ArcGIS Enterprise

There also exist two *levels* of ArcGIS Enterprise--ArcGIS Enterprise and ArcGIS Enterprise Workgroup.

ArcGIS Enterprise level

The ArcGIS Enterprise level is designed for medium to large-sized teams. At this level, enterprise geodatabases are utilized with ArcGIS Enterprise allowing an unlimited number of simultaneous connections to the database. This level comes with one four-core processor license and is scalable with additional two-core add-on packs.

ArcGIS Enterprise Workgroup level

The ArcGIS Enterprise Workgroup level is designed for smaller teams and organizations, allowing a maximum of 10 simultaneous connections to workgroup and file geodatabases; enterprise geodatabases are not supported. The base ArcGIS Enterprise deployment (Server, Portal, Web Adaptor, or Data Store) must be deployed all in one on a single machine with up to four cores. Server roles have a maximum of four cores--no add-on two-core packs are available.

Named user entitlements

Licensing for ArcGIS Enterprise 10.5.1 is like licensing at 10.4. With your purchase of ArcGIS Enterprise is included a set of named user entitlements to be used within Portal for ArcGIS. A named user is a specified user for running ArcGIS Pro or a Premium App through ArcGIS Online of Portal for ArcGIS. The number of entitlements you receive depends on the edition and level of ArcGIS Enterprise purchased by your organization. Named user entitlements also differ for licensing under an **Enterprise Licensing Agreement** (**ELA**), education site license, or any other special licensing agreement with Esri.

The following are the named user entitlements:

ArcGIS Enterprise Advanced Edition		ArcGIS Enterprise Workgroup Advanced Edition	
30 L1	50 L2	0 L1	10 L2
ArcGIS Enterprise Standard Edition		ArcGIS Enterprise Workgroup Standard Edition	
30 L1	5 L2	0 L1	5 L2

ArcGIS Enterprise with GIS Server Basic cannot be deployed with Portal for ArcGIS; therefore, named users are not available in this edition.

Level 1 (**L1**) users are content viewers who can only view content shared with them through the organization. L1 users cannot own items or edit items. **Level 2** (**L2**) users can view, create, edit, and share content and can be assigned into the Portal roles of **User**, **Publisher**, and **Administrator**. L1 access is no different than public anonymous (*Share with Everyone*), but allows named users to participate in focused sharing through groups.

Installing ArcGIS Server

ArcGIS Server installation at 10.5.1 is very similar to installation at 10.4 and will be a familiar process for many.

System and hardware requirements

The following is a high-level overview of some of the more important system and hardware requirements of ArcGIS Server 10.5. Consult the official ArcGIS Server 10.5 online documentation for further information and an exhaustive list of all requirements.

Operating systems

ArcGIS Server is supported on Windows Server 2012 R2 Standard and Datacenter; Windows Server 2012 Standard and Datacenter; Windows Server 2008 R2 Standard, Enterprise, and Datacenter; and Windows Server 2008 Standard, Enterprise, and Datacenter. Flavors of Windows 10, 8.1, and 7 are also supported *for basic testing and application development only, not for production environments*. Throughout this book, we will focus on ArcGIS Server on Windows.

ArcGIS GIS Server, GeoEvent Server, Image Server, or Business Analyst for Server are recommended to have 8 GB of RAM *per unique license role* in a production environment. ArcGIS Server requires a minimum of 10 GB of available disk space.

Ports

ArcGIS Server requires several ports be open to allow communication with machines both externally on the internet and internally on an intranet. The following ports need to be allowed on your firewall:

- HTTP port `6080`.
- HTTPS port `6443`: If HTTPS is enabled, ArcGIS Server uses port `6443` by default.
- Ports `4000-4002`: These ports are used for communication between ArcGIS Servers.
- Internally used ports: Other ports such as `1098`, `6006`, `6099`, and others are used by ArcGIS Server to start processes with each ArcGIS Server machine. These ports do not have to be open for access by other machines.

Secure socket layer

ArcGIS Server comes preconfigured with a self-signed **secure socket layer** (**SSL**) certificate. Although not required, it is **highly recommended** that you purchase and install an SSL certificate from a trusted **certificate authority** (**CA**) or a local domain CA. SSL provides encryption of sensitive information (such as usernames and passwords for logging in to ArcGIS Server and Portal) and authentication to ensure that information is being sent where it is intended to go, and not to an imposter. The downside to SSL is that certificates do cost money, but at the time of this writing, SSL certificates can be purchased for around $70 USD, a small price to pay for peace of mind for you and your end users. See the *SSL certificate installation* section, discussed later in this chapter, for more information.

For the latest system requirements, please consult the ArcGIS Enterprise online help.

Hardware scenarios

There are several ways that ArcGIS Enterprise can be deployed. These range from simple single-machine deployments to more complex multi-machine scenarios. Prior to ArcGIS Enterprise 10.5, a base deployment consisted primarily of ArcGIS Server and the ArcGIS Web Adaptor. At 10.5, a base deployment consists of the four main components of ArcGIS Enterprise--ArcGIS Server, Portal for ArcGIS, ArcGIS Data Store, and ArcGIS Web Adaptor, all working together.

Single-machine deployment

In a single-machine deployment, all components of ArcGIS Enterprise are installed in one single machine, either physically or virtually. This means the one machine acts as a database server, application server, and web server. This is a minimalist configuration that *can* be used in a production environment, but it is better suited for a testing or development environment. For the purposes of this book, we will use a single-machine deployment in Amazon Web Services. In a minimalist, conceptual form, a single-machine deployment would look like the following diagram:

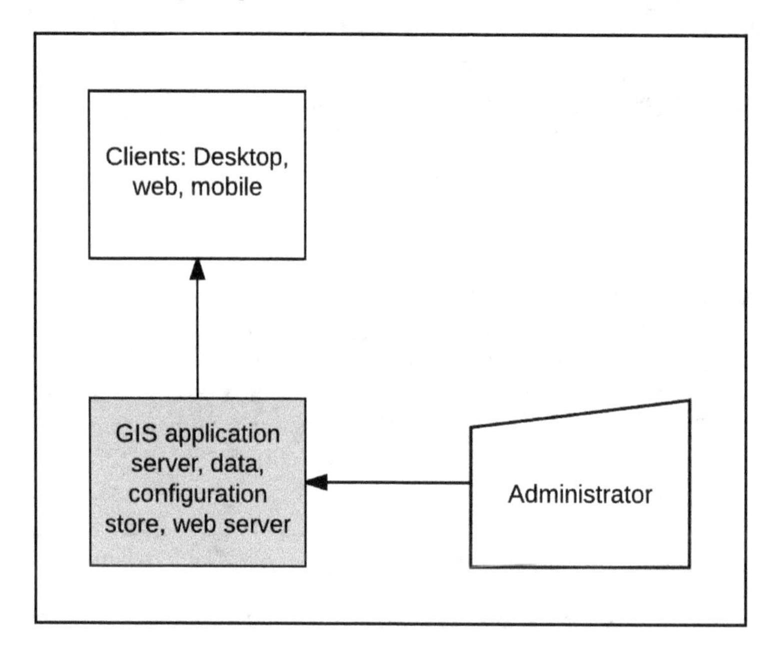

Esri recently released ArcGIS Enterprise Builder, which provides a simple installation and configuration experience for a base ArcGIS Enterprise single-machine deployment.

Multi-machine (multi-tiered) deployment

The multi-machine, or multi-tiered (where each machine is a **tier**), is the most common deployment scenario. Here, each component of ArcGIS Enterprise is installed on a separate virtual or physical machine. This means that there is a separate machine for each of the following:

- ArcGIS Web Adaptor (web server)
- Portal for ArcGIS

- ArcGIS Server
- ArcGIS Data Store
- Enterprise geodatabase

If absolutely necessary, Portal and the Web Adaptor can reside on one server, with ArcGIS Server and Data Store on another. Bear in mind that, for performance reasons, this is not what is recommended for production environments.

Although more complex than the single-machine deployment, the multi-tiered deployment allows for isolation of the different components and distribution of the workload. A multi-machine configuration would conceptually look like the following diagram:

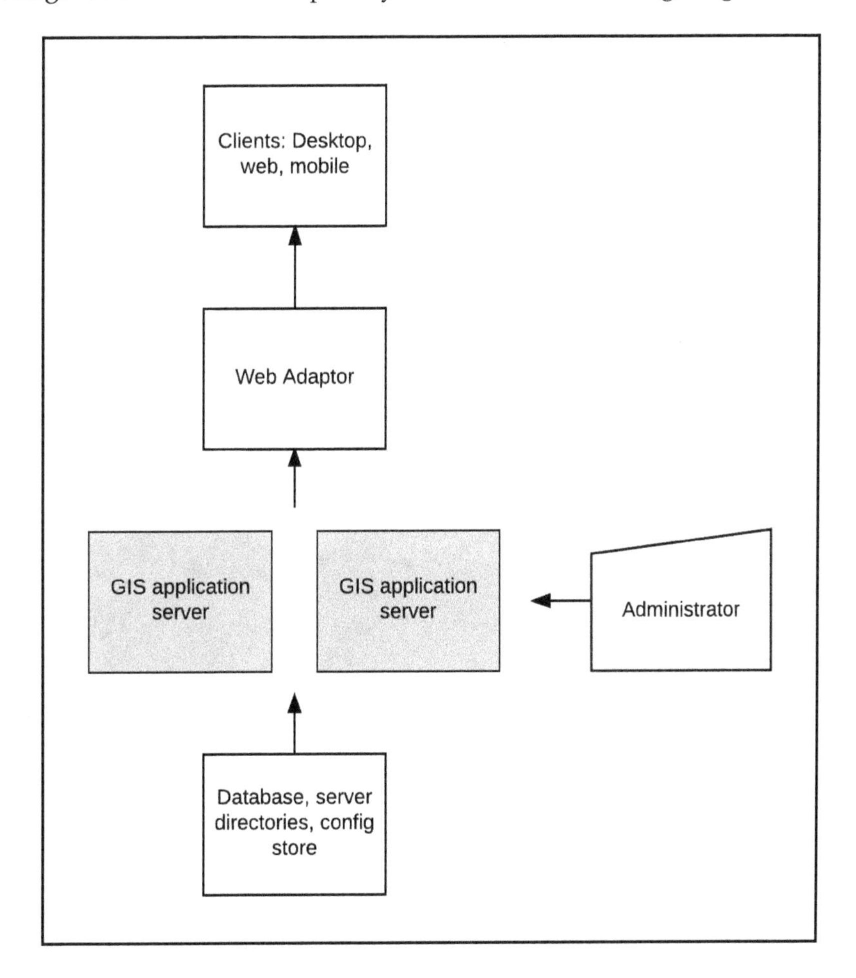

Hardware virtualization, utilized today by even the smallest of organizations, makes having and utilizing a multi-tiered deployment feasible.

Within the multi-tiered deployment, it is possible to have multiple ArcGIS Server machines functioning as a single logical unit. These servers operate in conjunction with the ArcGIS Web Adaptor to form a collective unit referred to as an ArcGIS Server **site**. Within a site, all ArcGIS Servers share the same configuration store and ArcGIS Server directories. Once configured, the site can be administered from any of the servers within it. For more information on ArcGIS Enterprise deployment scenarios, consult the online documentation.

ArcGIS Enterprise in the cloud

In addition to hosting ArcGIS Enterprise within your own infrastructure, whether it is on physical or virtual hardware, ArcGIS Enterprise can also run in the cloud. Esri supports ArcGIS Enterprise deployments on Amazon Web Services and Microsoft Azure. Standing up your ArcGIS Enterprise instance in the cloud offers several advantages to traditional on-premise deployments, such as:

- **Ease of setup**: Get an account set up and you can have a server up and running in just a few minutes.
- **Maintenance**: You don't have to maintain hardware infrastructure.
- **Scalability**: Machines can be added and removed as necessary, allowing you to distribute workloads for increased performance. Resources such as hard drives, CPUs, and memory can be easily scaled up as needed. Adding machines may require additional licensing depending on your licensing terms.

Amazon Web Services

With **Amazon Web Services** (**AWS**), there are several options available for launching ArcGIS Enterprise architectures.

AWS Marketplace

Through the AWS Marketplace (`https://aws.amazon.com/marketplace`), you can purchase an **Amazon Machine Image** (**AMI**) with ArcGIS Enterprise that can be easily deployed from your AWS account. Using the Marketplace, you purchase an AMI and then launch it as a virtual machine through the AWS Management Console:

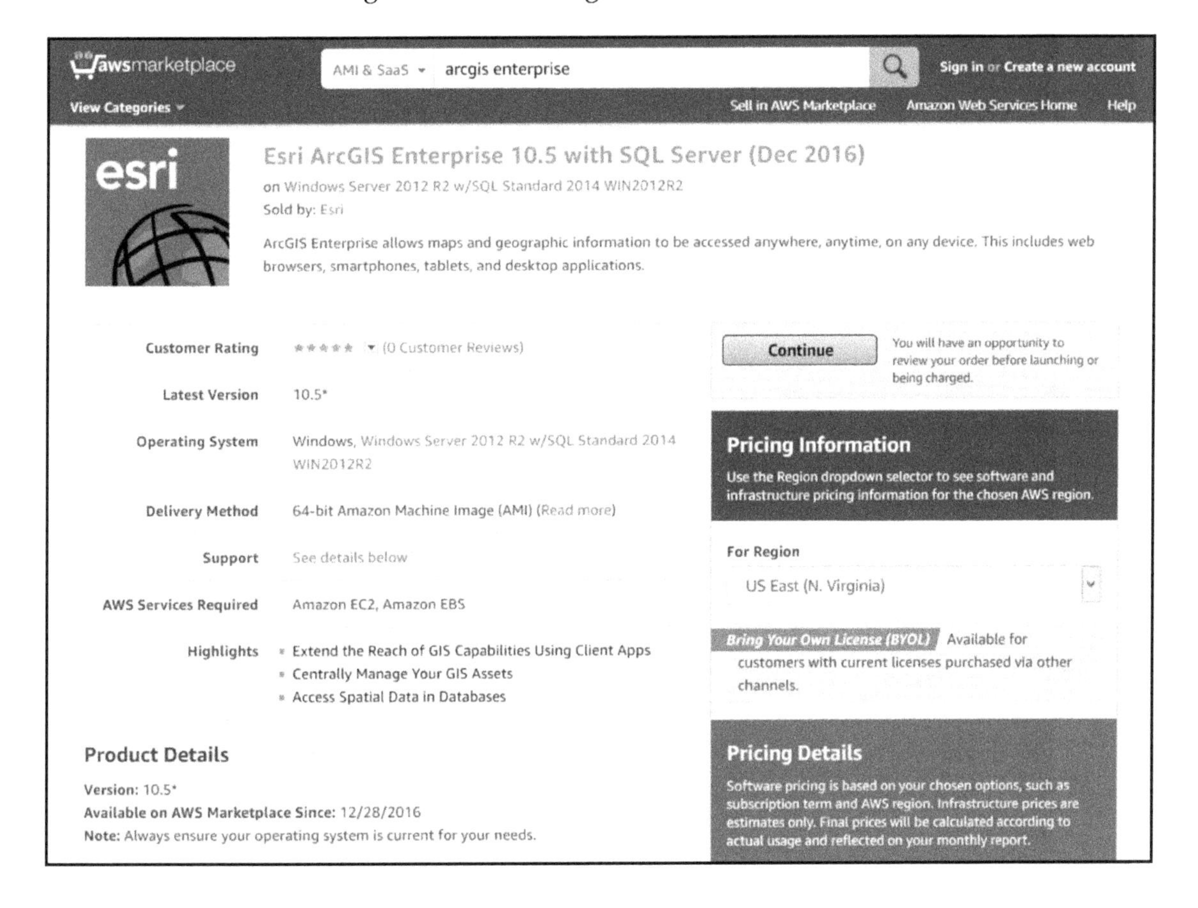

CloudFormation

CloudFormation is an AWS service that utilizes infrastructure as code to let you define architectures for the services you want to set up and utilize. Esri provides sample AWS CloudFormation templates that you can use to configure ArcGIS Server or ArcGIS Enterprise deployments for AWS. These template architectures vary in complexity, ranging from a simple single machine, ArcGIS Enterprise Deployment to a disaster recovery-ready configuration of multiple ArcGIS Enterprise deployments in two different AWS regions. See the ArcGIS Enterprise online documentation on *AWS CloudFormation and ArcGIS* for more information.

Cloud Builder

ArcGIS Server Cloud Builder is an Esri application that allows you to build and maintain a simple to complex ArcGIS Server site on AWS. With Cloud Builder, you can build, maintain, access, and backup your site, all from the Cloud Builder interface. It is perfect for those without cloud experience wanting to stand up infrastructure on AWS.

See the ArcGIS Enterprise online documentation on *ArcGIS Server Cloud Builder* for more information.

Manual deployment using the AWS Management Console

For the adventurous and those preferably with AWS experience, the **AWS Management Console** (**AWS Console**) can be used to administer any facet of the entire AWS ecosystem. From the AWS Console, you can stand up servers, manage security, view billing information, and add or remove any piece of AWS architecture to or from your system. With a manual deployment, you are responsible for planning, creating, and deploying all the machines in your site; setting up storage; configuring and managing security; and installing and configuring all components of ArcGIS Enterprise. For the purpose of this book, a single-machine deployment will be utilized in AWS, configured completely manually.

Microsoft Azure

As with AWS, there are options for using Azure to deploy ArcGIS Enterprise.

Azure Marketplace

Much like the AWS Marketplace, in the Azure Marketplace, you can search for a wide variety of preconfigured, readily available machines ready to be purchased and easily launched in the Azure cloud. The following is an example of an ArcGIS Enterprise machine available for purchase in the Azure Marketplace:

Cloud Builder

ArcGIS Enterprise Cloud Builder for Microsoft Azure is an application provided by Esri that you can use to deploy ArcGIS Enterprise and ArcGIS Server standalone sites on the Azure platform. With Cloud Builder for Azure, you can complete tasks, such as deploying ArcGIS Enterprise, adding sites to your deployment, installing an SSL certificate, adding a data store, and managing machines in your deployment.

ArcGIS Server installation

The ArcGIS Server installation process is straightforward. With a little planning and preparation, things can go smoothly.

Before you get started

Before starting the installation of ArcGIS Server, there are a few items to acquire:

- An authorization file for ArcGIS Server. Get this from `https://my.esri.com`.
- ArcGIS Server setup program. Get this from `https://my.esri.com`.

ArcGIS Server account

ArcGIS Server runs as a Windows service on the application server. All Windows services have an operating system *service account* that they run under; the ArcGIS Server default service account is a local account called `arcgis`, and it is commonly referred to as the *ArcGIS Server account*. The default local `arcgis` account is sufficient for development or testing environments, but Esri recommends using a domain account for production environments. If your organization uses a domain account, try to get the account set so that the password never expires. If your organization has security policies in place that require password expirations, determine when your ArcGIS Server account password will expire, and set a calendar reminder in advance. Once the password expires, the ArcGIS Server service will not be able to start and your ArcGIS Server site will be down. Always use a strong password, such as one generated at `https://xkpasswd.net`. To update the (expired) password, run the **Configure ArcGIS Server Account Utility** located in the Windows Start menu. See `Chapter 10`, *Troubleshooting ArcGIS Enterprise Issues and Errors* for more information on troubleshooting and issues with permissions and the ArcGIS Server account.

SSL certificate installation

If you will be utilizing an SSL certificate with your ArcGIS Server site, which is the recommended practice, Esri recommends installing this first before the installation of ArcGIS Server. The acquisition and installation of SSL certificates are quite often not well understood by GIS professionals. This is understandable, as SSL certificates are usually handled by systems administrators. That said, your systems administrator may indeed handle all aspects of SSL certificates within your organization, so contact them first before proceeding with purchasing one yourself. Regardless, let's demystify the process of acquiring and installing SSL certificates.

Acquiring an SSL certificate

Requesting and purchasing an SSL certificate is not as scary as it may seem. Armed with the knowledge of the process, it can be done in a few hours spread out over a few days in most cases.

Requirements

To acquire a basic SSL certificate, a few items are necessary:

- Web server access
- An account with a certificate authority
- A domain name and unique IP address

First, you will need administrative access to the web server that the ArcGIS Web Adaptor will be installed on. For our purposes here, we will be using IIS 8.5 on Windows Server 2012 R2. SSL certificates can, of course, be installed on any flavor of web server. See your web server's documentation for details on SSL certificate installation. Secondly, you, or someone in your organization, will need an account with a certificate authority, such as Digicert, GoDaddy, or Entrust, through which you will apply for and purchase the certificate. Again, check with your systems administrator before proceeding with the purchase of any SSL certificates. Finally, you will need a unique IP address and domain name to go along with it.

Getting the certificate

The first step in acquiring an SSL certificate is the generation of a certificate signing request or CSR. A CSR is a block of encoded text generated on the server where the certificate will be installed; it contains information that will be included in the certificate, such as the organization and domain name. Think of CSR as a digital signature for your server. To generate a CSR in IIS, follow these steps:

1. Launch IIS, select the machine name in the left **Connections** menu, then double-click on **Server Certificates** in **Features View**:

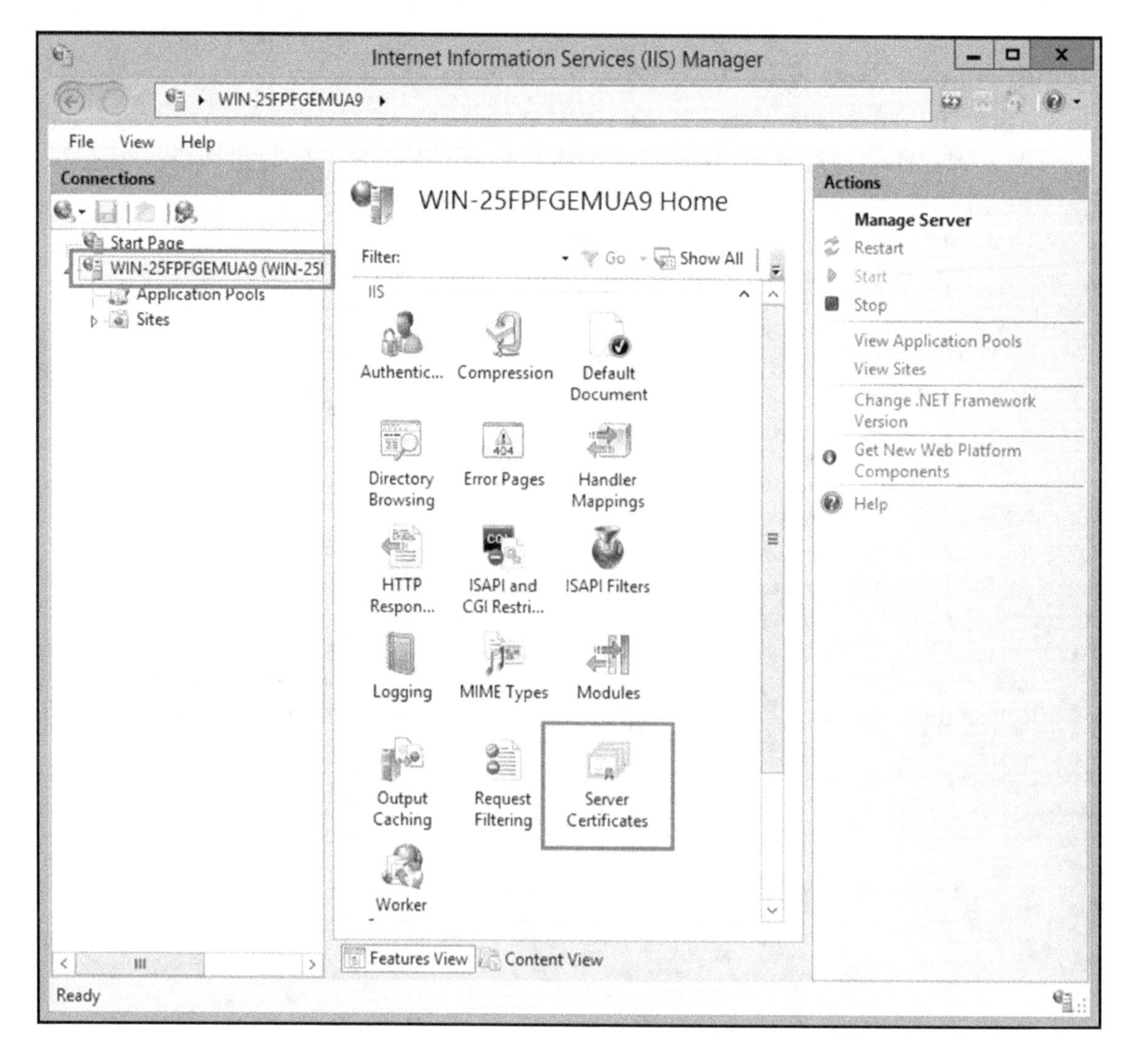

2. In the right **Actions** menu, click on **Create Certificate Request...**:

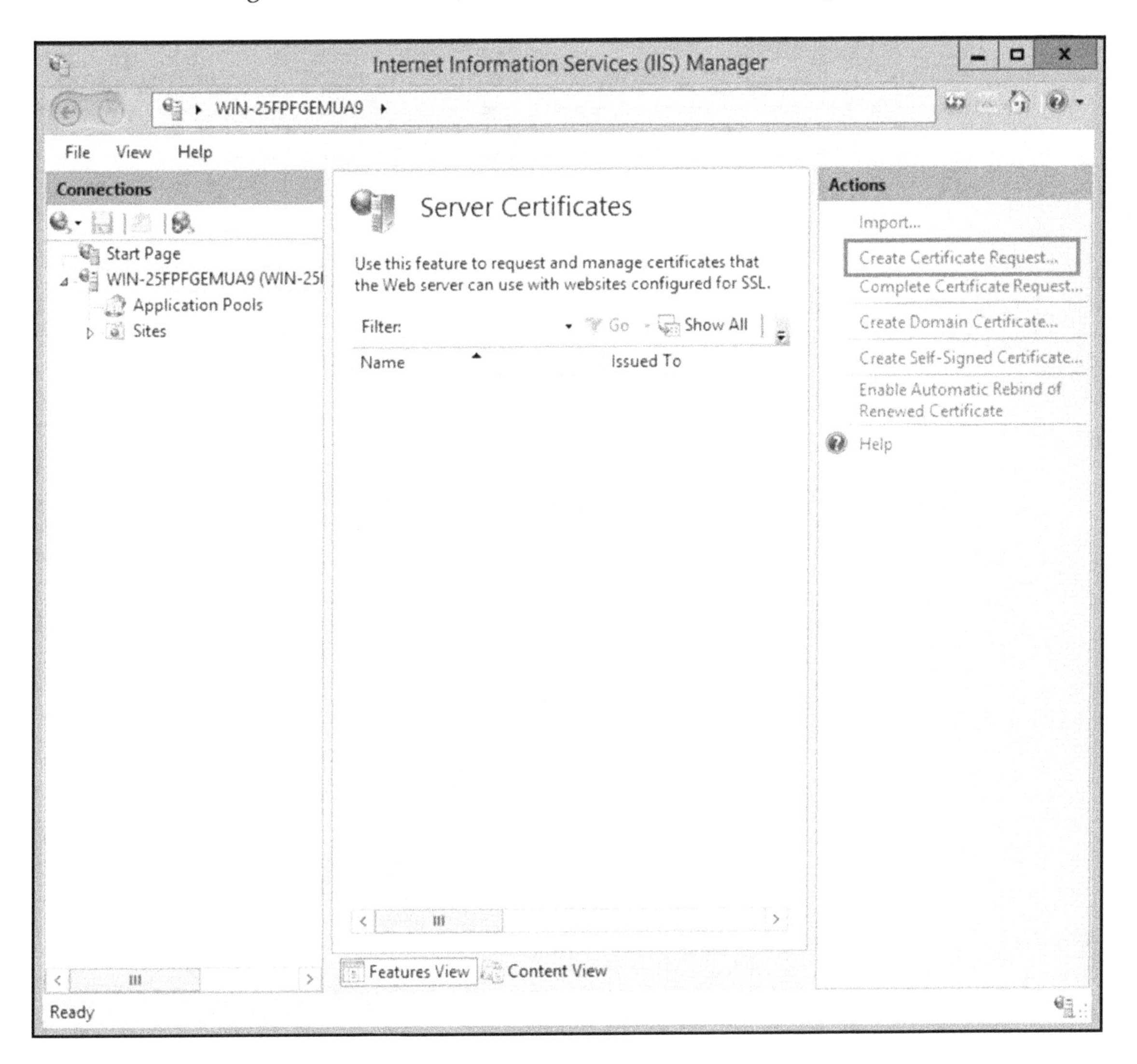

3. Fill out the **Distinguished Name Properties**, being careful to match these items (especially the **Organization** name) to those of the `WHOIS` record for your domain name. Click on **Next**:

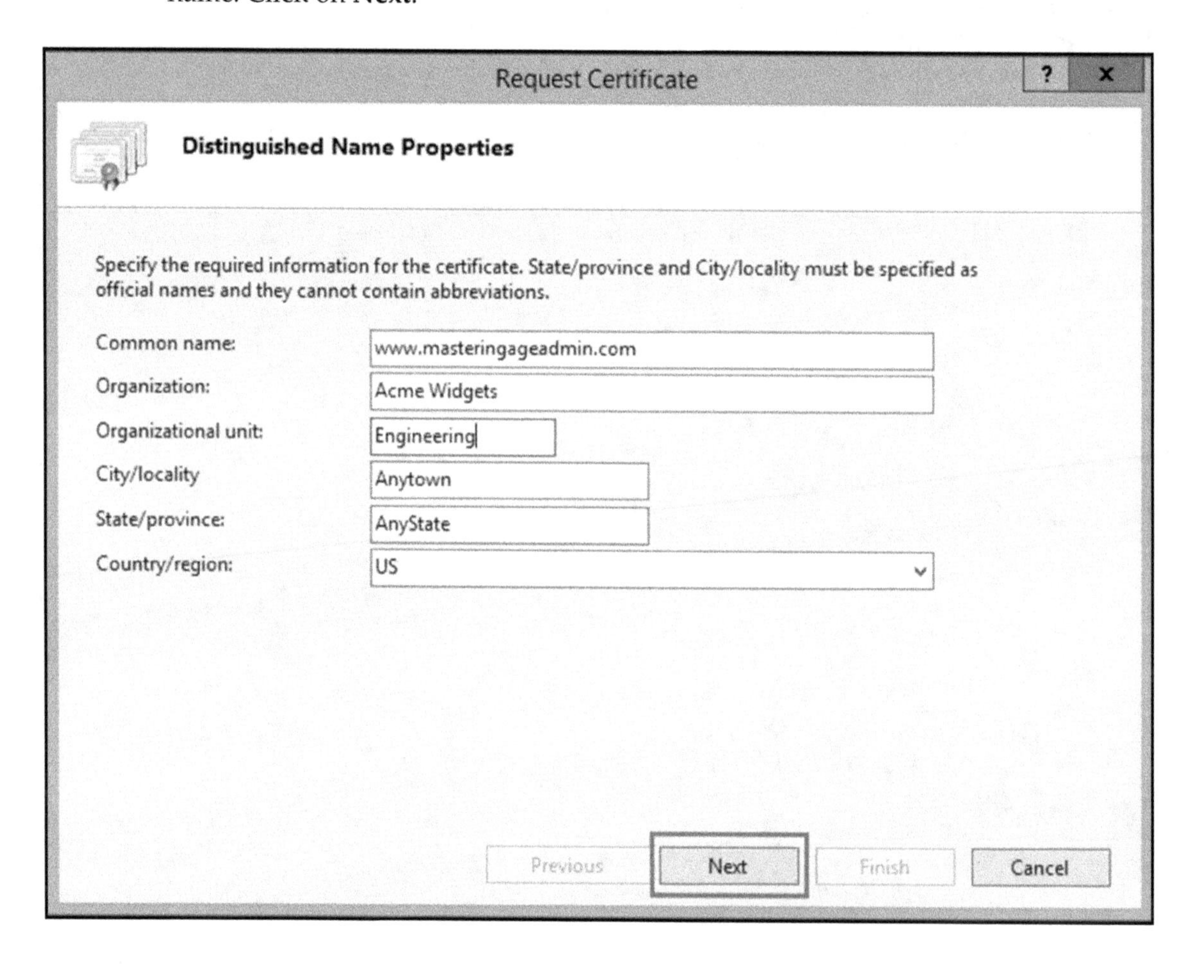

4. For **Cryptographic Service Provider Properties**, select **Microsoft RSA SChannel Cryptographic Provider** with a **Bit length** of **2048**; these are typical industry standards:

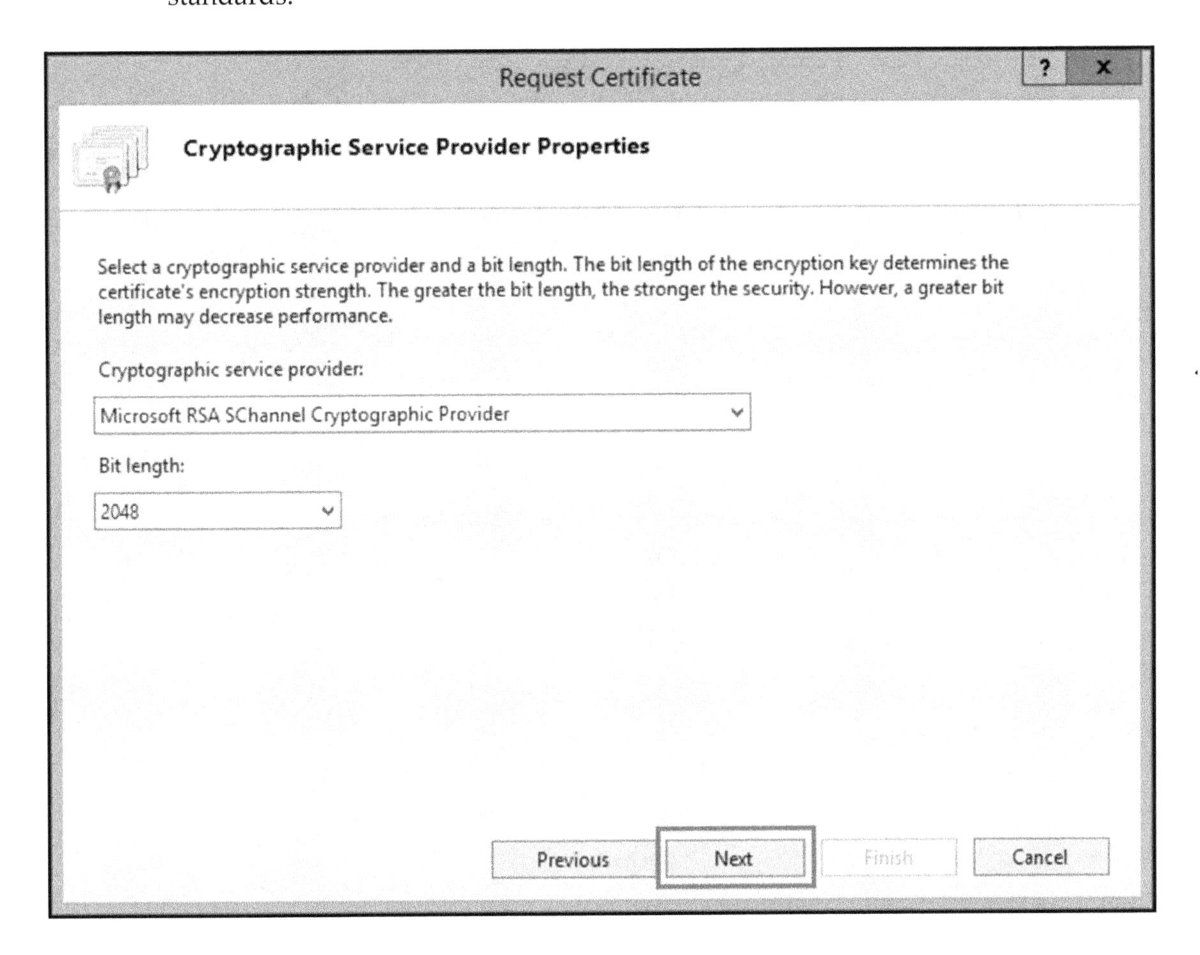

5. Specify a name and location for your CSR text file, as shown in the following screenshot:

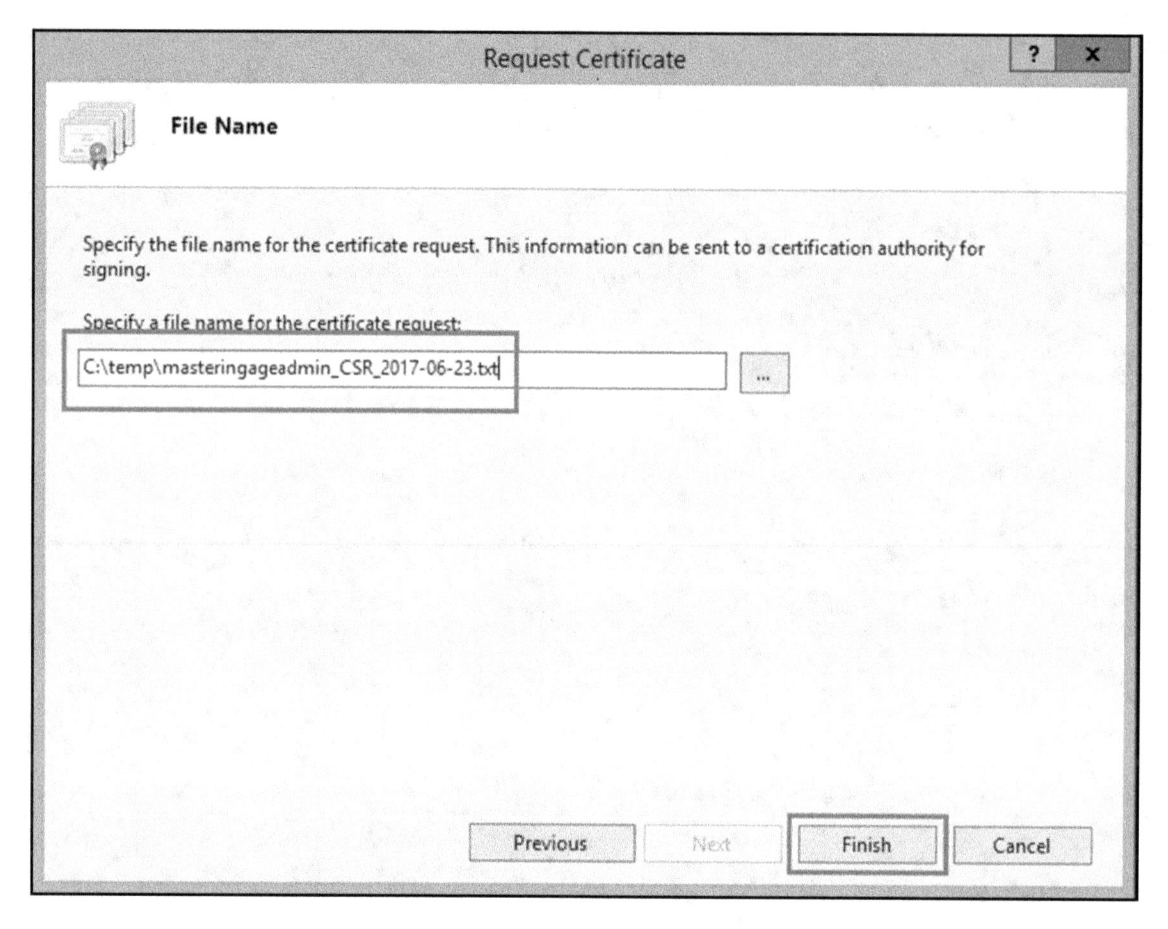

6. Open your CSR in a text editor; it will look like the following screenshot:

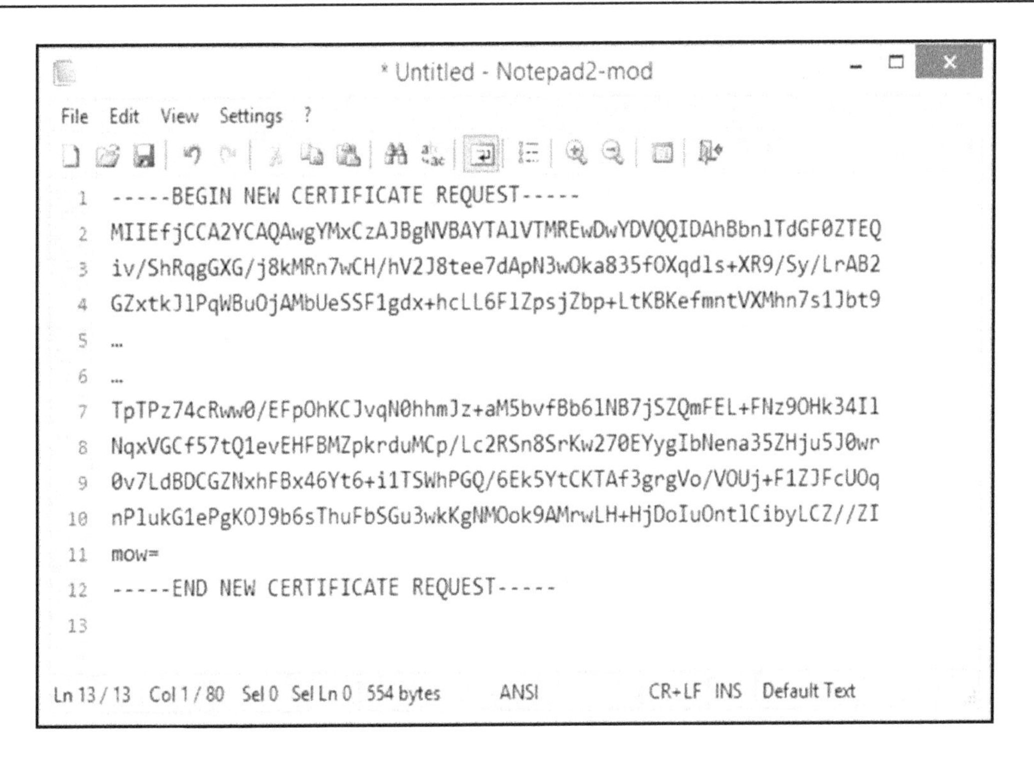

The second step in acquiring an SSL certificate is to purchase the certificate from the certificate authority, or CA. All CAs are different, but the process is the same in principle. First, log in to your account and purchase your SSL certificate. There are different options, so research them and find out which is best for your needs. Next, purchase your certificate. After you make the purchase, it will be available to you in your account.

The final step in this process is to apply your CSR to the certificate in your account. Here, you are *requesting* the certificate with the certificate signing *request* from your web server--this will bind the SSL certificate to your server, ensuring your end users that the site they are going to is indeed your site. After a successful request of the certificate from the CA, you will be able to download the certificate as a ZIP file.

Installing the SSL certificate

On your web server, you are now ready to install your SSL certificate. Launch IIS and complete the following steps:

1. In the **Connections** pane, select your server. Next, in **Features View**, double-click on **Server Certificates**. Finally, in the **Actions** pane, click on **Complete Certificate Request...**:

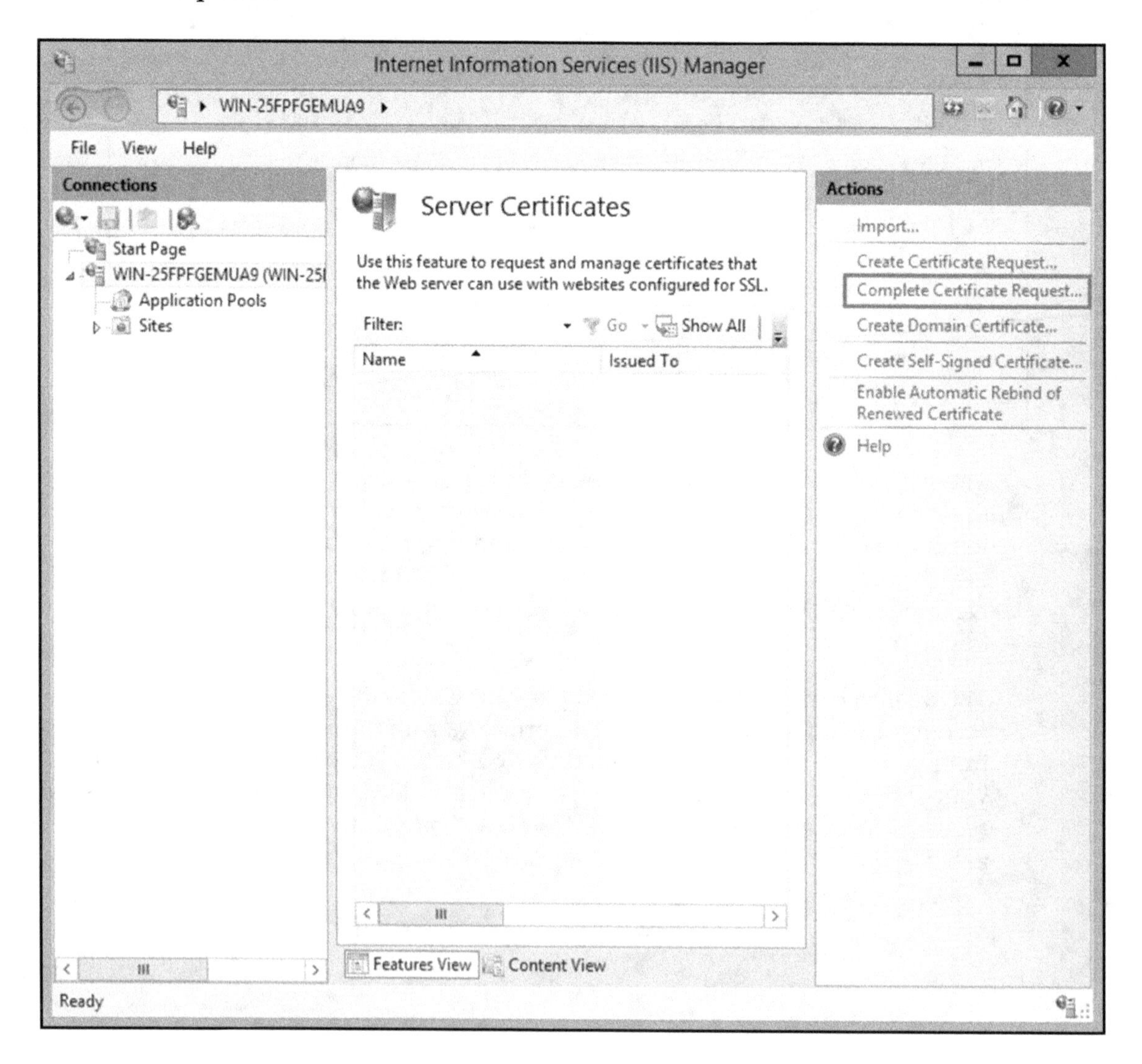

2. Enter the path to your `.crt` SSL certificate, then enter the friendly name (your domain name) and select the **Personal** certificate store:

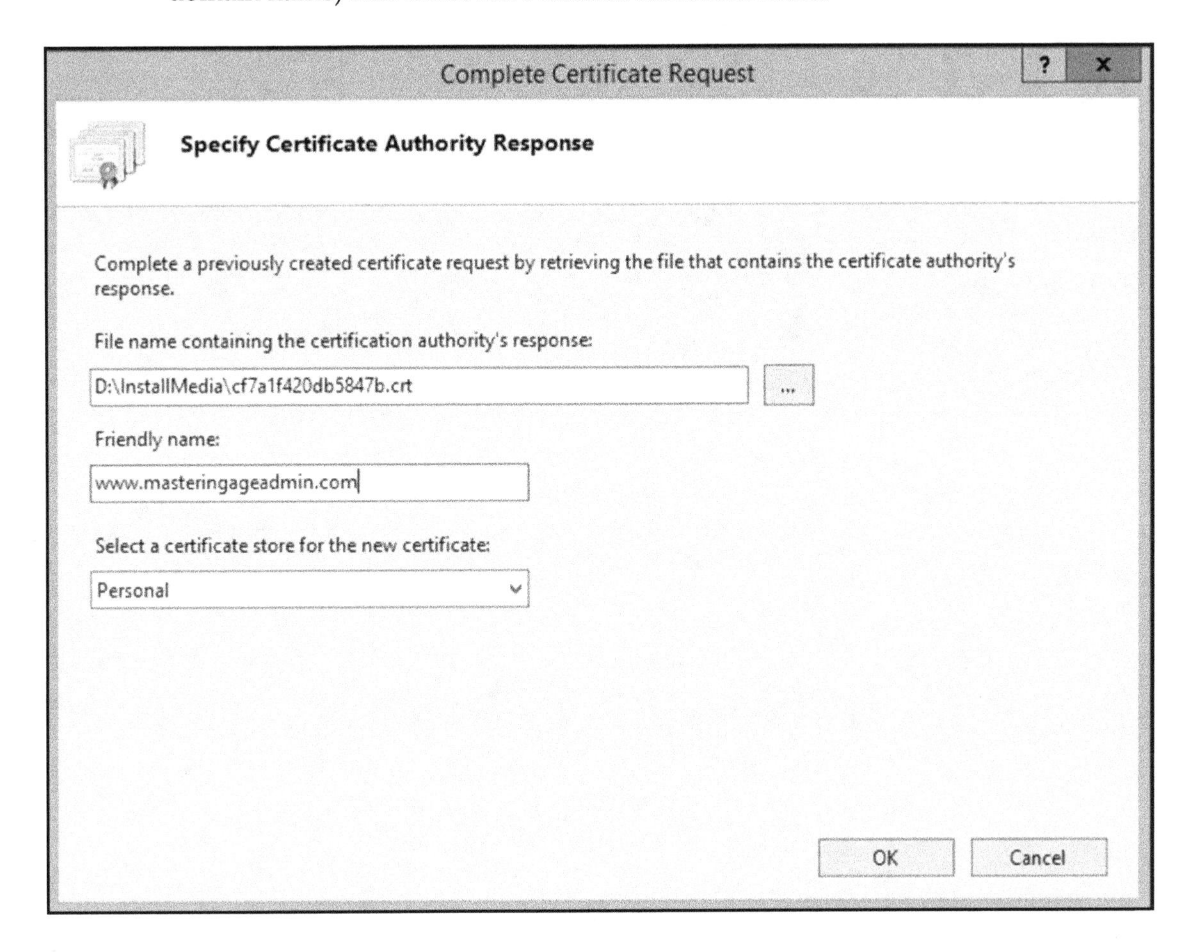

3. Your SSL certificate is now installed on your web server and should be listed in the **Server Certificates** pane, as shown here:

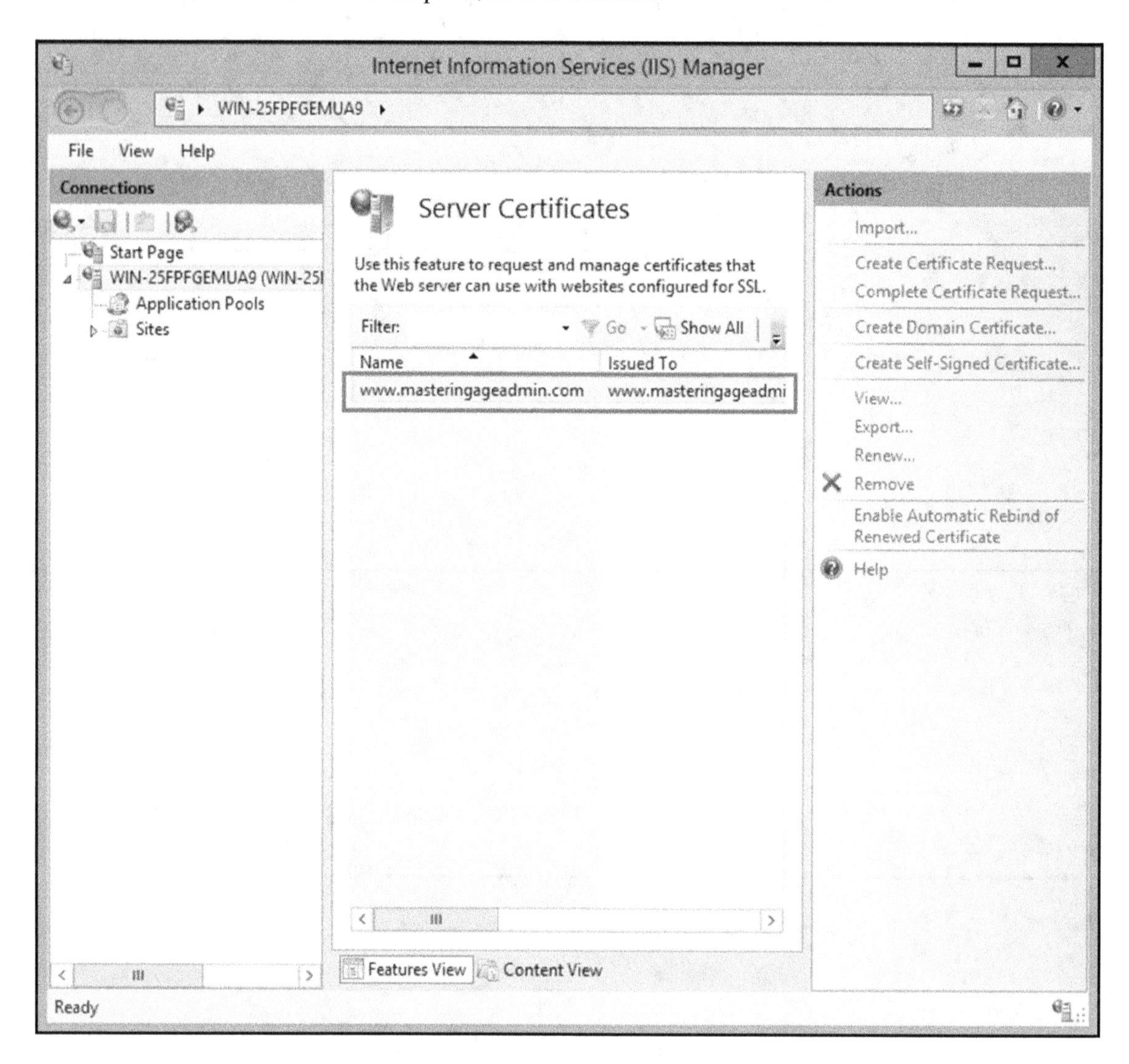

Setting your site bindings

Next, you need to *bind* your server's IP address and host header to port `443` with your SSL certificate. This is done through the **Site Bindings** settings in IIS. Again, open IIS and complete the following steps:

1. In the **Connections** left pane, select your website. In the right **Actions** pane, select **Bindings...**:

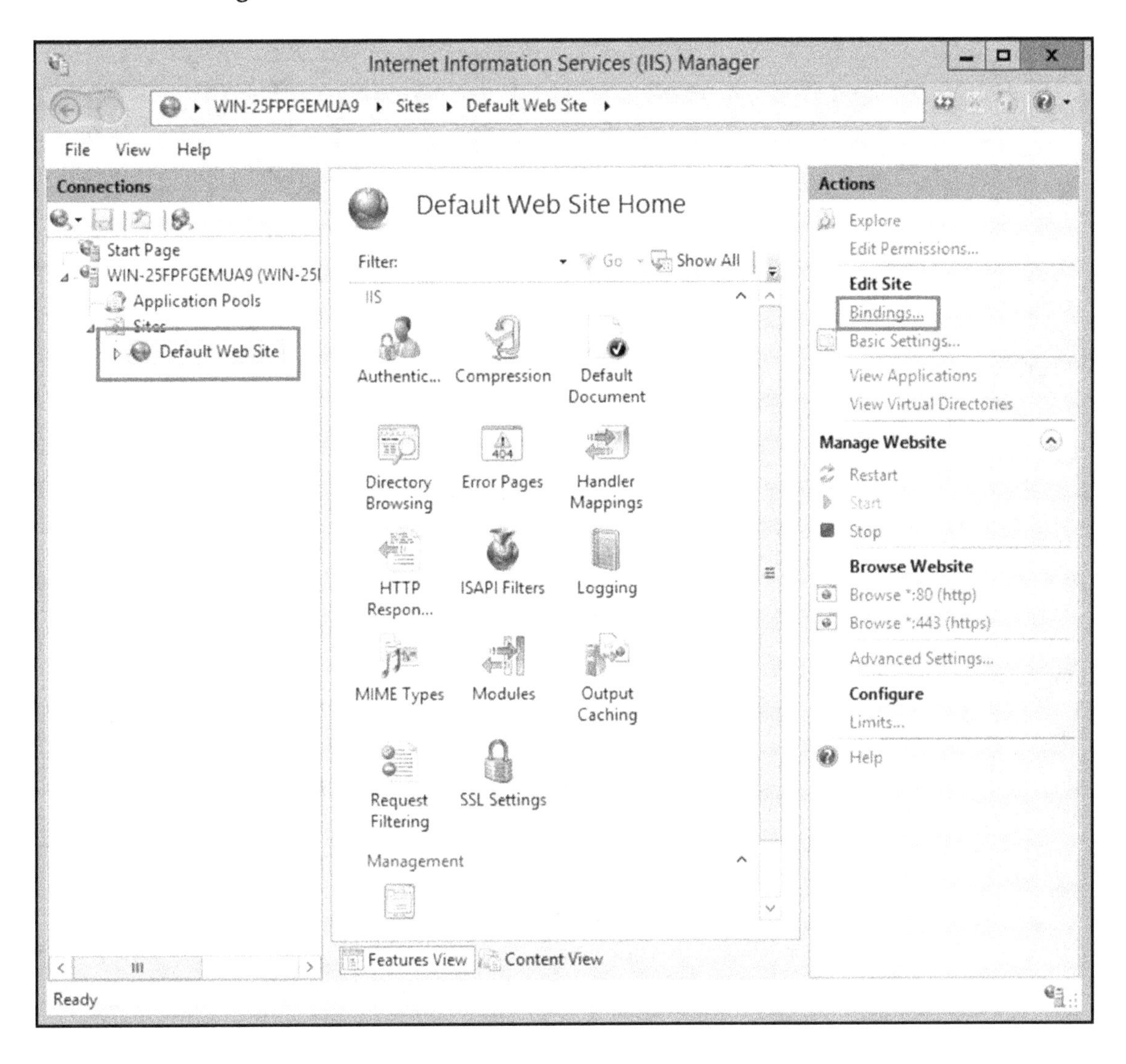

2. In the **Site Bindings** window, you will more than likely only have one binding for port `80` on `http`. Click on **Add**:

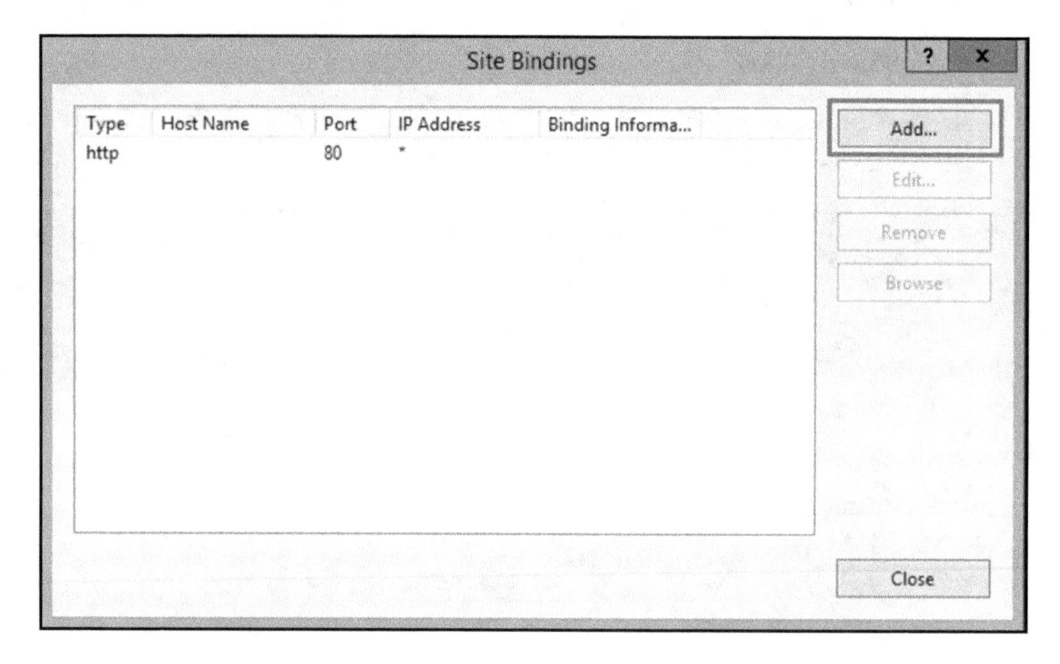

3. Add a binding for **Type:https**, **IPAddress:AllUnassigned**, **Port:**`443`. Select your **SSLcertificate** from the SSL certificate dropdown, as shown in the following screenshot, and then click on **OK**:

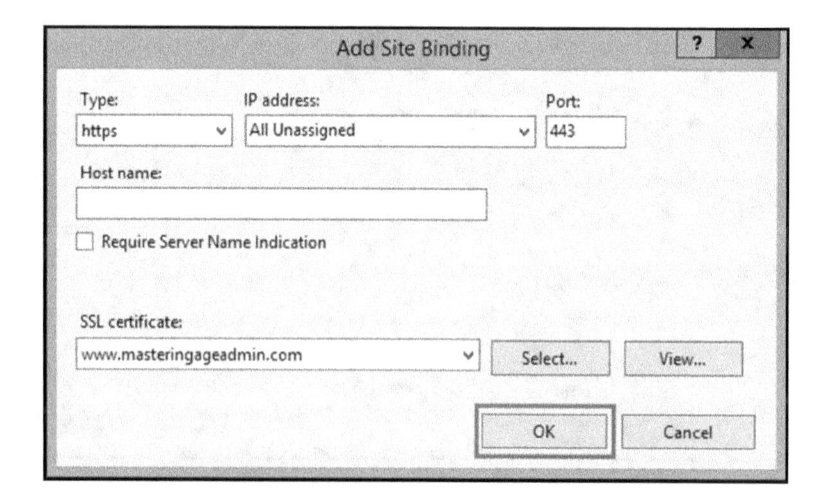

Your SSL certificate is now bound to port `443`. In a browser, navigate to your site over `https`; in my case, it is `https://www.masteringageadmin.com`:

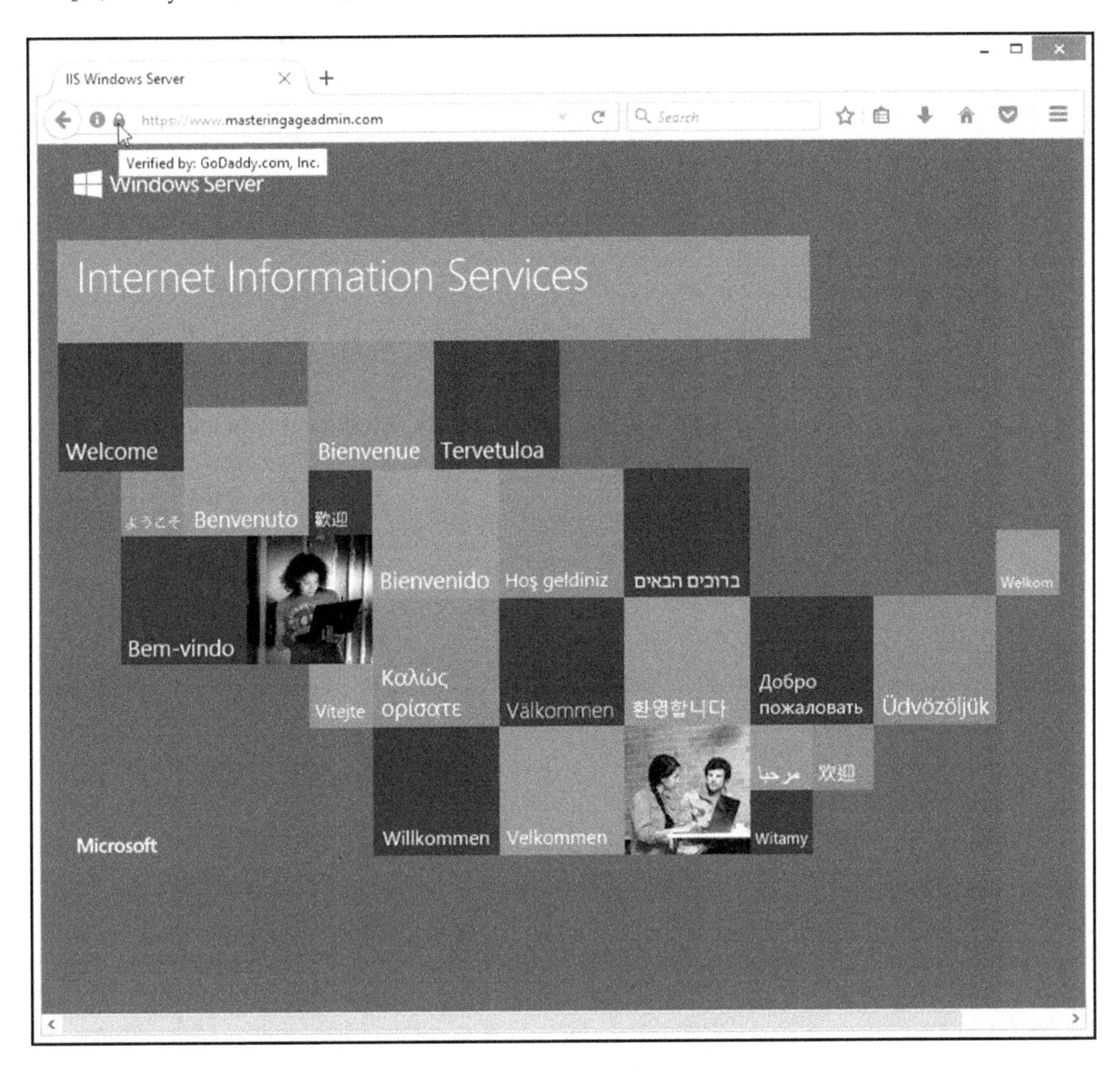

Running the installation program

Now that your SSL certificate is in place and you have your software authorizations and installers from Esri, it is finally time to install ArcGIS Server. The installation of ArcGIS Server is a straightforward process; as such, we will walk through the process at a high level, while highlighting some of the more important sections:

1. Double-click on the setup executable to launch it. The first step is to choose a location for setup installation files. This typically defaults to a path such as `C:\Users\Administrator\Documents\ArcGIS 10.5`. It is good practice to change this to a temp directory such as `C:\temp\ags`. Why, you ask? During this step of the installation, several very large `.cab` files (compressed files containing the installation pieces for ArcGIS Server), totaling almost 2 GB in size, are extracted to setup installation location. If you leave this location as the default, you will be placing almost 2 GB of files in the profile directories of the user running the installation. By storing them in a known location such as `C:\temp`, these files are more likely to get cleaned up and not be left sitting around needlessly on your system:

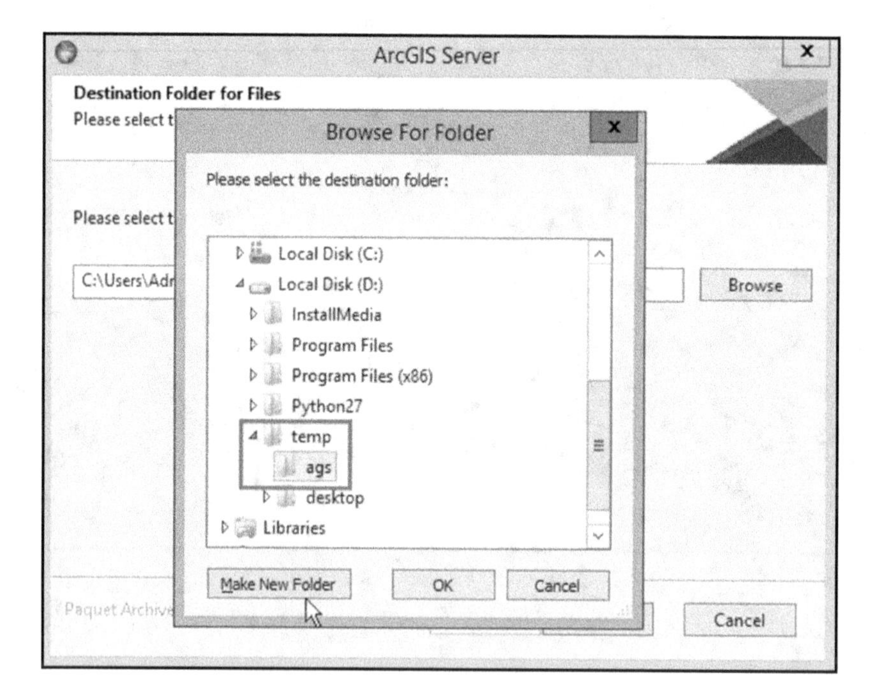

2. After the extraction of the installation files, check the checkbox to launch the setup program and click on **Close**.
3. Accept the license agreement and click on **Next**.
4. Choose features and the location in which to install them. Select all features (the default). The default installation directory for ArcGIS Server is `C:\Program Files\ArcGIS\Server`. However, some organizations, as a best practice, often have additional drives on servers to house application installs and data. If you have the option, it is a best practice to keep the ArcGIS Server installation off the operating system drive and installed on a secondary drive. This helps mitigate risks such as having an oftentimes relatively small operating system drive fill up and cause performance issues. To change the installation location, click on the **Change...** button and simply change the drive letter in the folder name path:

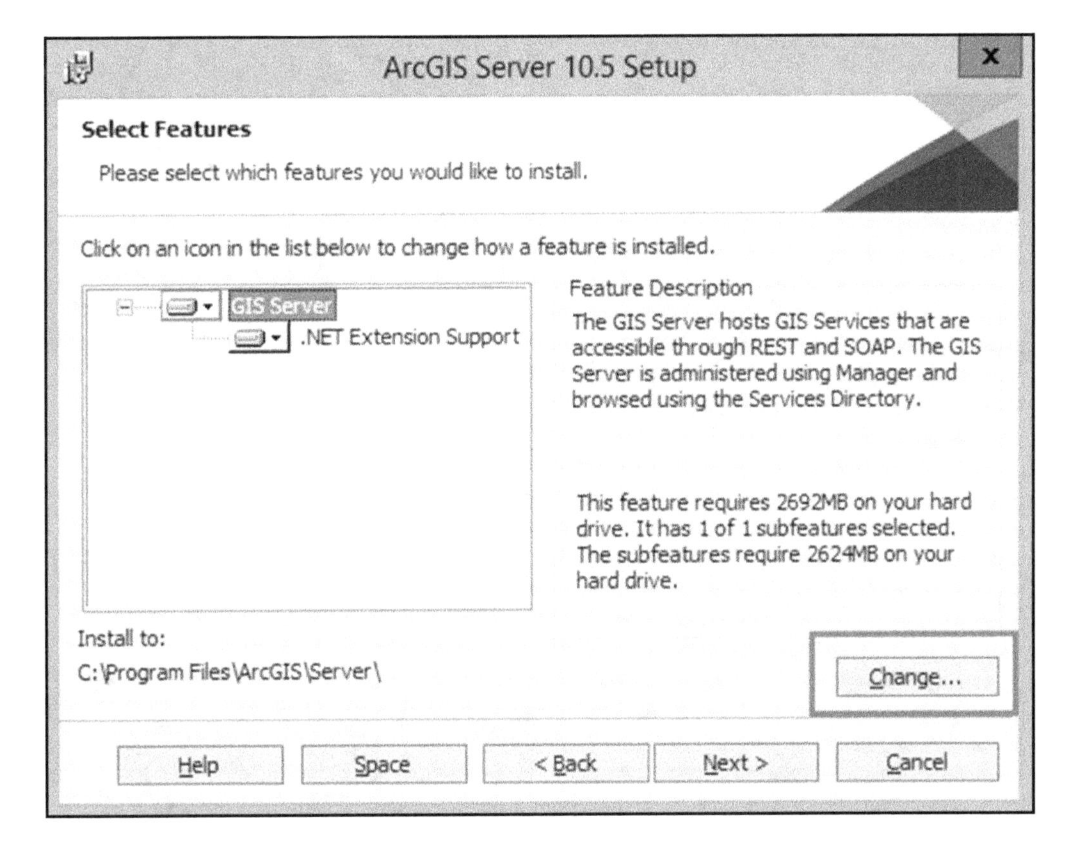

5. If you changed the installation location to another drive for ArcGIS Server, do the same for the Python installation by simply changing the drive letter.

Next, we come to the setup of the ever-important ArcGIS Server account. The ArcGIS Server account was discussed earlier in this chapter. If you will be using an existing domain account, enter it as domain\user along with the proper password. If you will be using a local account, you can stick with the default name of `arcgis` or change it. Remember to use a strong password, and it must meet the Windows password requirements. If you have previously saved a configuration file during a previous ArcGIS Server installation and would like to use the same account to run ArcGIS Server, you can use that here to avoid entering the ArcGIS Server account information:

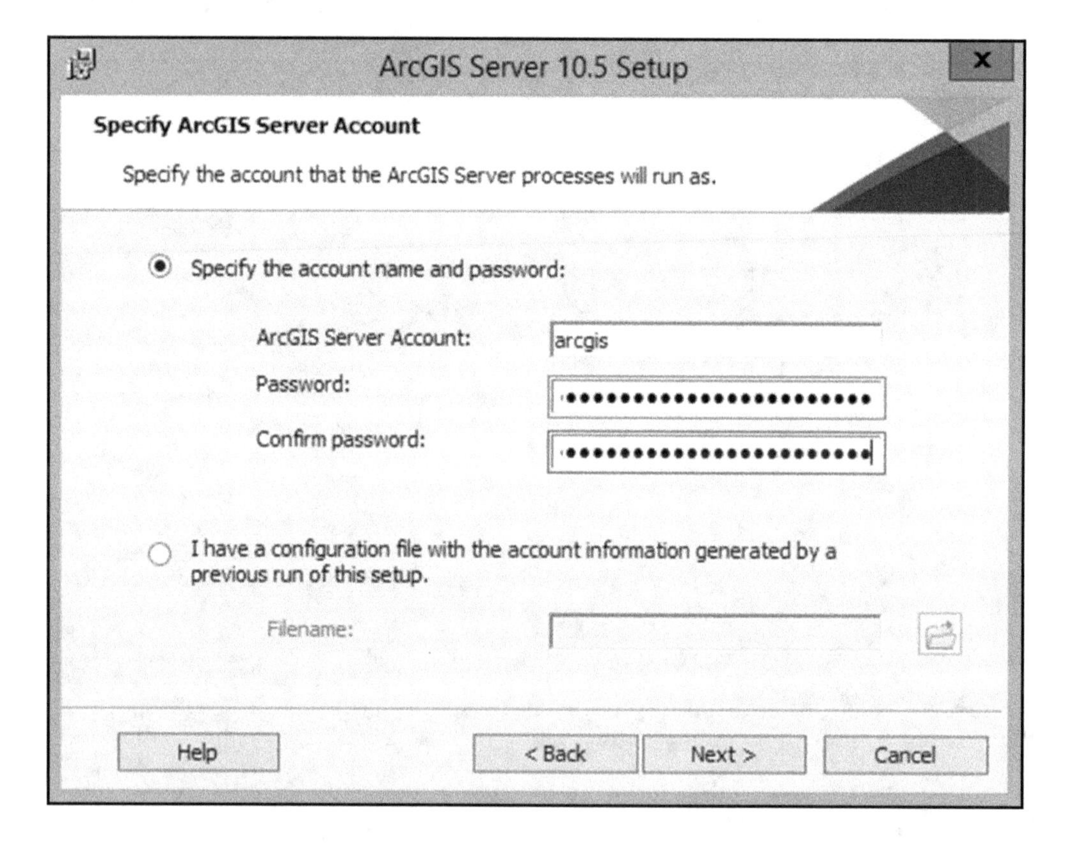

6. Next, you can optionally save a configuration file to use in later installations of ArcGIS Server. These configuration files can be useful to allow someone to do an installation without needing to know the ArcGIS Server account credentials, performing multiple installations of ArcGIS Server in a multi-server environment, or just to keep on hand in case of disaster recovery.
7. After you have specified ArcGIS Server account, the process will continue and install ArcGIS Server. Once finished, you will need to authorize your software.

Authorizing the software

Once the ArcGIS Server installation is completed, the Software Authorization Wizard launches. You can authorize with an authorization file you downloaded from `https://my.esri.com`, or authorize by email or additional extensions through the wizard.

You can also launch the Software Authorization Wizard manually from the Windows Start menu.

Using an authorization file from `https://my.esri.com` is usually the easiest and most common method of authorization. To authorize with an authorization file, follow these steps:

1. Select the **I have received an authorization file and am now ready to finish the authorization process** radio box. Navigate to and select your provisioning file (`.prvc`), and then click on **Next**:

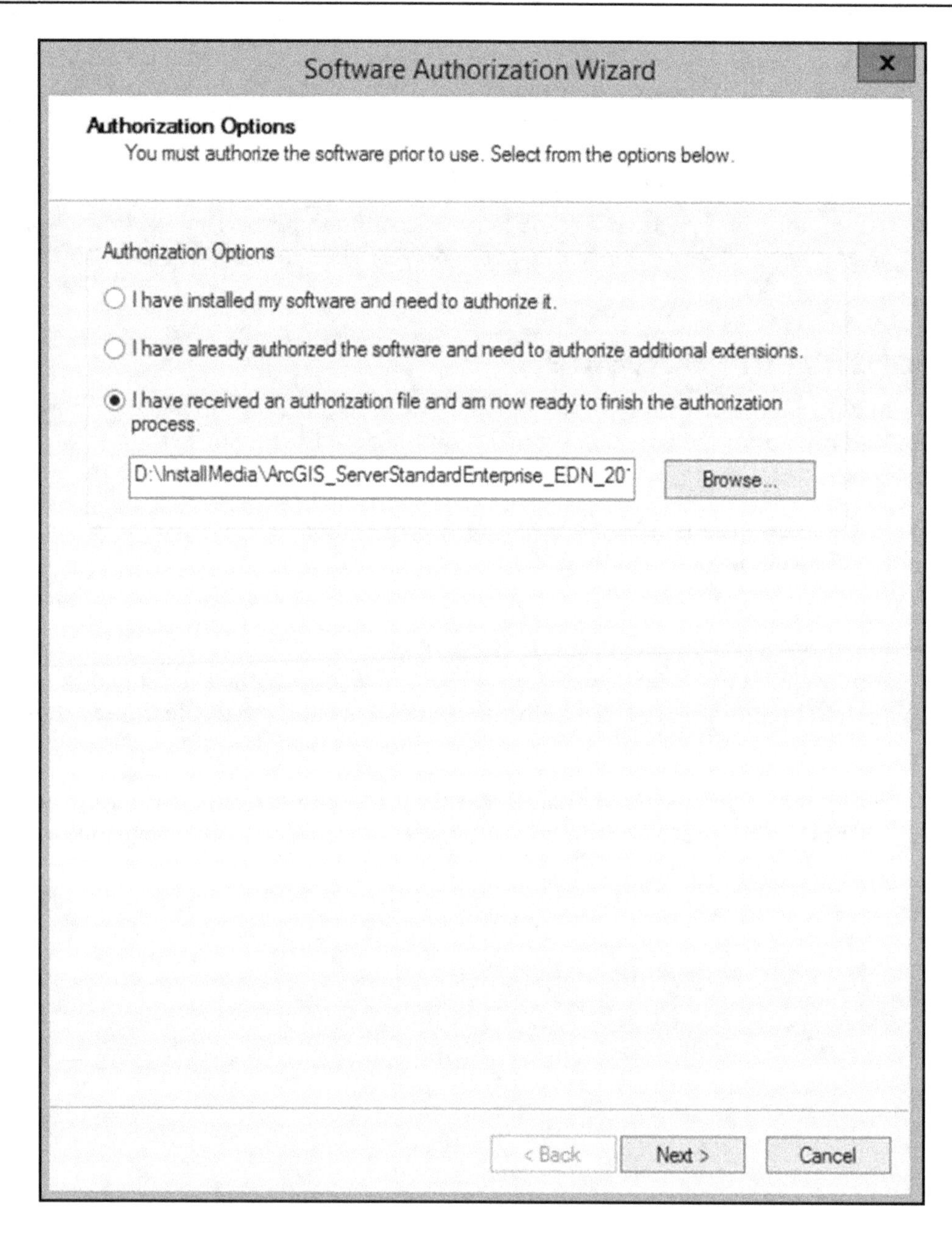

2. Select **Authorize with Esri now using the Internet**.

3. When using a provisioning file, your **Authorization Information** should fill in for you, as this was entered when the license was provisioned on `https://my.esri.com`. If not, fill in the contact and organizational details. Click on **Next**.
4. Continue by entering the organization information.
5. Your software authorization number, commonly referred to as an ECP number, will get populated from your provisioning file. Click on **Next**.
6. Next, you can authorize extensions for which you have licensing or authorize trial extensions for evaluation copies of several ArcGIS Server extensions.
7. Finally, your authorization information is sent to Esri and your software is authorized. Click on **Finish**.

ArcGIS Server initial configuration

Once ArcGIS Server is installed and authorized, you need to either create a new ArcGIS Server site or join an existing ArcGIS Server site. When you open ArcGIS Server Manager, the web-based ArcGIS Server administration panel, for the first time, you will be prompted to create a new site or join an existing site. An ArcGIS Server site is a deployment of ArcGIS Server.

Creating a new ArcGIS Server site

If you are installing ArcGIS Server on a single application server, or you are doing the first of several installations in a multi-machine environment, then you will create a new site:

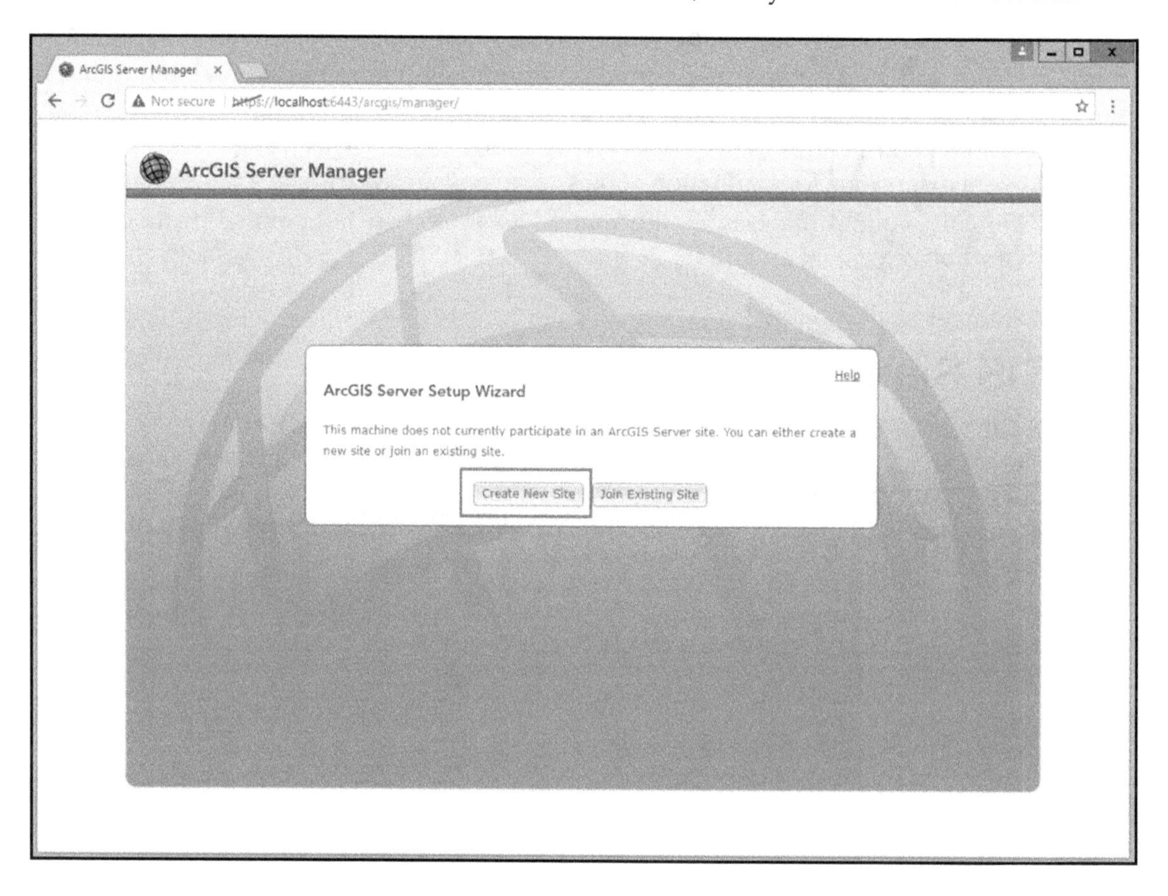

To begin, go to `https://localhost:6443/arcgis/manager` in a web browser. There is also an installed shortcut on the Start menu called **ArcGIS Server Manager**:

1. Step one of setting up an ArcGIS Server site is to create the **Primary Site Administrator** (**PSA**) account. This account is often referred to as the `siteadmin` account, as that is the default username, which most people utilize. This is not an operating system account, nor is it the same as the ArcGIS Server account (this is often a point of confusion). The `siteadmin` account has unrestricted access to the ArcGIS Server site. You can name this account differently, or you can disable it once you have configured other administrative accounts. Regardless, choose a very strong password for this account, enter it, and click on **Next**.

2. Specify your root server directory and **configuration store (config store)** locations. These will default to `D:\arcgisserver\directories` and `D:\arcgisserver\config-store` respectively, with the drive letter matching the drive you installed ArcGIS Server onto. In single-machine deployments, it is common to keep the config store and root directory on a local drive. With a multi-machine ArcGIS Server setup, all machines in the site share a configuration store, so it needs to be accessible by all machines. This is typically accomplished by using a network share for the config store.
3. Click on **Finish** to create your ArcGIS Server site. You may now login to your new ArcGIS Server site with your `siteadmin` credentials.

Joining to an existing ArcGIS Server site

In a multi-machine ArcGIS Server configuration, once the first server in the site is stood up and an ArcGIS Server site is configured there, you will add subsequent ArcGIS Server machines to that first site. This process is referred to as *joining to an existing site*. Before joining an ArcGIS Server machine to an existing site, make sure it meets the following criteria:

- Ensure that the machine to join is running the same operating system as the other machines in the site. It is best practice to have all site machines running on the same hardware and operating system.
- The ArcGIS Server version of the joining machine must match that of the other site machines, and it must be running under the same license.
- The joining machine must be able to read and write to the site's configuration store and server directories, and it must be running ArcGIS Server under the same ArcGIS Server account as all other machines on the site.

In this scenario, the ArcGIS Server account can be a local account with the exact same name and password on all site machines, but it is **highly** recommended to use a domain account for the ArcGIS Server account in a multi-machine setup.

- The joining machine must be able to communicate with all other site machines through the required ArcGIS Server ports.
- The joining site must be able to read any data referenced by any machines in the site.

To join a new ArcGIS Server machine to an existing ArcGIS Server site, follow these steps:

1. Open ArcGIS Server Manager by going to `https://localhost:6443/arcgis/manager` from the machine to be joined, or use the Start menu shortcut called **ArcGIS Server Manager**.

If you are prompted to login instead of being presented with the option to create or join an existing site, then this machine is already either its own ArcGIS Server site or it has been joined to another site.

2. Click on **Join an existing site**.
3. Enter the **fully qualified domain name** (**FQDN**) to the ArcGIS Server site you want to join this machine to. This should follow the format of `https://machinename.domain.com:6443/`.
4. Enter in administrator credentials of the site you are joining to. This is typically `siteadmin`, but could be any other administrator credentials as well.
5. If you have more than one cluster on your main site, then choose the cluster to join to. Otherwise, you will join the new machine to the default cluster.
6. Review your selected configuration and click on **Finish** to join the machine to the site.

ArcGIS Web Adaptor for ArcGIS Server

The ArcGIS Web Adaptor is one of the four components of ArcGIS Enterprise. Running in your existing website, the Web Adaptor forwards requests to your ArcGIS Server machines, typically forwarding incoming traffic on port `443` to `6443` and `80` to `6080`. In addition, the Web Adaptor keeps track of the ArcGIS Server machines on your site and forwards and distributes traffic to only currently participating machines. The Web Adaptor also allows you to do the following:

- Expose your ArcGIS Server through your standard website and port by leaving off the default port `6080` (or `6443`, if using SSL)
- Block the ArcGIS Server Administrator Directory and ArcGIS Server Manager from external viewers outside of your network
- Use web-tier authentication, such as Integrated Windows Authentication, to secure your ArcGIS Server

The Web Adaptor can be installed in your ArcGIS Server application machine, but is often put in an existing web server or a web server dedicated to GIS services.

Installing the ArcGIS Web Adaptor for ArcGIS Server

The ArcGIS Web Adaptor comes as a separate installer that you can download from `http://my.esri.com`.

Requirements

The Web Adaptor for ArcGIS is supported on IIS 10 on Windows Server 2016 Standard and Datacenter 64-bit and Windows 10 Pro and Enterprise; IIS 8.5 on Windows Server 2012 R2 Standard and Datacenter and Windows 8.1 Pro and Enterprise; IIS 8 on Windows Server 2012 Standard and Datacenter; and IIS 7.5 on Windows Server 2008 R2 Standard, Enterprise, and Datacenter and Windows 7 Ultimate and Professional. Windows 10, 8.1, and 7 are also supported *for basic testing and application development only, not for production environments.*

Microsoft .NET Framework 4.5 is required, as are specific IIS components. See the online help for *ArcGIS Web Adaptor 10.5.x system requirements* for an exhaustive list of these requirements.

Web Adaptor for ArcGIS Server installation

The ArcGIS Web Adaptor installation process is quick and easy:

1. Double-click on the installation executable to launch it.
2. Select a location to unpack the installation files to; a known temporary location is best for easy cleanup later.
3. After the installation files are unpacked, launch the setup program.

4. The ArcGIS Web Adapter for IIS requires certain components of IIS to be installed. At 10.5, there is a verification step in the installer that will detect what components are missing, and it will install them for you. Click on **I Agree** to install the missing IIS components, if any:

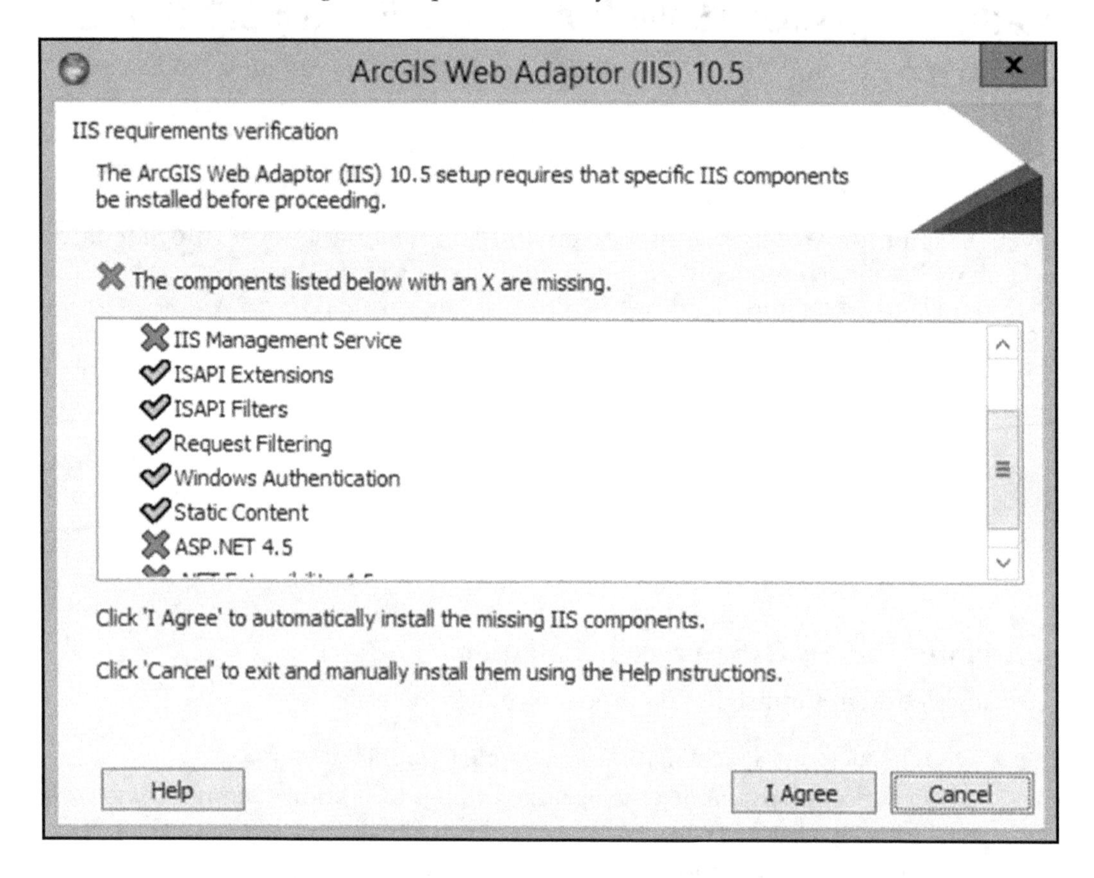

5. Click on **Next** and agree to the license agreement.
6. Select a port to install the Adaptor to. Since we installed and configured our SSL certificate already, port `443` is available to us. Select port `443` and click on **Next**:

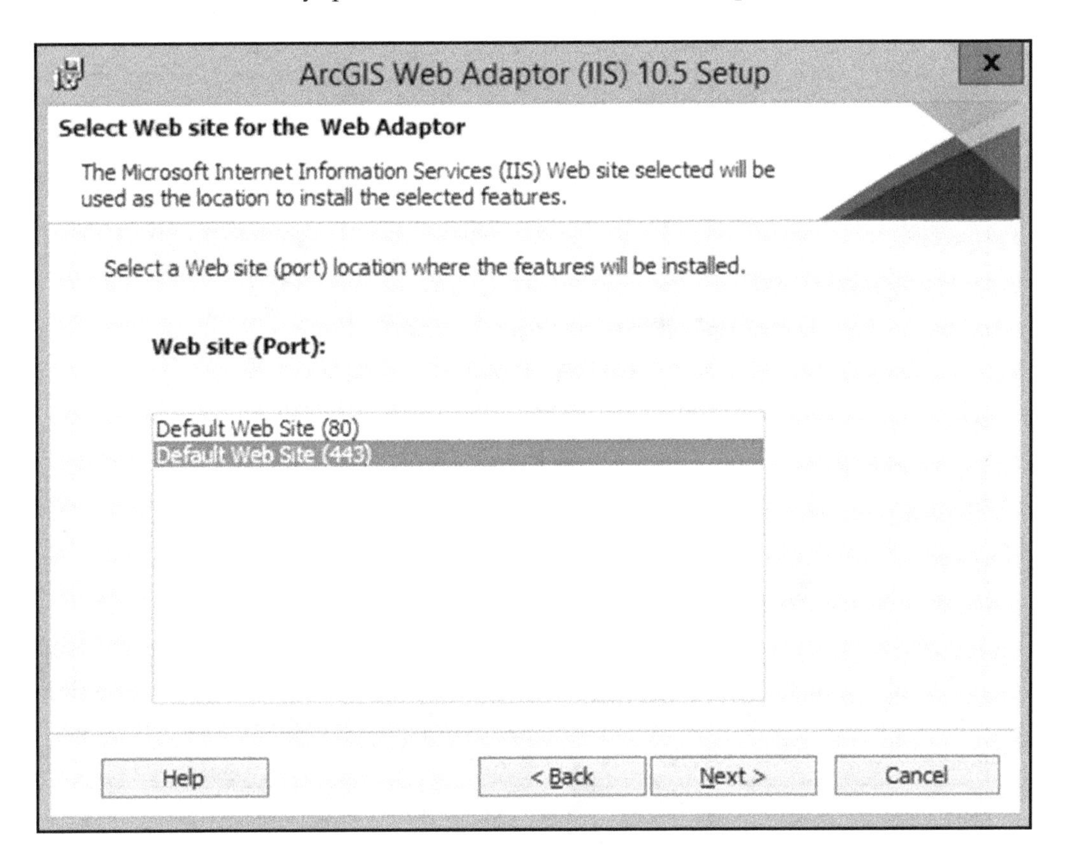

7. Specify the name of the ArcGIS Web Adaptor for your ArcGIS Server instance. The default here is arcgis. This is an important step in the process, as the Adaptor name will be in your services URL; for example, https://www.masteringageadmin.com/arcgis/rest:

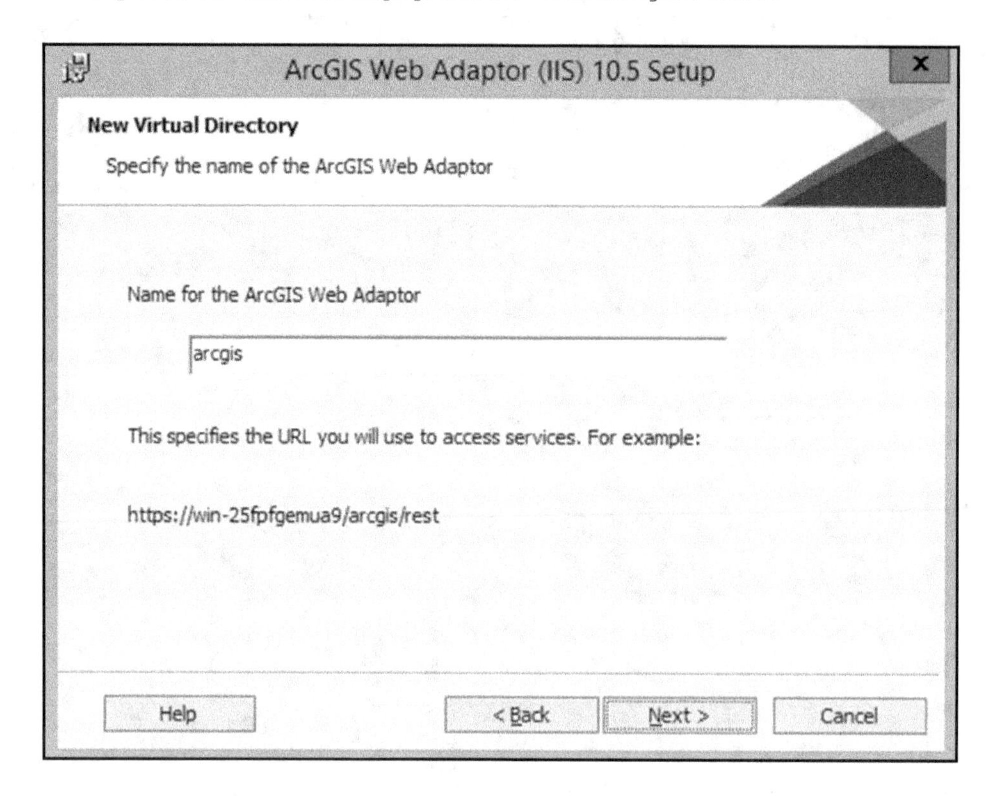

8. Click on **Install** to begin the installation and click on **Finish** when done.

Web Adaptor for ArcGIS Server configuration

Once the Web Adaptor for the ArcGIS installation process is complete, the configuration page should open in your default web browser (https://localhost/arcgis/webadaptor). To configure the Web Adaptor for ArcGIS Server, do the following:

There is also a Web Adaptor shortcut in the Windows Start menu named ArcGIS Web Adaptor - <web-adaptor name> (port), such as ArcGIS Web Adaptor - arcgis (443).

1. First, select the product to configure with the Web Adaptor. Here, we are configuring the Web Adaptor for ArcGIS Server. Later, we will also configure a Web Adaptor for Portal for ArcGIS. When we get to the Portal Web Adaptor configuration later, this configuration page will tell us that a server IS configured with our Web Adaptor. Here, select **ArcGIS Server** and click on **Next**:

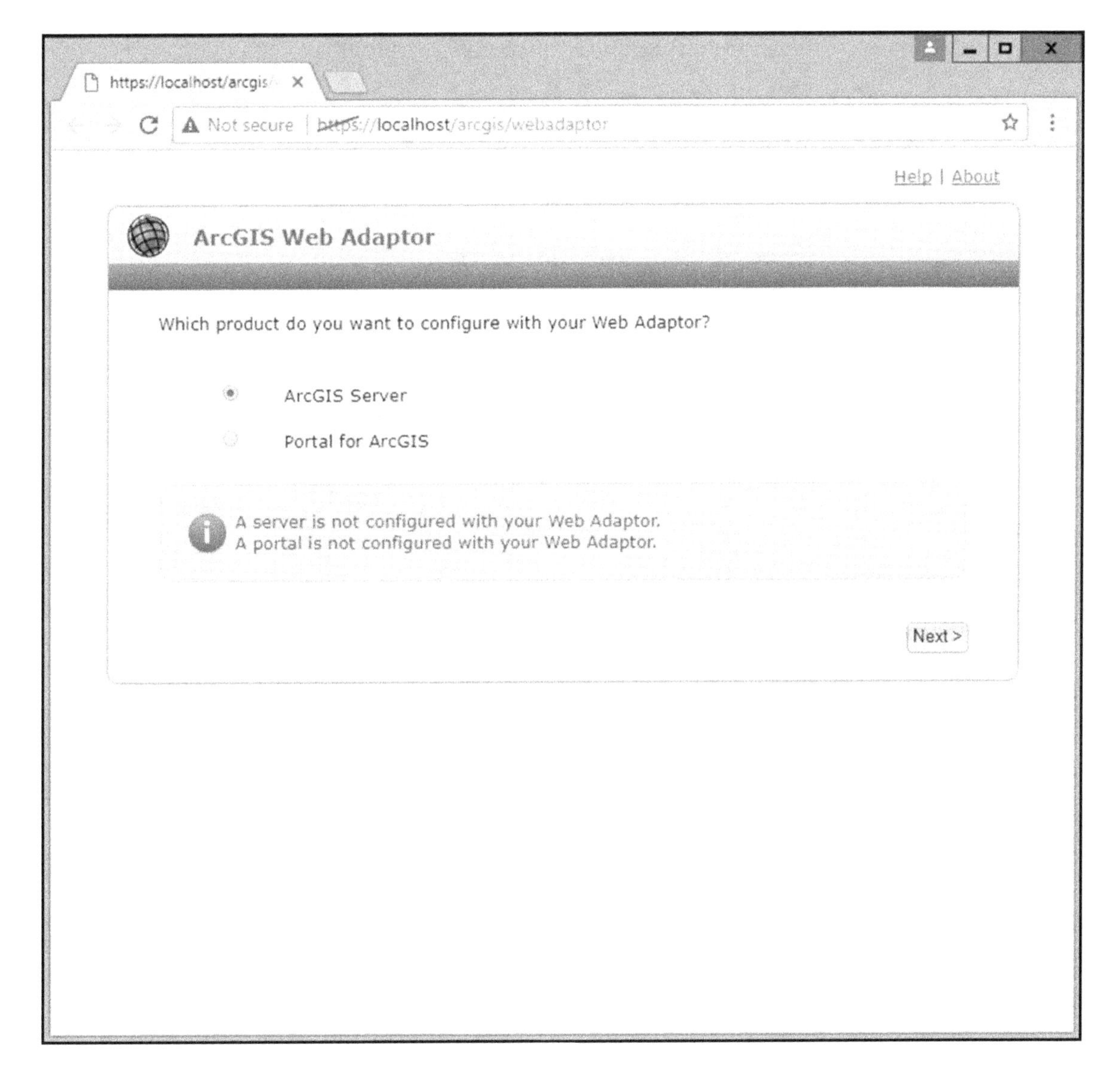

Next is the final and main configuration page.

2. Enter your **ArcGIS Server URL**. This is the URL to any one of the ArcGIS Server machines in your ArcGIS Server site (remember that all the machines in a site function together as one). The URL should take the form of `https://gisserver.domain.com:6443` or `http://gisserver.domain.com:6080` if you do not have SSL in place. Here, my ArcGIS Web Adaptor server is not on a domain, so my URL takes the form of my machine name, that is, `https://WIN-25FPFGEMUA9:6443`:

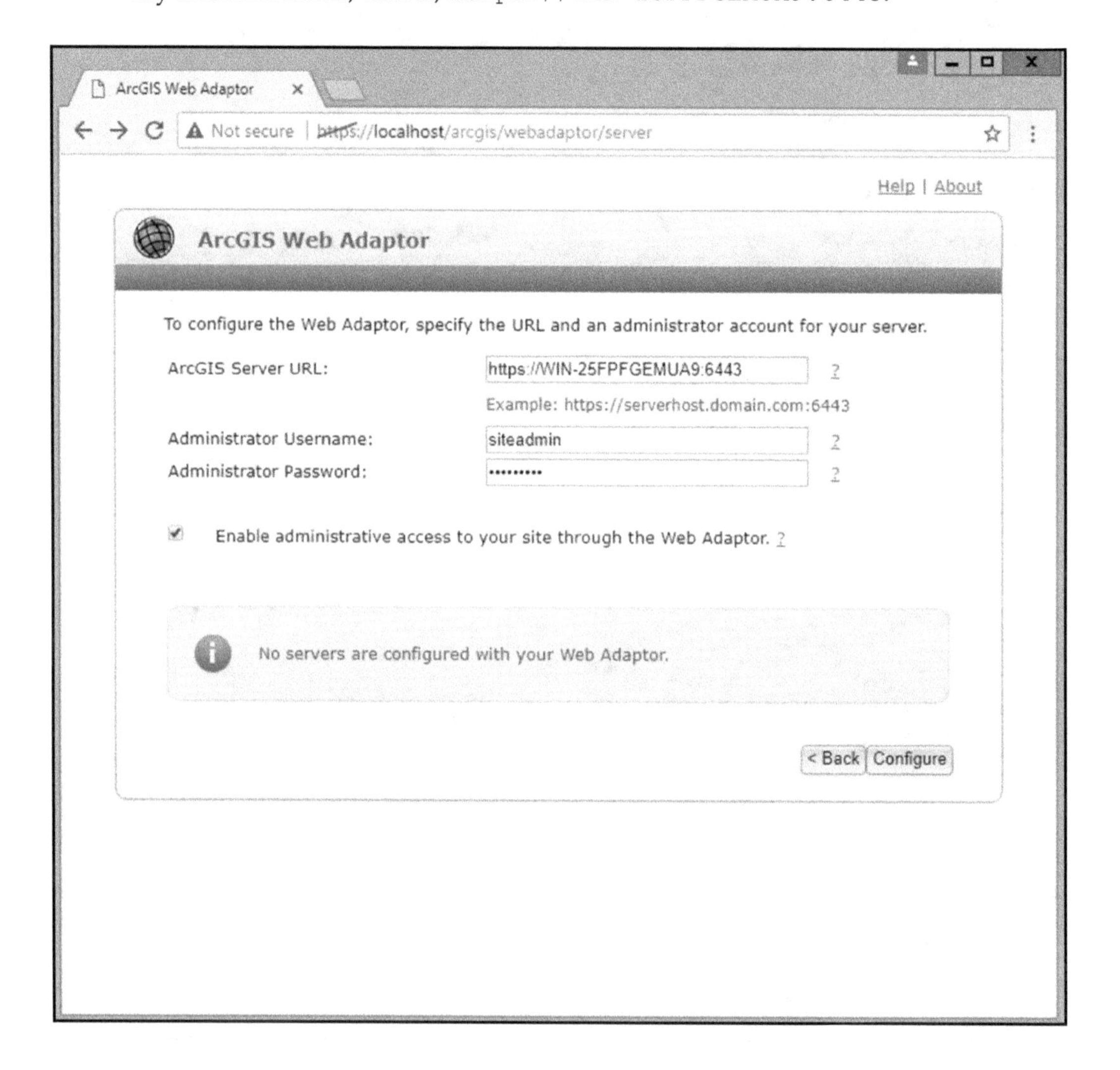

3. Enter your primary site administrator account credentials or the credentials of another administrative account.
4. Finally, choose whether or not to allow administrative access to your ArcGIS Server site through the Web Adaptor. Esri recommends disabling administrative access, but there are considerations, which are as follows:

- If disabled, administrators cannot access ArcGIS Server Manager and the ArcGIS Server Administrator Directory through the Web Adaptor URL. More importantly, ArcGIS Desktop users cannot establish administrative or publisher connections to ArcGIS Server, meaning publishers cannot publish services directly from their desktops (user connections can still be made regardless of this setting). However, if ArcGIS Server's internal URL is accessible, these connections can be made from there.
- If your ArcGIS Server will be configured with web-tier authentication (more on that later), you must enable administration through the Web Adaptor, allowing administrative and publisher users in the enterprise identity store to publish services from ArcGIS Desktop.

5. Click on **Configure** to continue. When the Web Adaptor configuration is successful, you will be presented with the following message telling you that your server is successfully configured with the Web Adaptor:

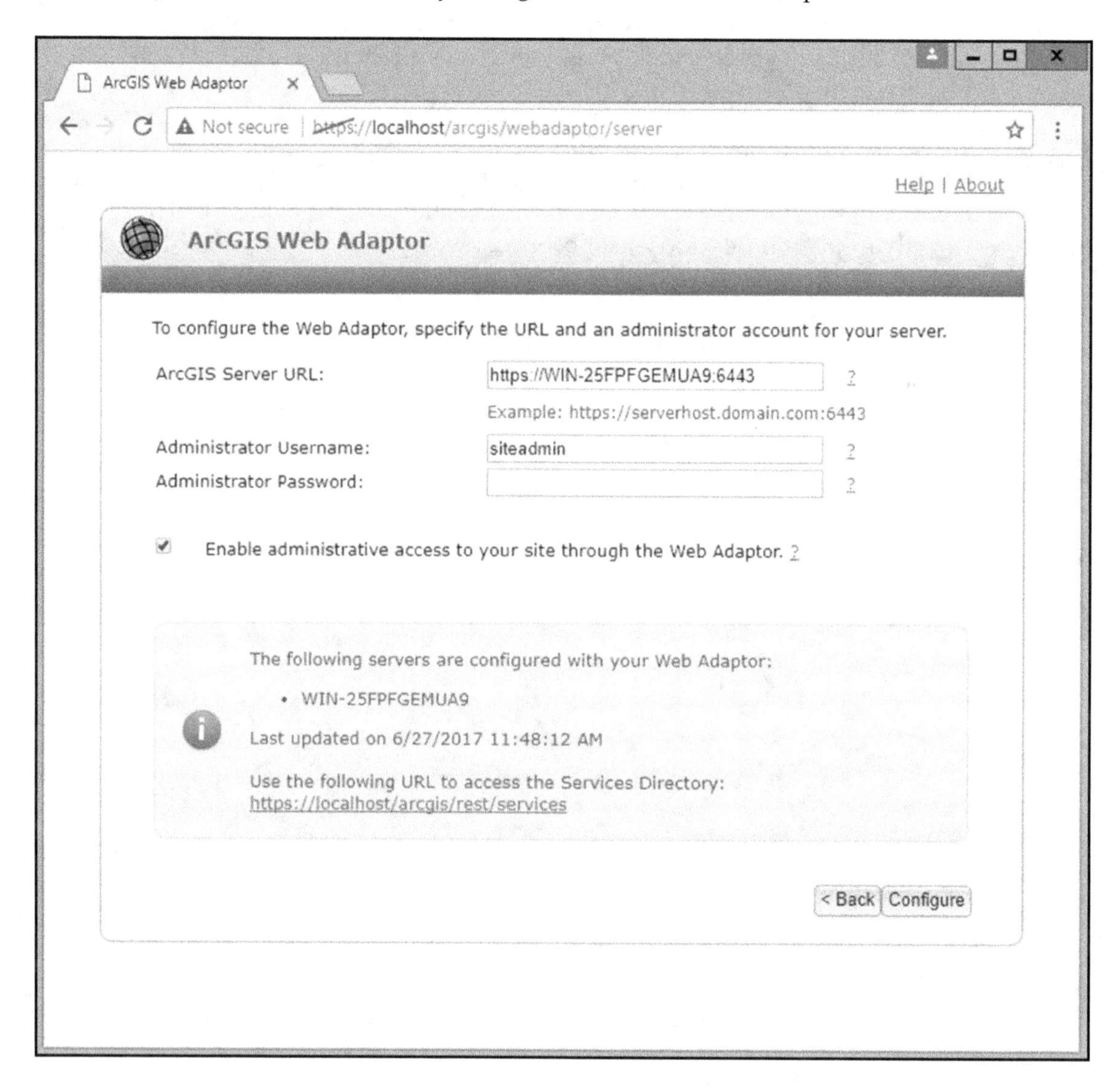

For a secure production environment, it is not recommended to allow administrative access through the same Web Adaptor used to host the REST services. Rather, install a second Web Adaptor with administrative access enabled through the Web Adaptor, possibly on an internal server that is only accessible to local users. This configuration ensures that public users are not presented with the option to access the ArcGIS Server Manager application. If an internal server is not available, a second Web Adaptor with additional security applied to it (Integrated Windows Authentication) that only publishers/administrators have access to can be installed on the same server.

Once the Web Adaptor is successfully configured, you can access your ArcGIS Server site without the port number, such as `http://www.masteringageadmin.com/arcgis/rest/services`.

Installing Portal for ArcGIS

As stated earlier, you can think of Portal for ArcGIS as being like an on-premise version of ArcGIS Online. Portal for ArcGIS is a website hosted on your network that serves as a repository for and gateway to your GIS data and content.

System and hardware requirements

Before diving into installation, let's first talk about system requirements. These have changed since earlier versions, so refer to the online documentation carefully for details and ensure that your hardware meets the minimum requirements.

Operating systems

Portal for ArcGIS is supported on Windows Server 2016 Standard and Datacenter 64-bit; Windows Server 2012 R2 Standard and Datacenter 64-bit; Windows Server 2012 Standard and Datacenter 64-bit; Windows Server 2008 R2 Standard, Enterprise, and Datacenter 64 bit; and Windows Server 2008 Standard, Enterprise, and Datacenter 64 bit. Windows 10, 8.1, and 7 64-bit are also supported *for basic testing and application development only, not for production environments*.

Portal for ArcGIS is not supported on the 32-bit operating systems.

Hardware

Portal for ArcGIS 10.5 requires one four-core processor for every 100 concurrent users, 8 GB of RAM, and 10 GB of disk space minimum for installation.

If you plan on using Insights for ArcGIS with your ArcGIS Enterprise system, you will need between 16 and 32 GB of RAM on your Portal server.

Ports

Like ArcGIS Server, Portal communicates through several predetermined ports. You must ensure that your firewall allows traffic through these ports:

- HTTP port `7080`: This is the main HTTP communication port for Portal for ArcGIS
- HTTPS port `7443`: This is the default port used to send encrypted information, such as user credentials
- Intermachine communication ports: `5701`, `7005`, `7099`, `7199`, `7654`, `7120`, and `7220` are used by Portal for ArcGIS for intermachine communications and must be allowed by your firewall

SSL

Again, much like ArcGIS Server, Portal for ArcGIS comes preconfigured with a self-signed server certificate suitable for installations and initial testing. However, Portal for ArcGIS requires that you **must** request a SSL certificate from a trusted certificate authority and configure your Portal to use it. This is especially important if you will be federating your ArcGIS Server with your Portal.

ArcGIS Web Adaptor

The ArcGIS Web Adaptor is a **required** component of Portal for ArcGIS; Portal for ArcGIS cannot be deployed without the Web Adaptor, unless you are implementing Portal in a highly available configuration with a load balancer.

Note that for Portal for ArcGIS, you need to install the ArcGIS Web Adaptor *again* to create a Web Adaptor for Portal.

Portal for ArcGIS installation

Much like with ArcGIS Server, the Portal for ArcGIS installation process is a simple and straightforward one. Double-clicking on the installation executable launches the installation process:

1. Choose a well-known temporary location to extract the installation files to. After the extraction is complete, select **Launch the setup program**.
2. Accept the license agreement.
3. Change the installation destination folder, if necessary.
4. Change the Portal configuration store location, if necessary.
5. Specify the Portal for the ArcGIS service account. This is the Windows account that the Portal for ArcGIS Windows service runs under. This can be either a local account or a domain account. For production systems, it is recommended to use a Windows domain account.
6. Optionally, save an installation configuration file.
7. Install the Portal for the ArcGIS software.

After installation of the Portal for the ArcGIS software is complete, it must be authorized for use. The Software Authorization Wizard launches automatically after the installation is complete, but can also be launched from the Start menu as **Software Authorization for Portal for ArcGIS**. As with the authorization of ArcGIS Server, there are several ways to complete the authorization, but, for Portal for ArcGIS, a common method is to enter ECP numbers for your Level 1 and Level 2 Portal for ArcGIS entitlements. In the Software Authorization Wizard, do the following:

1. For the authorization option, select **I have installed my software and need to authorize it**.
2. Select **Authorize with Esri now using the Internet**.

3. Enter in pertinent organizational information related to licensing.
4. Enter in your ECP numbers for Level 1 and 2 users:

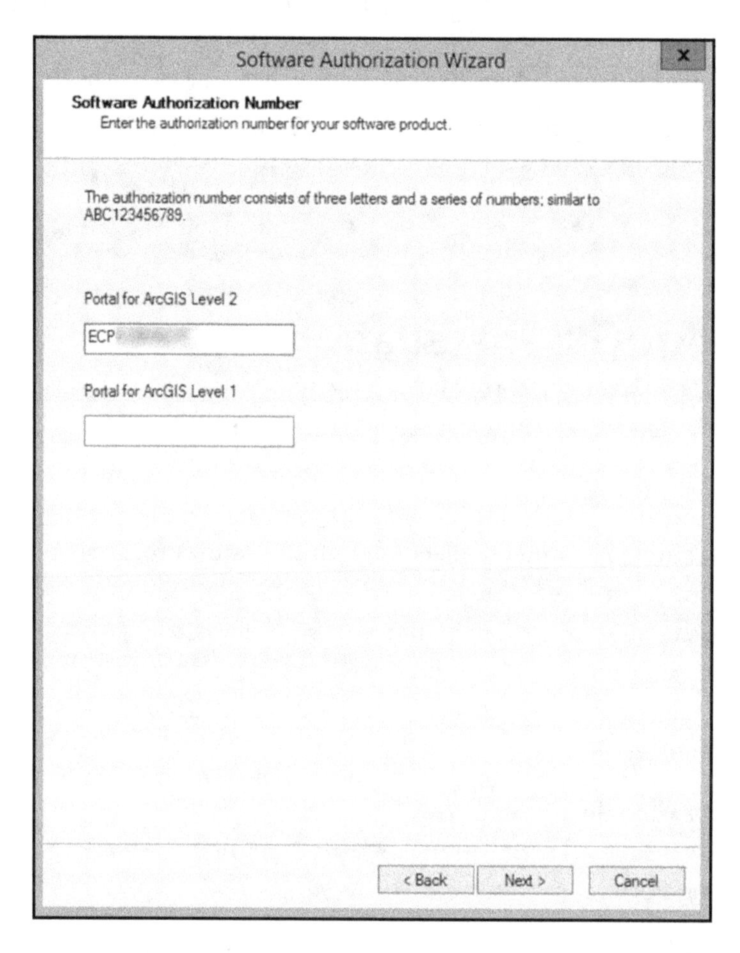

Your Portal for the ArcGIS software is now installed and authorized.

Portal for ArcGIS initial configuration

The final step in the Portal for the ArcGIS installation process is to create your Portal. Once your Portal for ArcGIS software authorization completes, your default web browser will launch and prompt you to create or join a Portal. The URL will be in the form of `https://<machine name>:7443/arcgis/home/createadmin.html`. Perform the following steps to initially configure your portal:

Note that you can also get this URL from the Start menu through the **Portal for ArcGIS** app link.

1. If you are creating a new Portal instance, select **Create New Portal**.
2. Create the Portal for the ArcGIS **primary site administrator** (**PSA**) account. This account will be the first administrator account created for your Portal. You can add additional administrator accounts later, but this account will be used to initially log on to your Portal. Common practice is to name this account `portaladmin`. Your Portal content directory is automatically chosen for you based on your installation location:

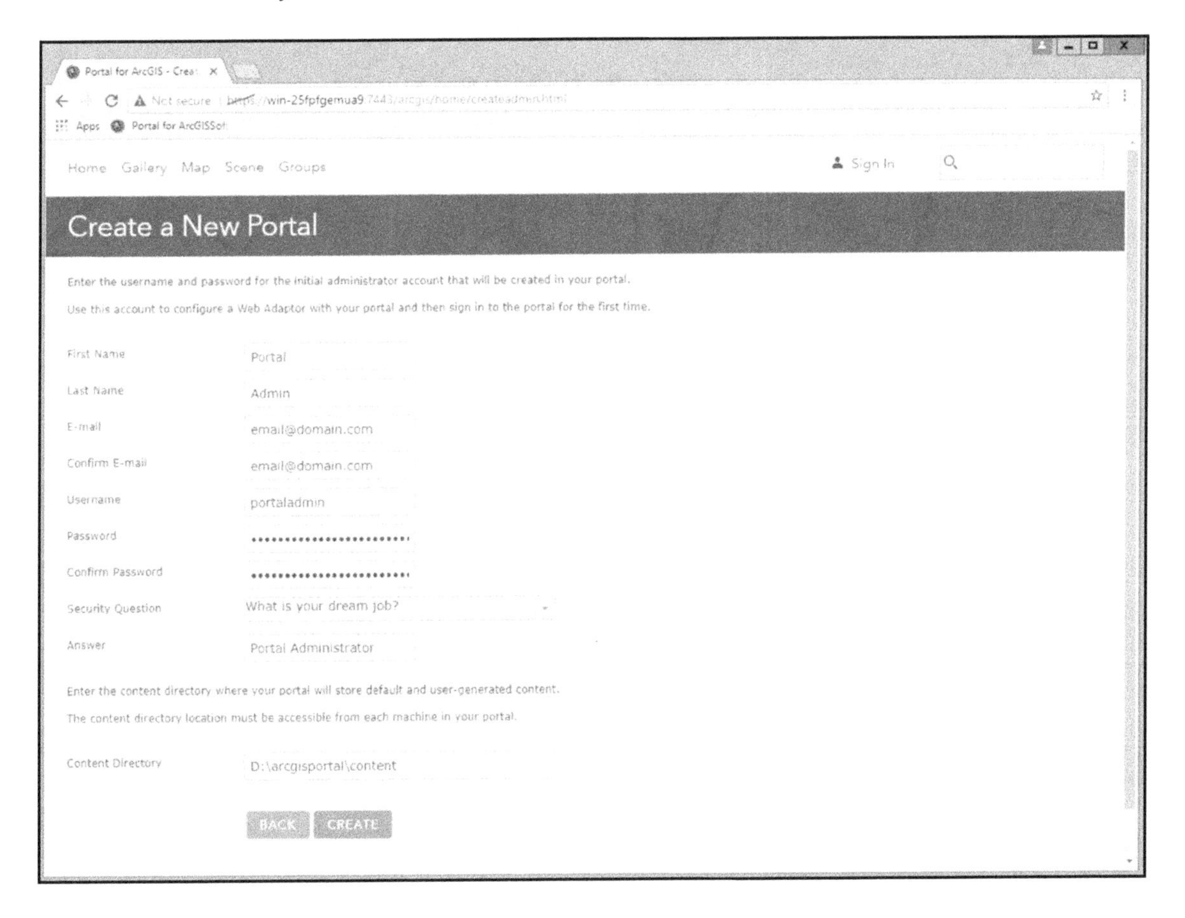

2. Click on **Create**. It may take several minutes to create your Portal and PSA account, after which you will be informed that you must install and configure the ArcGIS Web Adaptor for your Portal.

ArcGIS Web Adaptor for Portal for ArcGIS

Much like with the Web Adaptor for ArcGIS Server, a Web Adaptor is needed with Portal for ArcGIS to forward incoming traffic over port `443` to `7443`, the port which Portal listens on.

Installing the ArcGIS Web Adaptor for Portal for ArcGIS

The ArcGIS Web Adaptor comes as a separate installer you can download from `http://my.esri.com`. A *completely additional, separate installation* of the ArcGIS Web Adaptor is required for Portal for ArcGIS, in addition to any you already have installed for ArcGIS Server.

Requirements

For the requirements for the ArcGIS Web Adaptor, see the earlier section under *ArcGIS Web Adaptor for ArcGIS Server*.

Web Adaptor for Portal for ArcGIS installation

Installation of the Web Adaptor for Portal is identical to the installation done previously for ArcGIS Server (see the preceding section, *Web Adaptor for ArcGIS Server installation*), except for one step. Your Portal Web Adaptor must have a different name from your ArcGIS Server Web Adaptor (we named our `arcgis` earlier--standard practice). Standard practice is to name your Portal for ArcGIS Web Adaptor `portal`.

Portal for ArcGIS Web Adaptor configuration

See the earlier section, *Web Adaptor for ArcGIS Server configuration*, for more details on the Web Adaptor configuration parameters. To configure the Portal for ArcGIS Web Adaptor, do the following:

1. Select the **Portal for ArcGIS** radio box.

2. Your Portal URL must be the fully qualified domain name and port to your Portal. This URL must be reachable from the server you are installing your Portal Web Adaptor to. This means that all required ports for Portal must be open inbound on your Portal server. In our case here, our Portal URL is `https://www.masteringageadmin.com:7443`. The administrator username and password are your Portal PSA--typically, `portaladmin`:

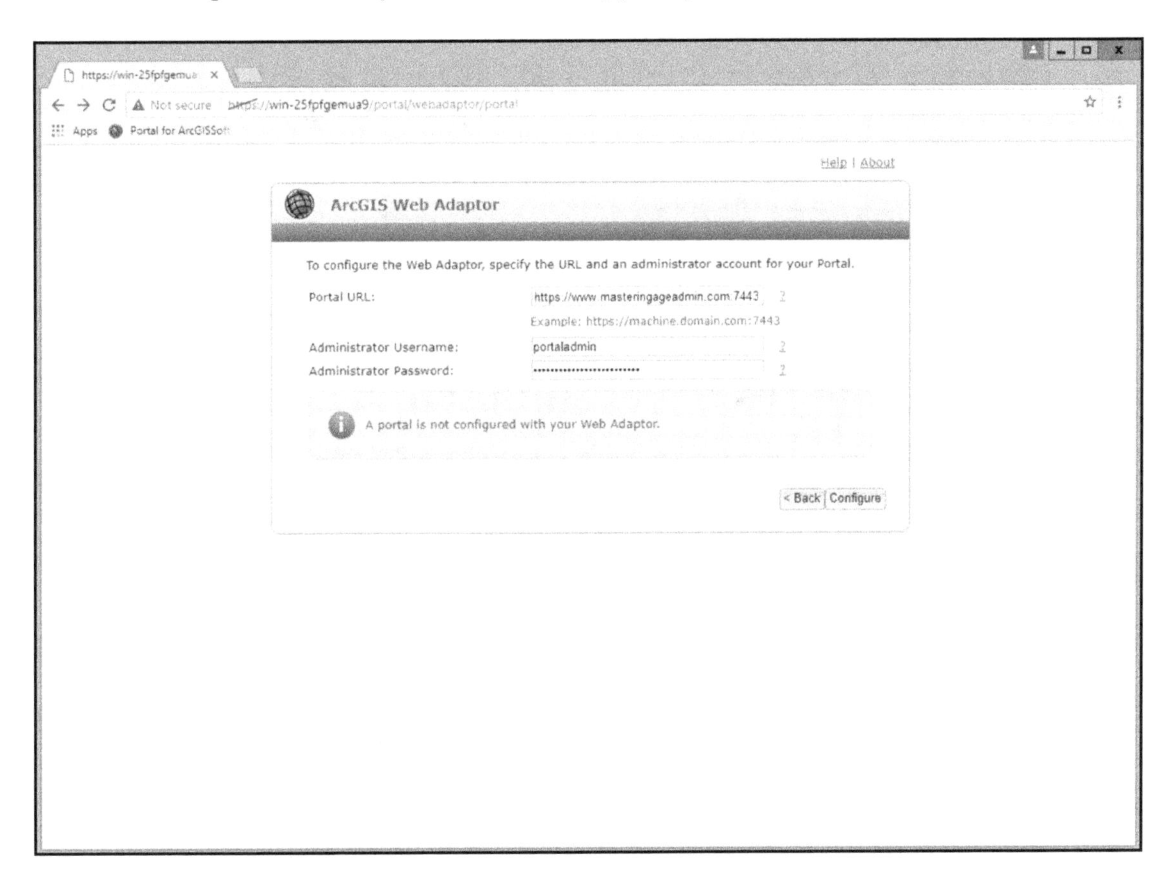

3. Upon successful configuration, you will be informed of the machine that has been configured with your Web Adaptor.

With the configuration of the ArcGIS Web Adaptor, your Portal for ArcGIS installation and initial configuration is now complete. You can proceed to your Portal from your Portal server at `https://<machine name>:7443/arcgis/home` or, externally, at your fully qualified URL, such as `https://www.masteringageadmin.com/portal`, and log in as your Portal PSA.

Installing ArcGIS Data Store

ArcGIS Data Store is an application to host data within your Portal. It provides a relational data store for your Portal's hosted feature data, a tile cache data store for storing your Portal's hosted scene layer caches, and a spatiotemporal big data store for storing observational data to use with ArcGIS GeoEvent Server and to store results generated from ArcGIS GeoAnalytics Server.

Some of the benefits of the ArcGIS Data Store include the following:

- **Publishing large numbers of hosted feature layers**: The ArcGIS Data Store relational data store can efficiently host thousands of feature layers with a smaller memory footprint, thus requiring less resources
- **Archiving high volume, real-time data**: With ArcGIS GeoEvent Server, you can use a spatiotemporal big data store to archive GeoEvent observation data

System and hardware requirements

As with the other components of ArcGIS Enterprise, system and hardware, minimum requirements must be met.

Operating systems

ArcGIS Data Store is supported on Windows Server 2016 Standard and Datacenter 64-bit; Windows Server 2012 R2 Standard and Datacenter 64-bit; Windows Server 2012 Standard and Datacenter 64-bit; Windows Server 2008 R2 Standard, Enterprise, and Datacenter 64-bit; and Windows Server 2008 Standard, Enterprise, and Datacenter 64-bit. Flavors of Windows 10, 8.1, and 7 64-bit are also supported *for basic testing and application development only, not for production environments.*

ArcGIS Data Store is not supported on 32-bit operating systems.

Hardware

Esri recommends installing ArcGIS Data Store on machines with large quantities of available disk space. The minimum amount of disk space required to install ArcGIS Data Store is 13 GB, but this does not include any data stores or backups. An *empty* relational data store alone uses up to 2.5 GB of disk space.

Ports

The ports used by ArcGIS data store are as follows:

- HTTPS port `2443`: Data Store is accessed over port `2443`
- Data store ports:
    - **Relational data stores**: Port `9876`
    - **Tile cache data store**: Ports `29080` and `29081`
    - **Spatiotemporal big data store**: Ports `9220` and `9320`
    - **Internal communication with Tomcat**: Port `9006`

ArcGIS Data Store installation

After ensuring that all the preceding requirements have been met, complete the following steps to install ArcGIS Data Store:

1. Double-click on the ArcGIS Data Store installer to begin.
2. As with all other ArcGIS Enterprise installations, choose a well-known temporary location to extract the installation files to, and then launch the setup program.
3. Accept the license agreement.
4. If you are installing to a drive other than C or to a non-default location, change the install directory accordingly.
5. Specify a Windows service account for Data Store to run under. As with ArcGIS Server and Portal for ArcGIS, this can be either a local account you create during this step of the installation process or a domain account. Best practice is to use a domain account for production systems. If using a local account, name it appropriately, such as `datastore`.
6. Continue with the installation process.

ArcGIS Data Store creation

Once the Data Store installation is complete, the ArcGIS Data Store Configuration Wizard will launch in your default web browser (`https://localhost:2443/arcgis/datastore`). Complete the following steps to configure your Data Store:

1. Enter the machine name and port to your GIS server; for example, in our case, `https://WIN-25FPFGEMUA9:6443`. Also, enter your ArcGIS Server PSA account credentials.

Do not use the Web Adaptor URL for the GIS Server URL.

2. Specify your Data Store content directory that will be used to store data, logs, and backup files. This directory should be located on the same machine that Data Store is installed on.
3. Choose the types of ArcGIS Data Stores to configure. Your choices are Relational (default), Tile Cache, and Spatiotemporal. See the preceding introductory section on *Installing ArcGIS Data Store* for more information on these Data Store types.
4. Review your configuration summary and click on **Finish**.

If your ArcGIS Server site is not federated with your Portal, you will need to do this and then set that ArcGIS Server site as your Portal's hosting server. See `Chapter 5`, *Portal for ArcGIS Administration* for more information on federation.

Summary

ArcGIS Enterprise 10.5 brings many changes to the world of ArcGIS Server and Portal for ArcGIS. Portal is now a core component along with ArcGIS Server, Data Store, and the Web Adaptor. The concept of server roles is introduced at 10.5, with former extensions now becoming added functionality to ArcGIS Enterprise as deployed in your own infrastructure. Installation of ArcGIS Enterprise consists of installing and configuring the core components. These components can live internally in your own infrastructure on physical or virtual hardware, in the cloud, or a combination of the two. Configuration options abound and it is important to find the optimal setup for your organization's needs. Now that core software is installed, next, in `Chapter 2`, *Enterprise Geodatabase Administration,* we will look at how to go about creating, configuring, loading data into, and maintaining an enterprise geodatabase.

2
Enterprise Geodatabase Administration

At the heart of any good enterprise GIS system lives a clean, tidy, and performant enterprise geodatabase. The geodatabase is the core of a strong GIS system; without data, you have nothing. Likewise, a poorly installed, configured, or maintained geodatabase leads to disappointing applications for end users. Proper installation, configuration, tuning, maintenance, and administration of the geodatabase is crucial to the health and usability of a GIS.

Before we can cover enterprise geodatabase administration, we first need to discuss what makes an enterprise geodatabase; how we install, create, or enable one; how we connect to it; and how we load data into it. Also, keep in mind that, in no way can we cover all aspects of Enterprise geodatabase administration in one chapter; an entire book could be dedicated just to this topic. Instead, this chapter will highlight several aspects of installing, configuring, and maintaining an SQL Server enterprise geodatabase. Keep in mind that many of the principles covered can be applied to other RDBMSs as well.

After the completion of this chapter, you will know how to install, utilize, manage, and maintain an ArcGIS enterprise geodatabase.

This chapter will cover the following topics:

- What exactly is an enterprise geodatabase?
- Installation and configuration of the RDBMS (SQL Server 2014)
- Creating or enabling an enterprise geodatabase
- Database connections
- Loading data
- Users, role, and privileges

- Database maintenance

What constitutes an enterprise geodatabase?

A geodatabase is a spatially-enabled database. Within the ArcGIS Enterprise framework, there are three types of geodatabases:

- **Personal geodatabase**: This uses Microsoft Access for data storage, and it has a size limit of 2GB.
- **File geodatabase**: This uses the file system folder for storage of GIS datasets; each dataset can be 1TB in size. If not using an enterprise geodatabase, this is the recommended file-based storage type.
- **Enterprise geodatabase**: This uses a **relational database management system (RDBMS)** for data storage, supports multiple simultaneous user connections, and is limited in size by the RDBMS.

Personal and file geodatabases are intended for single users and small workgroups with one writer and multiple readers, where concurrent user connections eventually degrade performance with more and more readers. File geodatabases can have only one editor per feature dataset, stand-alone feature class, or table. For medium to large organizations needing multiple writers and larger numbers of concurrent readers, an enterprise geodatabase is the optimal choice.

Using an enterprise geodatabase allows you to do the following:

- Bring your own Esri-supported RDBMS license for use in the ArcGIS Enterprise ecosystem
- Be limited on database size and number of connections only by your RDBMS
- Handle security within the RDBMS
- Utilize the RDBMS functionality such as versioning, backup and recovery, replication, SQL support, and high availability

Considering the preceding functionalities that an enterprise geodatabase offers, it is easy to see the advantages offered by an enterprise geodatabase. At 10.5, ArcGIS Enterprise supports the following relational database management systems and versions:

Database management system	**Version(s) supported**
Microsoft SQL Server	Microsoft SQL Server 2016 (64-bit) Microsoft SQL Server 2014 (64-bit) Microsoft SQL Server 2012 SP3 (64-bit)
Oracle	Oracle 11g R2 (64 bit) 11.2.0.4 Oracle 12c R1 (64 bit) 12.1.0.2
PostgreSQL	PostgreSQL 9.5.3 (64 bit) with PostGIS 2.2 PostgreSQL 9.4.8 (64 bit) with PostGIS 2.2 PostgreSQL 9.3.13 (64 bit) with PostGIS 2.2
IDM DB2	IBM DB2 Version 9.7 Fix Pack 4 IBM DB2 Version 10.1 Fix Pack 2 IBM DB2 Version 10.5 Fix Pack 5 IBM DB2 Version 11.1 IBM DB2 Version 10 for z/OS IBM DB2 Version 11 for z/OS
IBM Informix	IBM Informix Server 64 Bit 11.70.FC4 IBM Informix Server 64 Bit 12.10.FC3

Relational database management system installation and configuration

The first step in setting up an enterprise geodatabase is to install your RDBMS. In many organizations, this is handled by someone in the IT department, such as a database administrator or systems administrator. If so, try to work with them as much as possible for the setup of your RDBMS; not only will you build a stronger working relationship with them, you will further understand how your RDBMS was installed and configured.

For this book, we will be using SQL Server 2014 Standard Edition SP2 as our RDBMS. As always, consult the documentation of your RDBMS for detailed installation and configuration instructions.

RDBMS installation

RDBMS installations can be quite lengthy with dozens of steps. For this reason, we will not cover *every* step in the SQL Server 2014 installation, but instead touch on those points that might be tricky or otherwise important regarding our GIS system. Remember to check your RDBMS documentation regarding system requirements. Some items to ensure are completed during your SQL Server installation:

- If you have a non-OS drive to install on, change the installation location to that drive.
- Install only the SQL Server features you will need. Consult the SQL Server documentation for help on choosing features. Be sure to install Management Tools so you get SQL Server Management Studio.
- During the database configuration steps, ensure the following:
    - For authentication mode, do **Mixed Mode**, which will allow both the SQL and Windows accounts to log in to the database server. Also, add any Windows user accounts that you want to be a member of `sysadmin` to `sysadmin` as well.
    - Like you did for the installation location, set your data directories to your non-OS drive if you have one.

Creating or enabling an enterprise geodatabase

To create a geodatabase, you must use ArcGIS Desktop licensed at either Standard or Advanced level, ArcGIS Pro Standard or Advanced, or a Python script on a machine with the proper level of Desktop or Pro installed. There are two ways to create an enterprise geodatabase in SQL Server, depending upon your level of access to the database:

- You create the enterprise geodatabase using the Create Enterprise Geodatabase geoprocessing tool. Here, you are both the SQL Server database and geodatabase administrator. This option applies if you installed SQL Server and/or you have `sysadmin` access to the SQL Server instance.

In SQL Server, the database administrator owns everything in the entire SQL Server instance. The geodatabase administrator, on the other hand, owns only the objects within a geodatabase.

- Your SQL Server administrator creates the database and the geodatabase administrator (you) creates the geodatabase.

Let's break down these options in further detail.

For either of the following scenarios, you will need an ArcGIS Server (Enterprise Advanced or Enterprise Standard) keycodes file to authorize your geodatabase. Your **keycodes** file will be located at **C:\Program Files\ESRI\License10.5\sysgen**, even if you installed ArcGIS Server to a drive other than C. You may have to copy the **keycodes** file from your ArcGIS Server machine to somewhere you can access it from the toolbox tools. Feel free to copy it somewhere with your software installers or keep it in your secure password manager of choice. (You *do* use a password manager, *don't you?*)

Creating an enterprise geodatabase

If you are both the SQL Server administrator (or you have been placed in a sysadmin role) and the geodatabase administrator and have access to ArcGIS Desktop Standard or Advanced on the database server, you can easily use the Create Enterprise Geodatabase tool to create a geodatabase from scratch. This is *by far* the easiest and quickest method of creating an enterprise geodatabase.

If you do not have the Desktop client installed on your database server, you can still create an enterprise geodatabase from another one of your servers that does have the proper Desktop client. You may, however, need to install the proper database drivers to connect to the database. See the following *Connecting to the geodatabase* section for more details.

SDE versus Dbo schema

Before proceeding, let's discuss schema ownership in your soon-to-be enterprise geodatabase. We've been discussing the geodatabase administrator, which is the user that owns the geodatabase. In SQL Server, the geodatabase administrator can be either a database user named `sde` or the `dbo` database user. The user you connect with to create your enterprise geodatabase is the owner of your geodatabase. A full list of the differences between these two ownership models can be found at the ArcGIS Enterprise online docs by searching for **comparison of geodatabase owners in SQL Server**. In the following section, we will summarize some of the major differences and advantages/disadvantages of these two schemas.

Dbo schema

To create an enterprise geodatabase with a `dbo` schema, you will configure and execute the **Create Enterprise Geodatabase** tool as shown in the following screenshot:

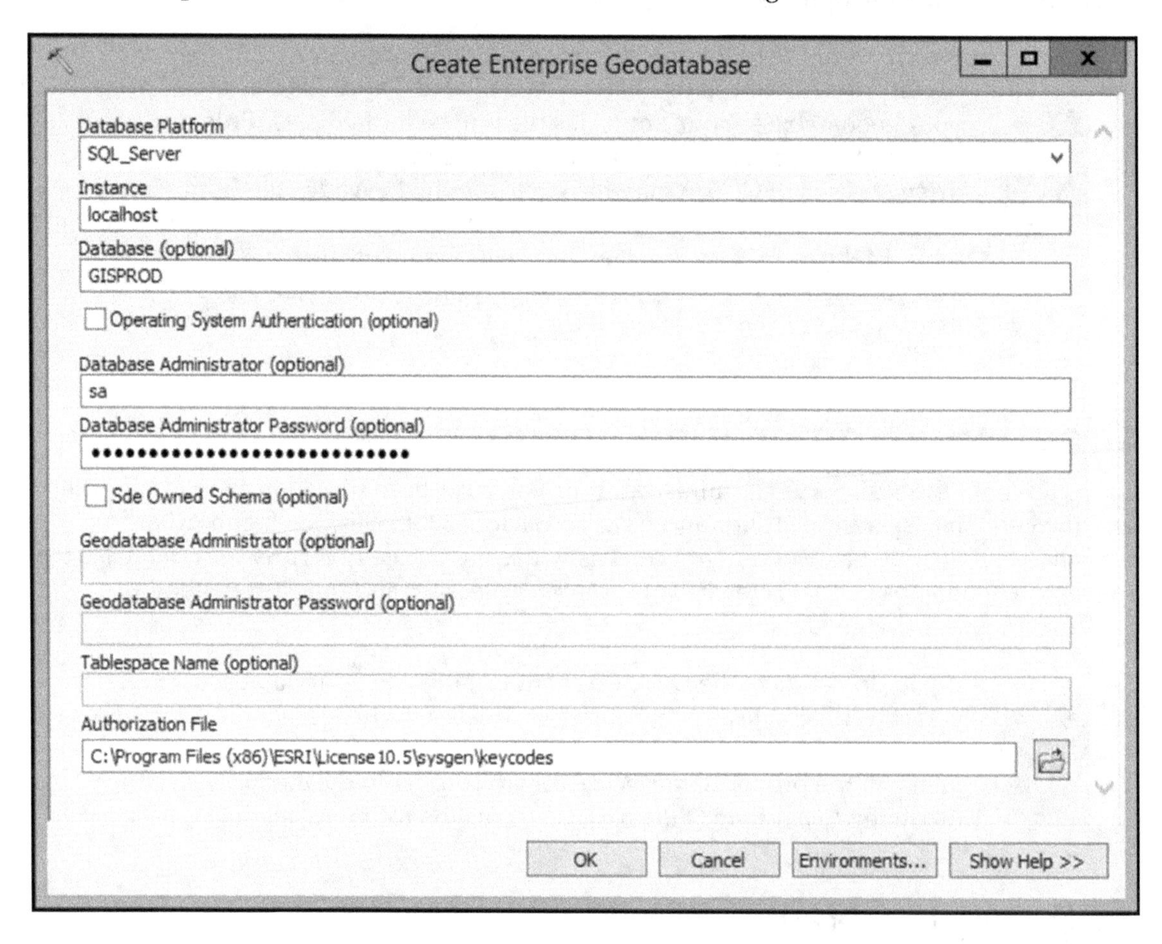

Our sa database admin user is a member of sysadmin, and we didn't specify to use an SDE owned schema. We end up with the following setup:

- Our enterprise geodatabase is owned by the `dbo` user. An account specifically for owning data can also be created.
- A `dbo` schema for our database, meaning all SDE system tables, will be prefixed with `dbo`.
- Any user who is a member of the `sysadmin` fixed server role can create data such as feature classes, feature datasets, or tables, in the `dbo` schema
- Since all it takes to become a geodatabase administrator is membership to the `sysadmin` fixed server role, it is possible to have multiple geodatabase administrators.

You have probably figured out by now that the `dbo` schema allows for a loose management model; you can have many geodatabase administrators, either by design or by accident. If you have a small department or group of users, then using the `dbo` schema poses little risk. Do not confuse "loose" with insecure; in many large organizations, the database administrators would have `sysadmin` privileges and would therefore be geodatabase administrators. There are pros and cons to this. Not all of these DBAs might be in charge of the geodatabase, but the ones that are would not need to have a second account (`sde`) to manage it. I have SQL Server 2014 Standard Edition in a local development environment. My Windows login is a member of sysadmin at the database-server level. I routinely standup quick enterprise geodatabases for client projects and use the `dbo` schema with a data loader account. With this setup, it is quick and easy to setup and administer enterprise geodatabases for development and testing purposes where I will be the only person using them.

SDE schema

To create an enterprise geodatabase with an sde schema, you will configure and execute the Create Enterprise Geodatabase tool like the following:

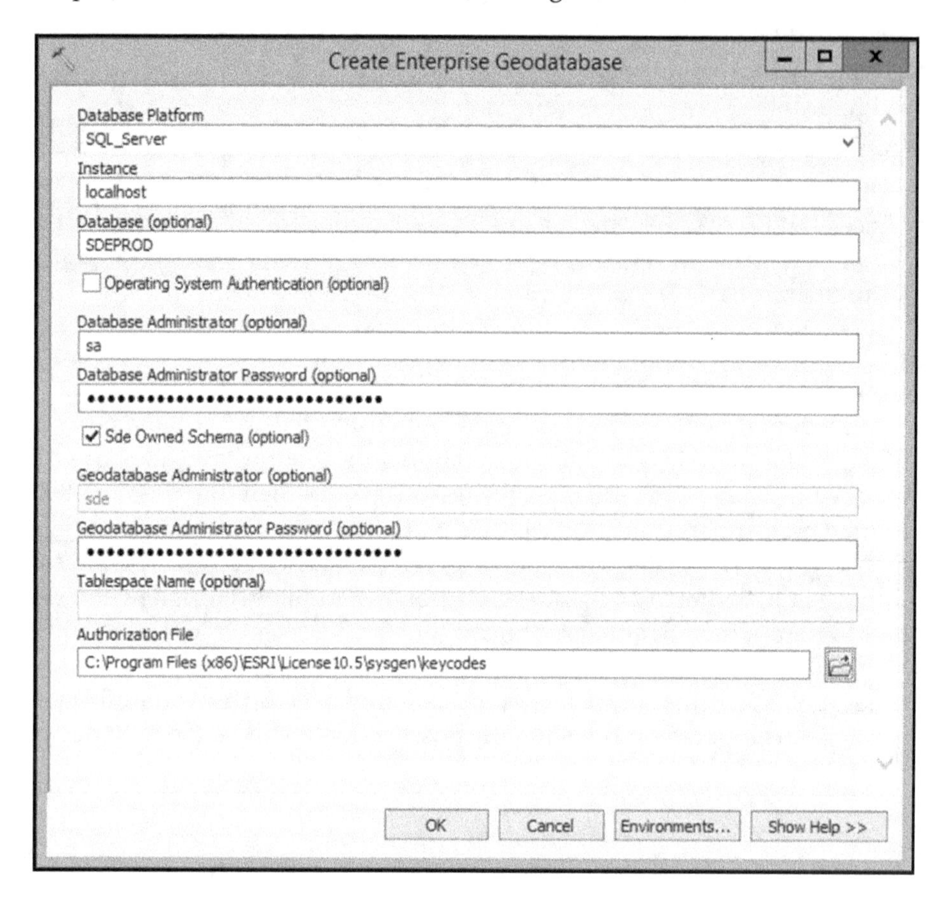

Here, we will specify the `sa` account and the `sde` account. When we specify that we want an sde-owned schema, the Geodatabase Administrator textbox gets prepopulated with `sde`. If your database does not already have an sde account, specify a strong password and the sde user and schema will be created for you. It is also possible to connect to the database with the **operating system** (**OS**) authentication to create the geodatabase. To do so, your current login must be in the database sysadmin fixed-server role. After you have all parameters entered, hit **OK**, and the tool will create your enterprise geodatabase for you as follows:

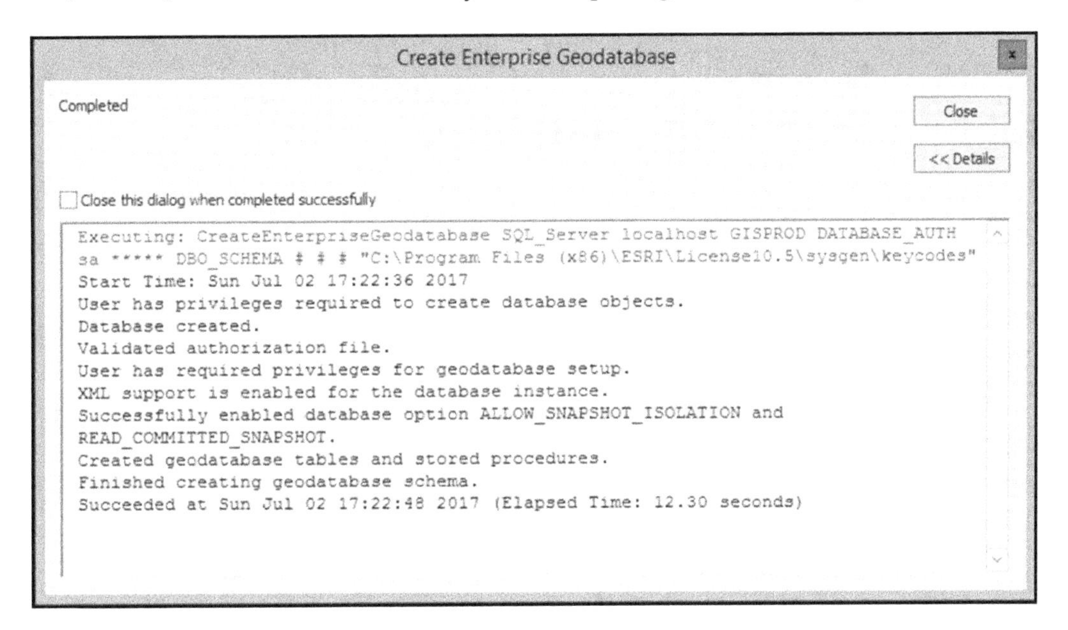

With this configuration, we get the following enterprise geodatabase setup:

- Our enterprise geodatabase is owned by the `sde` user
- We are free to create multiple owner accounts, each with their own schema, for data ownership (remember that the `sde` user should **never** own your GIS data, but since the schema is owned by `sde`, all SDE system tables are owned by the `sde` user)
- Users can only load data by logging in with a data owner account

As you can see, there are many factors to consider when deciding which schema to use. If databases administrators will be managing the geodatabase, then a `dbo` schema may be better as it simplifies the requirements for a being a geodatabase administrator; the user must be either a member of `sysadmin` or database owner. This method also allows for the use of existing accounts as owner only. On the flip side, if the GIS team will be managing the geodatabase, it may be better to use an `sde` schema. This provides for one account (`sde`) with elevated privileges and keeps from needing to grant elevated privileges to other accounts.

Enabling an existing database

In our previous geodatabase creation scenario, the geodatabase administrator is also the database administrator, allowing for the ease of use of the Create Enterprise Geodatabase tool. However, what if you are the geodatabase administrator but *not* the database administrator? In other words, you have the credentials to create the geodatabase, but not the initial *database* in the RDBMS. When this is the case, your database administrator will first have to create the database and geodatabase administrator's login, user, and schema.

For a full list of steps to be carried out by your database administrator in this scenario, search the ArcGIS Enterprise online help for **Create an enterprise geodatabase in SQL Server**.

After your database administrator has created your database for you, you can proceed with *enabling* the geodatabase functionality within it with the Enable Enterprise Geodatabase tool. Now that your database administrator has done all of the heavy lifting with the creation of the database, geodatabase administrator's login, user, and schema (things that in our previous scenario were done with the Create Enterprise Geodatabase tool), you can enable the database by following these steps:

1. In ArcMap, ArcCatalog, or ArcGIS Pro, connect to the database as the geodatabase administrator.
2. Search for and find the **Enable Enterprise Geodatabase** tool.
3. Drag your geodatabase administrator connection to the **Input Database Connection** field.

You can also right-click on the geodatabase administrator connection and select **Enable Geodatabase**.

4. Browse to your ArcGIS Server authorization (keycodes) file to add it to the **Authorization File** parameter:

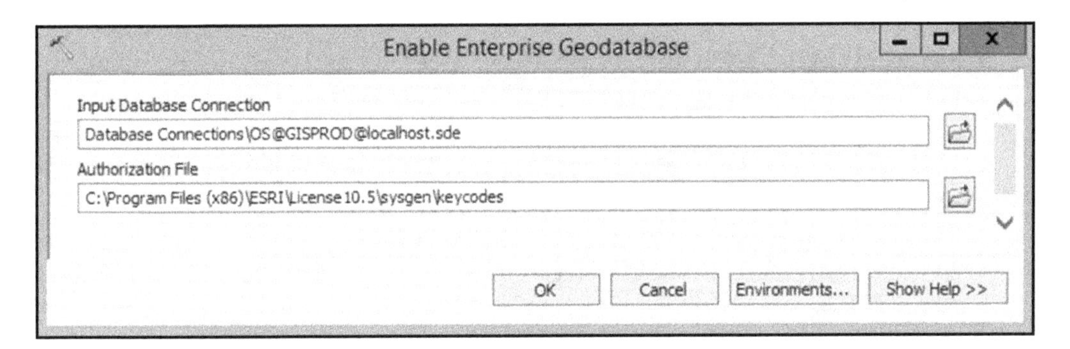

5. Click on **OK** to execute the tool and enable geodatabase functionality within the database.

Connecting to the geodatabase

Now that we have a shiny new enterprise geodatabase, we need to connect to it. A connection allows us to use, manage, and administer the geodatabase. Before we can connect, there are a few items to configure.

To allow connections from machines other than the SQL Server machine itself, we must ensure that remote connections to the database server are allowed. To do this, first open SQL Server Management Studio on your database server and log in with the `sysadmin` credentials. In the **Object Explorer** pane, right-click on the database server and go to **Properties**. In the **Properties** window, select the **Connections** page. Under remote server connections, ensure that **Allow remote connections to this server** are checked. Click on **OK**.

Next, open SQL Server Configuration Manager and under **SQL Server Network Connection**, select **Protocols for <your server>**. Ensure that **TCP/IP** is enabled. Right-click on **TCP/IP** and go to **Properties**. Select the **IP Address** tab and scroll down to the **IPALL** section. Note the TCP port listed here; `1433` is the default port for SQL Server. If your instance of SQL Server is running on a non-standard port, you will need the port number later, when we connect to the geodatabase.

For your PC to connect to the remote SQL Server instance, you will need a piece of software known as a **client**. The client contains drivers that allow your PC to connect to the enterprise database server. On a 64-bit operating system, you must install the 64-bit SQL Server native client.

Earlier, we discussed that ArcGIS Enterprise 10.5 supports Microsoft SQL Server 2012 SP3, SQL Server 2014, and SQL Server 2016. For these versions of SQL Server, there are three clients available:

- **Microsoft SQL Server 2012 SP3 Native Client (32 and 64-bit)**: This client will work for connecting to a SQL Server 2012 SP3 instance.
- **Microsoft ODBC Driver 11 for SQL Server**: This client will connect to Microsoft SQL Server 2005, 2008, 2008 R2; SQL Server 2012; and SQL Server 2014 databases.
- **Microsoft ODBC Driver 13.1 for SQL Server**: This client will connect to Microsoft SQL Server 2008, SQL Server 2008 R2, SQL Server 2012, SQL Server 2014, and SQL Server 2016 databases. Note that this client is only supported by ArcGIS 10.5.x and 10.4.1 clients. ArcGIS 10.4 clients must use Microsoft ODBC Driver 11.

These clients can be downloaded at `http://my.esri.com` or directly from Microsoft.

- Once you have your database open to connections and have your client drivers installed, you can add a database connection in ArcMap, ArcCatalog, or ArcGIS Pro. In the catalog tree of either ArcMap or ArcCatalog, expand Database Connections and double-click on **Add Database Connection**. To connect to an SQL Server geodatabase instance, do the following:

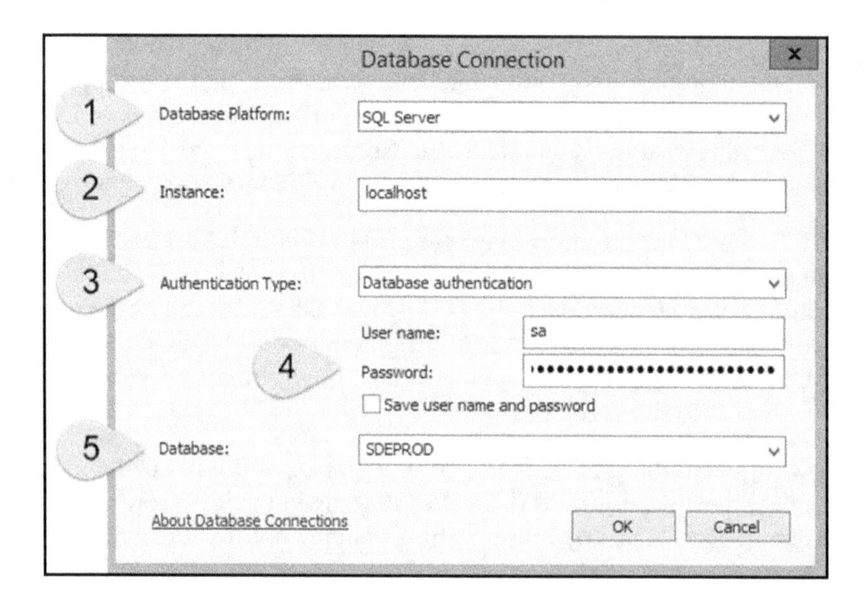

1. From the **Database Platform** dropdown, select **SQL Server**.
2. **Instance** is the name of your database instance. Here, I am connecting to a local SQL Server instance on the same machine, so I can simply use **localhost**. If your instance is a named instance other than the default SQL Server instance, connect to it as `<server name>\<instance name>`, such as `gisprod\gis`. If your database is listening on a port other than the default `1433` SQL Server port, include that in the instance name as well, separated from your instance name by a comma, for example, `gisprod\gis,15000`, for example.
3. There are two choices for **Authentication Type**: **Database authentication** and **Operating System** (**OS**) Authentication. In the preceding, we are connecting as the database `sysadmin`, so we will use **Database authentication**.
4. Enter the proper credentials you intend to connect with. Note that if you chose OS authentication, this option is greyed out and the credentials of the currently logged in Windows user are utilized to connect to the database. Here, you can also choose to save credentials with the connection you are creating. This is fine to do with viewer and editor accounts, but it is highly recommended not to save credentials for `sde` or `sysadmin`-level connections. Not only it is to protect your database from unintended elevated access by others, but it is a measure to keep you yourself from accidentally connecting as an elevated user and possibly carrying out unintended actions.
5. Once you have provided all the preceding parameters, a connection to the database instance is attempted. If the connection is successful, the **Database** dropdown will get populated with a list of databases available to the credentials provided. Choose the database you wish to connect to.

Users, roles, and privileges

Within the geodatabase, there is a hierarchy of users, with each level being based on what actions the user can perform. We have talked at length about the most powerful two of these users, the database administrator and geodatabase administrator. These users are vital to the creation, management, and maintenance of the enterprise geodatabase. As the following diagram shows, with great power, there must also come great responsibility. Database administrators and geodatabase administrators are both powerful accounts with far-reaching privileges. The following diagram shows that with increased privileges in the database, come increased responsibilities:

Remember that the geodatabase administrator account should never own data in the geodatabase.

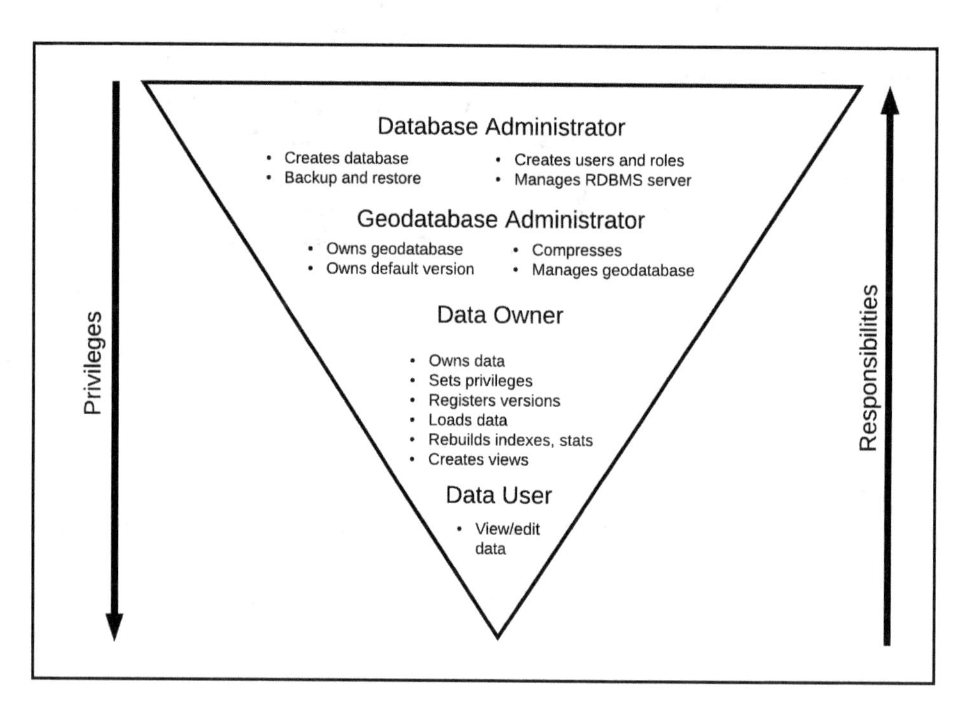

The data owner account

Another important account is the data owner; this account owns the schema and therefore the data, sets privileges, performs maintenance tasks, and probably most importantly, loads data into the geodatabase. For the data owner account, it is best to create a *headless* user account, an account that is not assigned to any person. If necessary, these credentials can be shared among those individuals trusted to create data in the geodatabase. If one staff member is unavailable, another has the necessary credentials to perform data maintenance.

Creating a data owner account

The data owner account is most commonly a database account. However, if your organization only allows OS authentication to the database, you can map an OS user login to a headless database user (the owner account). You could then log in to the database with the OS credentials, but keep the data owned by a headless database user. From the preceding diagram, note that only database administrators can create users. The database administrator can use the Create Database User tool to create users in the geodatabase without using any RDBMS tools. The Create Database User tool will handle granting the required privileges in the database and is the recommended method for creating database users. In SQL Server, the tool grants the following privileges:

- Create procedure
- Create view
- Create table

To create a data owner account, first create a connection to your geodatabase as `sysadmin`. Note that in the following connection, we are not saving **Password** for the `sysadmin` account:

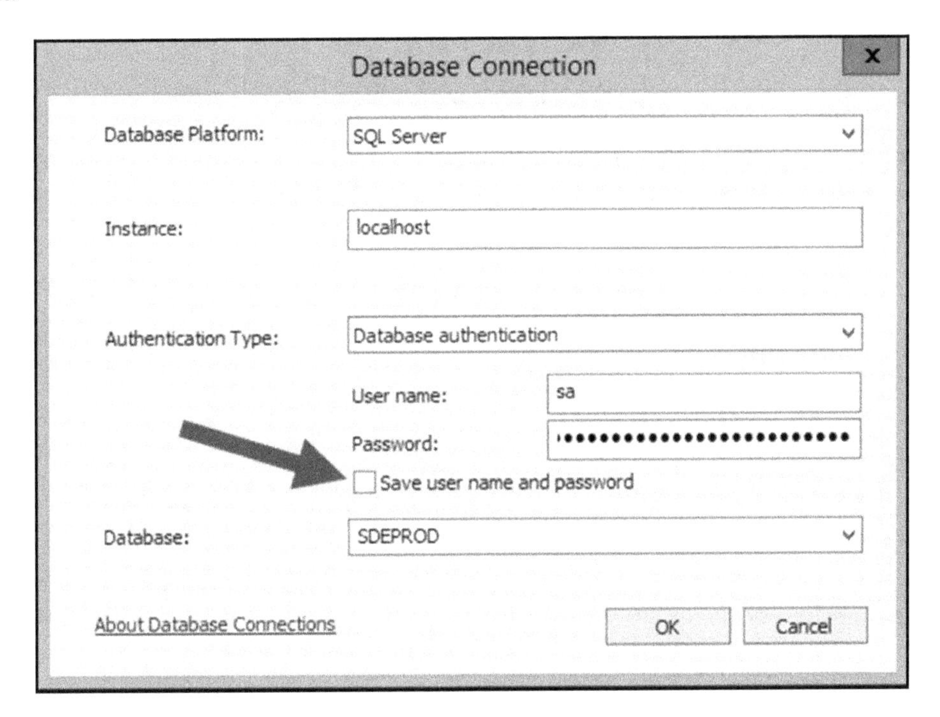

Not saving the username and password is a recommended security measure--remember that the `sysadmin` account has the highest of privileges. By not saving **Password** with the connection, you are protecting yourself from accidentally connecting as `sysadmin`.

Finding toolbox tools can get frustrating. As the screenshot below shows, you can use the Search tool to easily find the tool you are looking for. In either ArcMap or ArcCatalog, go to **Window** | **Search** and the **Search** window will get docked in the far-right panel. Select **Tools** and type part of the tool you are looking for. This can be a huge time saver:

Next, find and launch the **Create Database User** tool. Drag and drop your sysadmin `.sde` connection from **Database Connections** in the Catalog tree into the **Input Database Connection** input parameter:

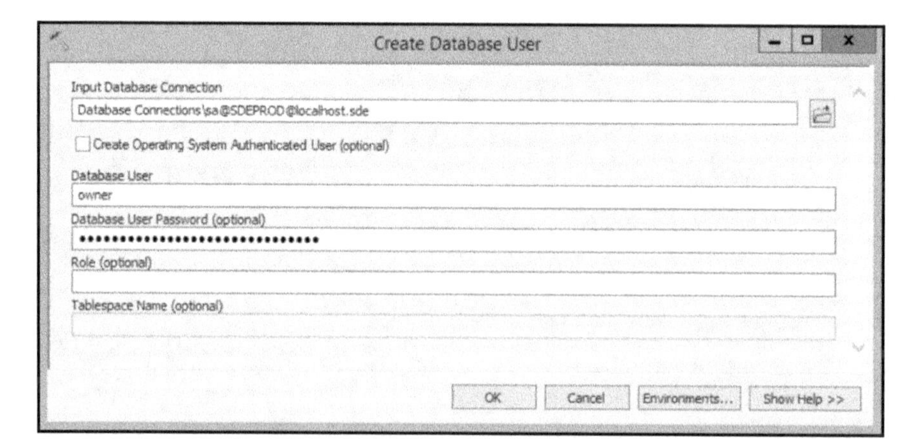

Next, enter the owner account name you would like to create. Choose the name wisely, as this account will own all data that is loaded with it, meaning that this account will own the data's schema as well. Common names for this account are *gis*, *owner*, or the organization or business function (*utilities*, *planning*, *fire*, and so on). For example, for a feature class named `wHydrants` in a database named `gisprod` loaded with a user named `gis`, the fully qualified table name would be `gisprod.gis.wHydrants`.

Data user accounts

The final level of database accounts is the user level. User level consists of viewers and editors in the geodatabase, where viewers have only `SELECT` privileges and editors can have `SELECT` privileges along with any combination of `UPDATE`, `INSERT`, or `DELETE` privileges. These accounts are indeed the users of the database--those creating and updating features and those utilizing your data for viewing and analysis. User accounts can be created using the **Create Database User** tool with a `sysadmin` connection, as was done by the data owner, or by a database administrator utilizing SQL Server tools.

Database versus operating system authentication

An often-deliberated topic when designing a GIS system is whether to utilize operating system or database authentication. In some organizations, this is a decision that is made outside the purview of the geodatabase administrator. OS and database-level authentication each have their own advantages, drawbacks, and use cases from organization to organization.

Database authentication

Much of what we have discussed has involved database authentication. With database authentication, the database administrator creates users in the database using either Esri tools (Create Database User tool) or database tools. The pros and cons of this method are as follows:

Pros

- It can connect from the same machine as multiple users
- Connection files can be saved with credentials and shared, allowing administrators to supply users connections without providing the actual credentials

Cons

- Connection files can be saved with credentials and shared. This is a good reason to not create `sysadmin` or `sde` connections with saved credentials.

Use cases

Database authentication is best for any organization with a small number of users that need to connect directly to the geodatabase. Maintaining and administering a small number of database user accounts by a database administrator is oftentimes easier and more convenient than adding/removing domain users from the database. Over time, database credentials can be passed on as staffing changes occur without requiring any access changes to the database.

OS authentication

With operating system authentication, logins are handled at the domain level, typically through Active Directory user management. The pros and cons of this method are as follows:

Pros

- More secure than database logins, as OS-authenticated logins pass an access token instead of a username and password.
- User management is already being handled at domain level, so OS authentication leverages what is currently in place
- Connections are simple for end users; they are not required to enter a username and password as their current logged on credentials grant them access to the database
- Change control in geodatabase items such as feature classes and tables can be monitored using Editor Tracking and user logins.

Cons

- User management is handled at domain level, a realm that database and geodatabase administrators usually have no control over
- It cannot connect as multiple users from the same machine.

Use cases

OS authentication is best for organizations with a solid domain membership management system (such as Active Directory) in place and the need for many users to connect directly to the geodatabase. This is often utilized in larger organizations where the cost of maintaining large numbers of database-level users is too high.

Managing user connections

Managing user connections to the enterprise geodatabase is a task that goes alongside data and database management. There will always be times when administrators need access to a dataset for an update or the entire database for a compress or other routine procedure, only to find out that the dataset is locked by another user. Luckily, these tasks of finding and dealing with locks can be easily accomplished through either the user interface or ArcPy and Python.

Determining who is connected to the geodatabase

In either ArcCatalog or ArcMap, connect as a geodatabase administrator. Right-click on your **connections**, go to **Administration** | **Administer Geodatabase**, then go to the **Connections** tab. Here, you will see a list of all current connections to your geodatabase. Your current connection will be the one listed in greyed-out italics:

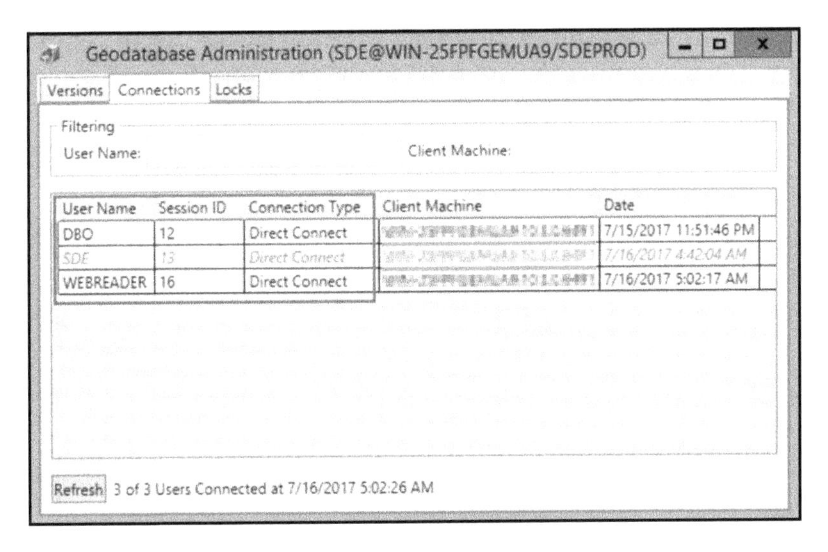

Connected users can also be shown in ArcGIS Pro. Right click the database connection, click **Properties**, expand **Connections** and then select **Show connected users and locks**.

Another way to get a list of your currently connected users is through a few quick lines of Python:

```
import arcpy
ws = r'Database Connections\sde@SDEPROD@localhost.sde'
users = arcpy.ListUsers(ws)
for user in users:
    print "{0}: {1}".format(user.ID, user.Name)
```

First, we will import the `arcpy` module. Second, we get a list of users by making a call to `arcpy.ListUsers` and passing in a geodatabase administrator connection. Finally, we loop through the list of `arcpy.utils.user` class instances returned from `arcpy.ListUsers` and print out the `Name` and `ID` of each. The preceding code lines executed against the database connection used in the preceding example returns:

```
12: DBO
13: SDE
16: WEBREADER
```

Disconnecting users

Oftentimes database and geodatabase administrators will need to disconnect users from the geodatabase in order to perform administrative functions. To disconnect users while in ArcCatalog, first connect as a geodatabase administrator user to your database. Next, right click on the **connection**, go to **Administration** | **Administer Geodatabase** and then the **Connections** tab. Finally, right-click any of the available connections and select **Disconnect** to disconnect them from the geodatabase:

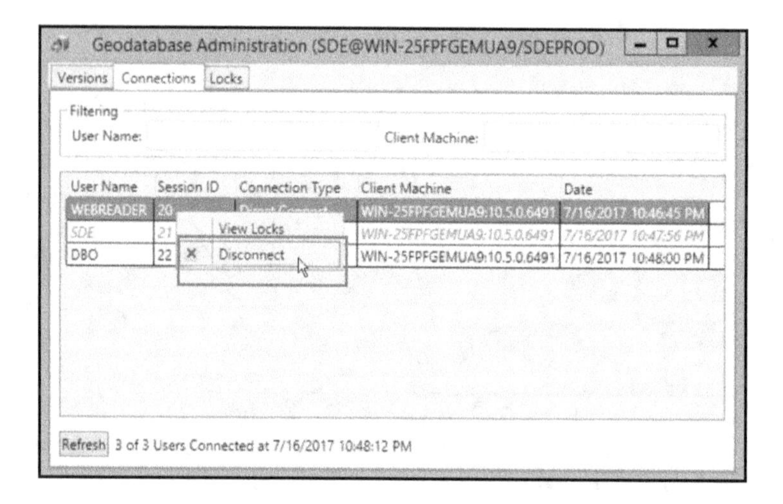

Disconnecting users programmatically is accomplished with the `arcpy.DisconnectUser` method. Continuing with our earlier example, let's use Python to disconnect our connected users:

```
import arcpy
ws = r'Database Connections\sde@SDEPROD@localhost.sde'
users = arcpy.ListUsers(ws)
for user in users:
    arcpy.DisconnectUser(ws, user.ID)
```

Like earlier, we import `arcpy` first. Next, we get a list of users. Finally, instead of simply printing out the user IDs and names, for each user, we call `arcpy.DisconnectUser`, passing in the workspace and user ID to disconnect the user.

Finding locks on datasets

Perhaps the database administrator or data loader only needs access to one dataset, but some users currently have the dataset open and therefore have locks on it. These users could be physical users that are viewing, editing, or analyzing the dataset, or they could be headless users such as ArcGIS Server admin or publishing accounts. Regardless, you can disconnect these users from individual datasets just like we did earlier to disconnect them from the entire geodatabase. In the **Geodatabase Administration** window under the **Locks** tab is a list of locks currently on all datasets in the geodatabase; to disconnect a user and effectively remove the lock, right-click on the object in the list and select **Disconnect User**.

For more information on locks in geodatabase, search **the ArcGIS Enterprise online help for schema locking**.

Preventing and allowing connections

In addition to disconnecting users from the geodatabase, it is often necessary to block connections during maintenance tasks and then allow them again once maintenance is complete. To do this in ArcCatalog, connect as a geodatabase administrator, right-click on **connection** and go to **Properties**. Next, on the **Connections** tab, uncheck the **Geodatabase is accepted connections** checkbox to keep users from creating new connections to the geodatabase.

Note that here you can also view connected users from this window with the **Show Connected Users** button:

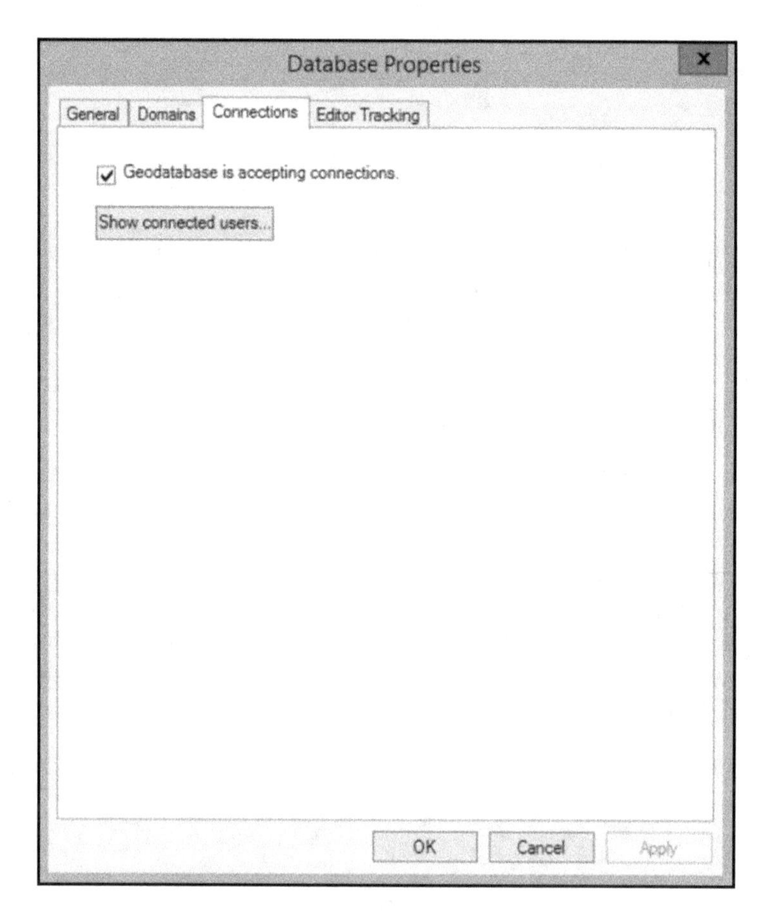

Blocking connections does not disconnect current users, it only blocks *new* connections from being made.

To block or allow users via Python, use the `arcpy.AcceptConnections` method. Blocking users is done by calling `arcpy.AcceptConnections` and passing in a geodatabase administrator connection and a Boolean `False` flag:

```
import arcpy
ws = r'Database Connections\sde@SDEPROD@localhost.sde'
arcpy.AcceptConnections(ws, False)
```

Likewise, to allow connections, the same code works with a Boolean `True` flag:

```
import arcpy
ws = r'Database Connections\sde@SDEPROD@localhost.sde'
arcpy.AcceptConnections(ws, True)
```

Loading data

For our purposes here, loading data refers to initially creating datasets in a geodatabase. Any user with proper edit privileges can insert data into an existing feature class or table. Loading data into the geodatabase is a task typically reserved for those with access to the data owner account credentials. Data should always be loaded (and feature classes and tables created) under the data owner account; in other words, the user loading the data must be connected to the geodatabase as the data owner.

Never load data while connected as the `sde` user; this will make the `sde` account the owner of that data.

There are many ways to load data into the geodatabase, and these vary based on the format of the source data. Let's discuss some of these methods where they are applicable, and their pros and cons.

Storage

Before discussing data loading, we should briefly touch on storage, the primary role of an enterprise database. Each RDBMS supported by ArcGIS Enterprise has its own mechanisms for storing the spatial component of geographic data (the geometries). For SQL Server, starting with ArcGIS 10.4, the default mechanism is the Microsoft Geometry spatial type. Other types available for use in SQL Server include ArcSDE compressed binary (the default prior to ArcGIS 10.4) and Microsoft Geography. These different storage types are each suitable for various environments and situations. The data storage type can be set when creating or loading/importing data. For example, in SQL Server, if you do not need SQL access to the spatial column of your data and you are more concerned with editing performance, you could use **SDEBINARY** as your data storage type. On the other hand, if you need SQL access to the spatial column and need to use latitude and longitude coordinates, you can specify the **GEOGRAPHY** configuration keyword when you create the feature class.

For more information on storage types and configuration keywords, search **the ArcGIS Enterprise online help for Configuration keywords for enterprise geodatabases**.

The easiest and quickest way to find out the storage type of a feature class is to connect to your geodatabase as any user, right-click on the feature class, and go to the **Properties** | **General** tab. Under the **Geometry Properties** section, **Storage** is listed:

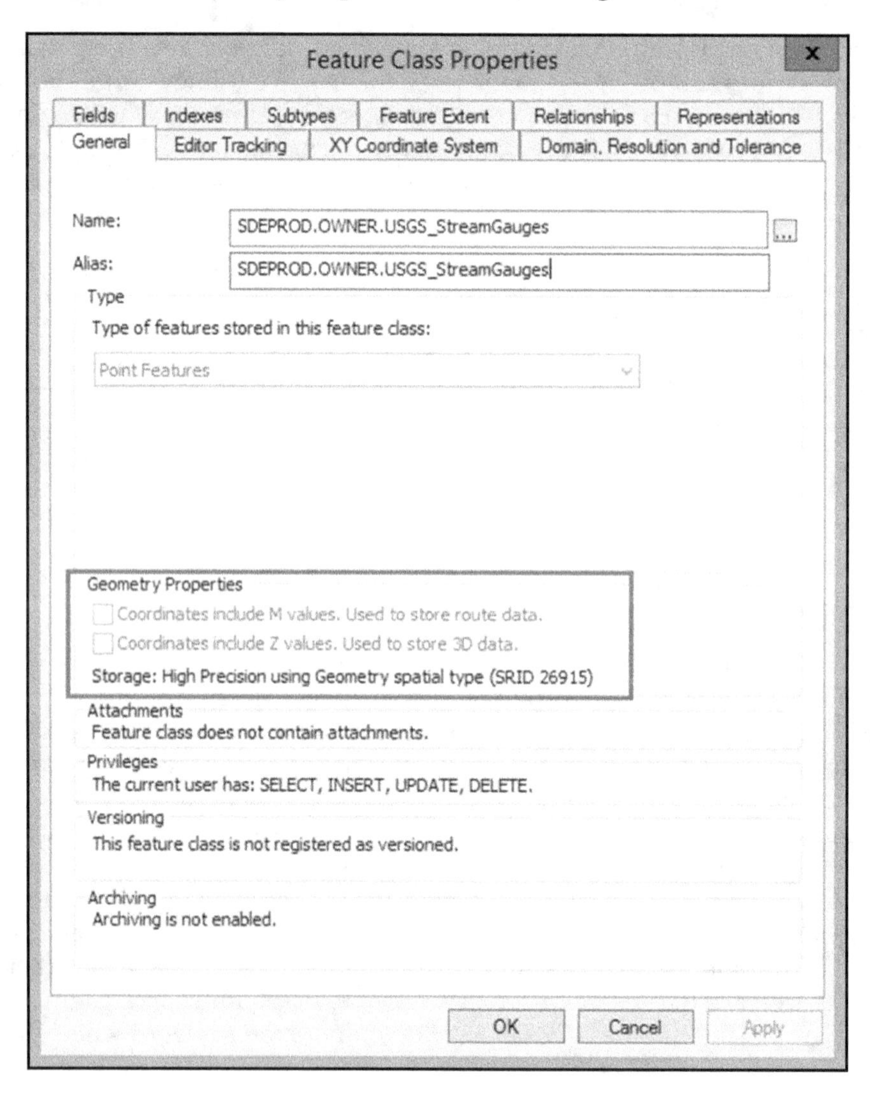

To set the storage type when creating a feature class, select the **Use configuration keyword** radio box in the **Configuration Keyword** step. From the drop-down, select the `storage` type you would like to use:

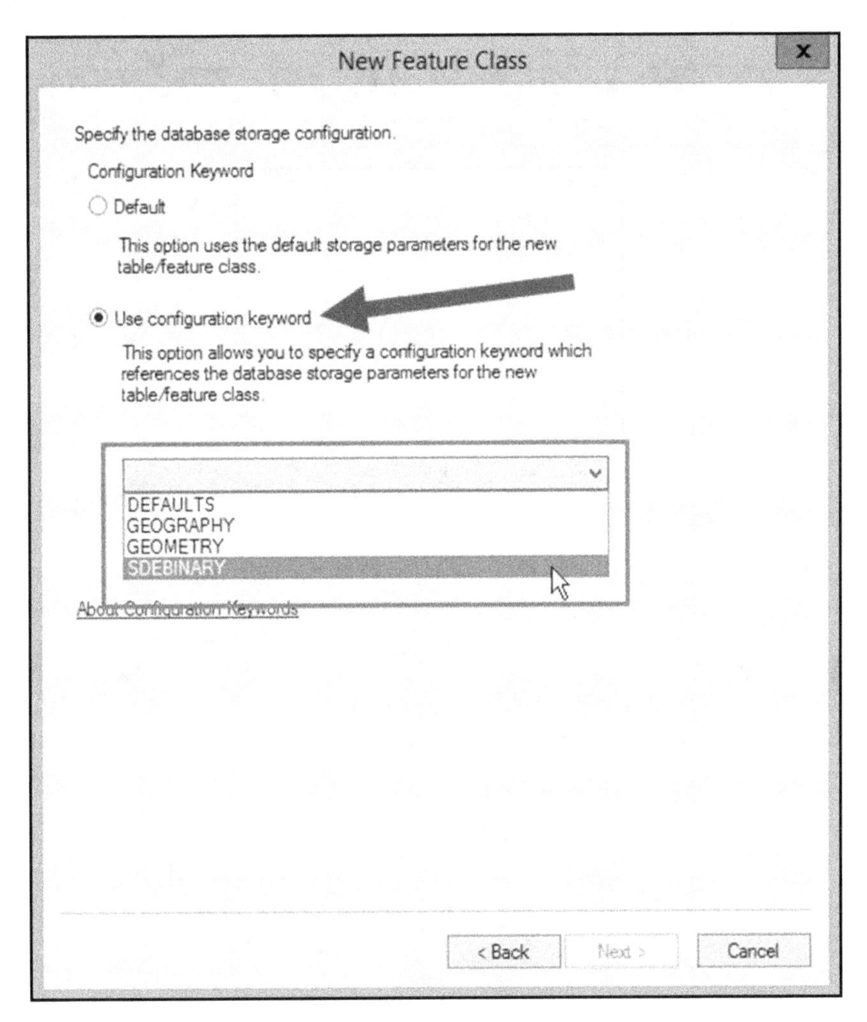

When using a geoprocessing tool to import or load data into the geodatabase, storage type can be specified under the **Geodatabase Settings** Configuration Keyword on the **Feature Class to Feature Class** tool. Likewise, when using copy/paste, the **Configuration Keyword** can be set in the **Data Transfer** window under the **Config. Keyword**'s drop-down:

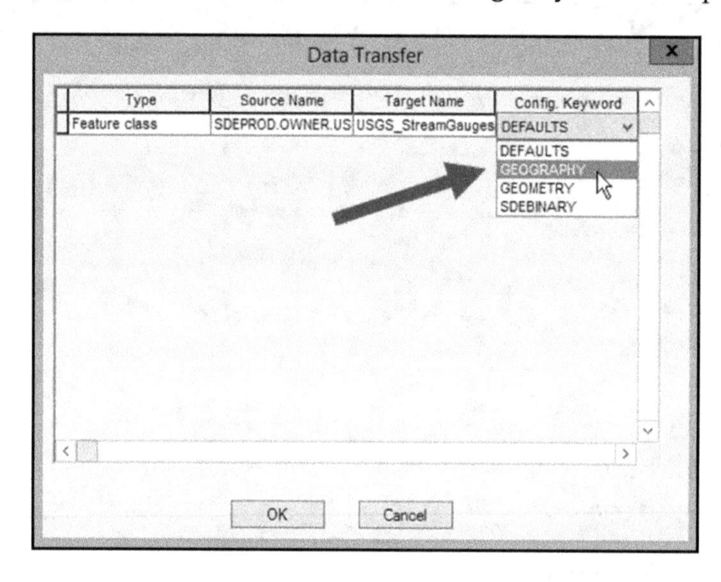

To change the default storage type in your enterprise geodatabase, edit the **GEOMETRY_STORAGE** parameter under the **DEFAULTS** configuration keyword in the **SDE_dbtune** table, from within SQL Server Management Studio. By default, it will be **GEOMETRY**; in the following screenshot, we have changed it to be **SDEBINARY**:

keyword	parameter_name	config_string
DEFAULTS	F_STORAGE	
DEFAULTS	GEOM_SRID_CHECK	1
DEFAULTS	GEOMETRY_STORAGE	SDEBINARY
DEFAULTS	GEOMTAB_OUT_OF_ROW	0
DEFAULTS	GEOMTAB_PK	WITH FILLFACT...
DEFAULTS	GEOMTAB_STORAGE	

Copy/paste

One of the fastest and easiest methods of loading data into an enterprise geodatabase is copy/paste. This method works best from within ArcCatalog and only if your source data is in a file geodatabase or another enterprise geodatabase. You simply right-click on the **dataset**, select **Copy**, and then, right-click in your target enterprise geodatabase location and select **Paste**. This action should only be carried out by a data owner. The pros and cons of this method are as follows:

Pros

- It is fast and easy.
- It handles related data. For example, if a feature class has attachments enabled and an additional related table, simply copying and pasting *only the feature class* will also copy over the relationship classes, attachments table, and the additional related table.
- It provides you with a list of which datasets and related datasets are to be copied:
    - This list will also inform you beforehand of issues such as naming conflicts or domain differences (domain with the same name exists in source and target, but the domain values are different, for example). You can then abort the copy, fix these issues, and try again.
- It provides progress bars during the paste operation.

Cons

- Limited data source options (only works from enterprise geodatabase to enterprise geodatabase or file/personal geodatabase to enterprise geodatabase).
- Target spatial reference will always be whatever source spatial reference was. It cannot overwrite target. If source exists in target, the pasted object will get renamed (with the _*n* format).
- The source and target spatial references must match when pasting a feature class into a feature dataset.

Use cases

- Data migrations from one enterprise geodatabase to another (development to production migration of data, for example).
- Just about any time you have file geodatabase data that needs to be loaded into an enterprise geodatabase.

Data Conversion tools

The Data Conversion toolbox contains a myriad of tools to convert data from one format into another, both spatial and non-spatial. Here again, these tools should only be used by a data owner to load data into the geodatabase. The pros and cons of this method are as follows:

Pros

- It has a wide variety of tools that cover just about any input format you would ever want to bring into the enterprise geodatabase
- Tools are input-focused (GPS, table, Excel, KML, and so on)
- Inputs fields can often be mapped to output fields

Cons

- It has many tools; you must find the right one for your source data (use the **Search**)
- Targets cannot already exist

Use cases

- Loading fresh complete datasets from scratch or doing complete data deletes and reloads.

Simple Data Loader

The Simple Data Loader has been a part of the core ArcGIS functionality for quite some time, allowing you to load data into several existing feature classes or tables that are either empty or already contain data. The Simple Data Loader is primarily used from within ArcCatalog and should only be used by a data owner or editor with proper permissions. The pros and cons of this method are as follows:

Pros

- Target can have data already or be empty
- It is fast as it performs no data validation

Cons

- Target must already exist
- Source and target schemas must match
- It does not handle geometric network feature classes, relationships with messaging, and feature-linked annotation

Use cases

- Quick loads of simple data into existing datasets

Object Loader

Like the Simple Data Loader, the Object Loader is functionality that has been around in ArcGIS for years. The Object Loader should only be used by a data owner or an editor with proper permissions.

Pros

- It loads during an edit session in ArcMap, so it provides undo capabilities
- It handles geometric network feature classes, relationships with messaging, and feature-linked annotation

- It allows for data validation

Cons

- It is slower, as it is performed during an edit session

Use cases

- More complex data loads into existing datasets where validation may be necessary

Truncate/load

Many of the methods we have discussed relate to loading new datasets into your enterprise geodatabase, but what do you do when you need to *reload existing* data? Truncate and load isn't anything new, it's been used in the database world for years. This process entails *truncating* the target feature class table, thus removing all features but leaving the schema intact, followed by *loading* (or *reloading*, as the case may be) the data back into the table. Tools here include:

- Truncate Table
- Delete Rows
- Append

The pros and cons of this method are as follows:

Pros

- It is a fast, efficient, tried, and proven method

Cons

- It removes all existing features from target prior to load
- Not a good choice if the feature class has Editor Tracking enabled

Use cases

- Fast, efficient, and simple full reloads of data from a source into the target

Managing user privileges

Once you have user accounts and data loaded into your geodatabase, you then need to grant users access to the proper datasets for them to be able to utilize them. For example, as shown in the following screenshot, we have loaded a feature class under the data owner connection (`owner` login), but have not yet applied any privileges to it. As seen in the following screenshot, when we connect with the webreader account, that account does not have access to the feature class:

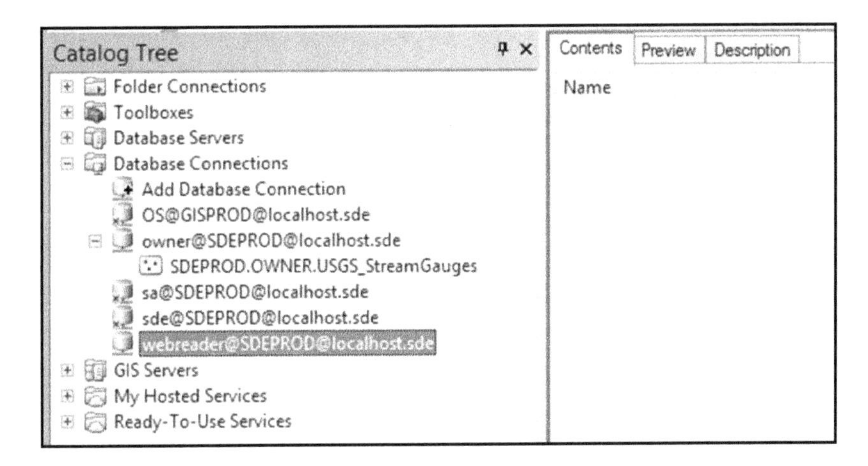

There are several ways to grant privileges on to tables through the user interface, geoprocessing, or Python.

When discussing privileges, the term "*table*" is used synonymously with more familiar terms such as feature class.

Before covering how to grant privileges, first, let's discuss some basics on privileges. There are four privileges that can be granted on tables: `SELECT`, `UPDATE`, `INSERT`, and `DELETE`. `SELECT` allows a user to read and select only on a dataset; `UPDATE`, `INSERT`, and `DELETE` allow you to modify a table by inserting new records, and updating and deleting existing records. Some general rules to keep in mind when granting and revoking privileges are as follows:

- Only the table owner can grant or revoke privileges to it.
- Only the table owner can alter its definition; in other words, only the owner can alter a table's schema, such as adding or removing a field.
- In a feature dataset, all feature classes must have the same privileges applied.

There are many other rules regarding privileges, search the Esri online documentation for **grant and revoke dataset privileges** for a full listing.

- In the Catalog tree of either ArcMap or ArcCatalog, the Privileges dialog box is a quick and easy way to grant or revoke privileges on datasets, but only one at a time. To alter privileges through the Privileges dialog box, first connect to your geodatabase as the data owner. Right-click on the dataset you wish to alter privileges on, go to **Manage** | **Privileges**. You will be presented with the **Privileges** dialog box and a list of users to whom you can grant or revoke privileges:

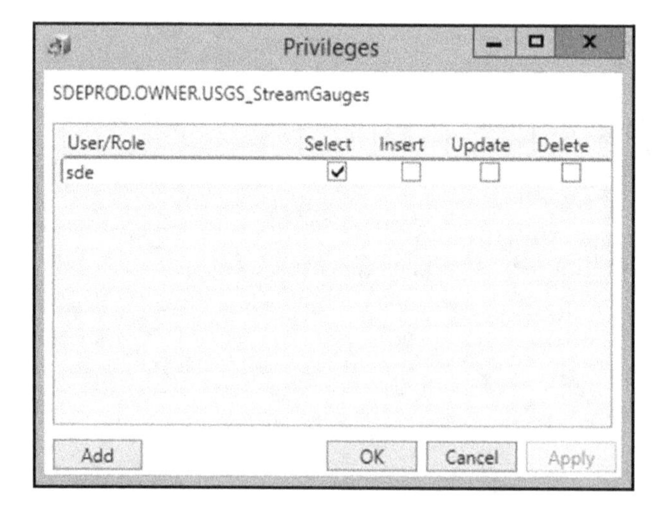

- You may find, as we do here, that not all users are always present in the user list. To add existing database users to this list and be able to grant them privileges, click on the **Add** button in the lower left of the **Privileges** window. We need to add the **webreader** account. Select the account you want to add and click on **OK**:

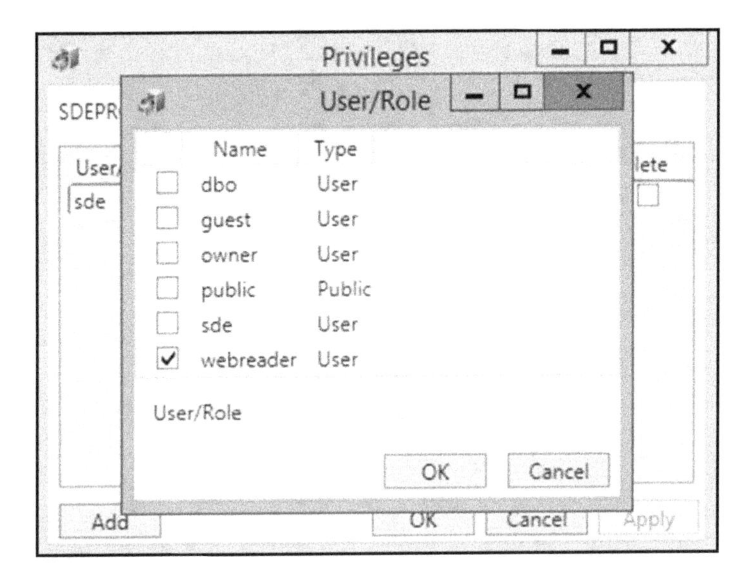

The account will now be available in the user list. `SELECT` privileges get checked and thus granted at a minimum by default. By granting `webreader` select privileges on **USGS_StreamGauges** and refreshing the `webreader` connection we looked at earlier, `webreader` can now view that feature class, as shown in the following screenshot:

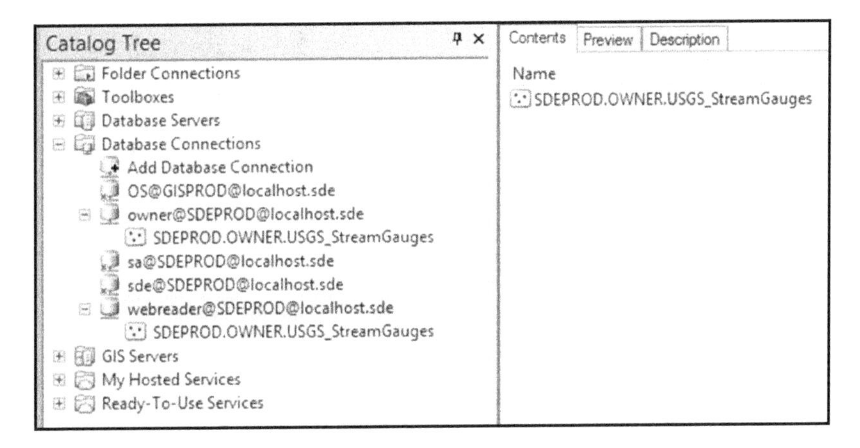

A second way to alter privileges is to use the Change Privileges geoprocessing tool. Search for and find this tool in ArcCatalog. Using an owner connection, drag the datasets whose permissions you wish to alter to the Input Dataset field. Enter the username and **VIEW** and **EDIT** privileges you wish to grant, as shown in the following screenshot:

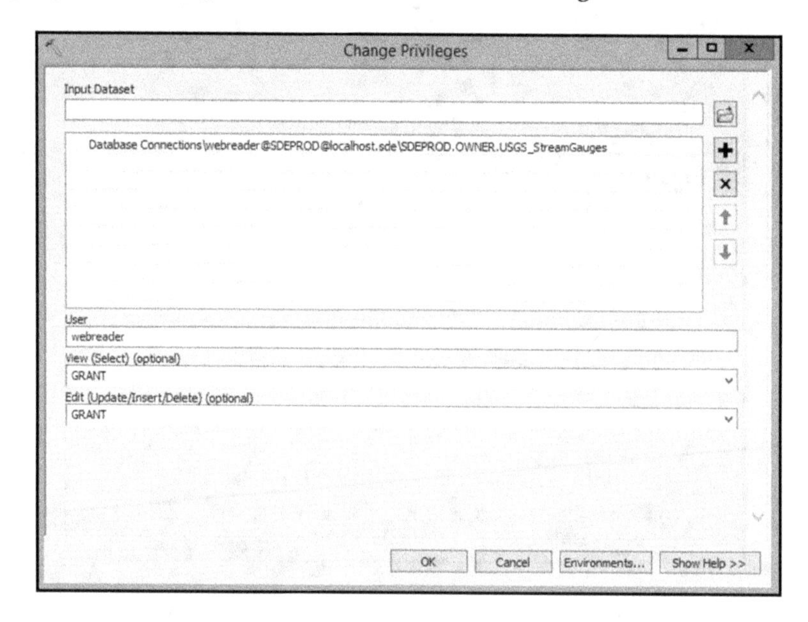

If we **GRANT View** and **Edit** on **USGS_StreamGauges**, refresh our connection, and view privileges now, we see that **webreader** now has full privileges on the feature class:

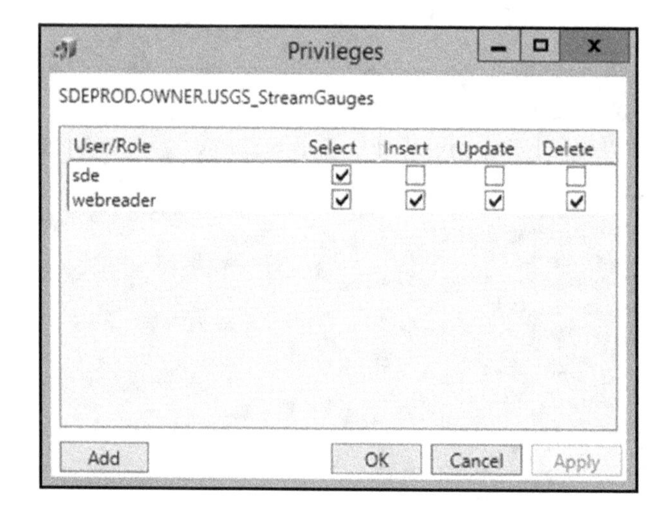

Regarding privileges, here are some brief definitions:

- **AS_IS**: This makes no changes and leaves permission as they currently are.
- **GRANT**: This grants the privilege.
- **REVOKE**: This revokes (removes) the privilege.
- A final method to modify user access on a dataset is to use Change Privileges in a Python script. The inputs are identical to what we used earlier when executing the tool through ArcCatalog. To grant **webreader** full privileges on **USGS_StreamGauges**, we will call the `arcpy` method as follows:

```
import arcpy
ds = r'Database
Connections\owner@SDEPROD@localhost.sde\SDEPROD.OWNER.USGS_StreamGauges"
arcpy.ChangePrivileges_management(ds, "webreader", "GRANT", "GRANT")
```

Database maintenance

A well-maintained geodatabase is a performant geodatabase. Database maintenance requirements vary from system to system, but there are several routine tasks that need to be carried out on all systems.

Backups

Although not necessary for performance, database backups taken on a routine schedule are crucial to the safety, integrity, and security of your system. Not only do database backups protect you from data loss in the event of system failure, they also protect you in the case of data corruption. Database backups are typically scheduled and handled by the database administrator, but it should also be the responsibility of the geodatabase administrator to ensure that this process is in place and carried through.

Just as important as taking the backups, is to routinely, yet randomly, *test* your database backups. This entails restoring backups to a different SQL Server instance to ensure that the backups are valid and current. This also keeps the staff current on the protocols for backup restoration in the case of an actual emergency.

Statistics

Keeping database statistics updated is crucial for maintaining query performance. Statistics should be updated after large data loads or large numbers of edits. For the latter reason, it is common practice to rebuild statistics on all layers after a reconcile/post/compress operation in a versioned database, oftentimes using the Analyze Datasets geoprocessing tool.

Indexes

Attribute indexes can speed up queries made on the geodatabase by users. Indexes should be maintained (not just created, but maintained) only on fields that are routinely used in queries as each index added to your feature class or table slows down editing on that item. Attribute indexes can be created and deleted in ArcCatalog by going into a feature class or table's **Properties** on the **Indexes** tab of a feature class or table's properties. When creating indexes in an enterprise geodatabase, there are several very important rules to keep in mind:

- Index names must be unique in the database.
- Index names cannot contain spaces and must start with a letter.
- Index names cannot contain reserved words.
- Index names cannot be over 16 characters in length; this is oftentimes the hardest of the rules to deal with, as index names must be unique, but short.

Regarding the naming of indexes, a common nomenclature is to prefix the name with "`IDX_`", followed by an abbreviation of the feature class or table name and then an abbreviation for the fields participating in the index. For example, `IDX_HYD_FID` would be an index on the `FID` field in a hydrants feature class. For a large database with many indexes, index names can get confusing and it is advisable to keep a document that defines the indexes. For example, something like the following table:

Index name	Featureclass/table name	Field name
`IDX_HYD_FID`	`wHydrant`	`FID`
`IDX_HYD_STR`	`wHydrant`	`STREET`

Considering that index names must be unique, in order to rebuild an existing index, it must first be deleted or removed. Luckily, `arcpy` and tools from the Indexes geoprocessing toolbox can perform these tasks. The following example takes an input list of indexes, deletes existing indexes if they exist by name, and then creates new indexes. This first section sets the workspace variable and defines a list of dictionaries containing key-value pairs of information needed to update indexes:

```
import arcpy
import os
ws = r'Database Connections\owner@SDEPROD@localhost.sde'
indexes = [{"fc": "wHydrant",
            "index_name": "IDX_HYD_FID",
            "field": "FID"},
           {"fc": "wHydrant",
            "index_name": "IDX_HYD_STR_STN",
            "field": ["STREETNO", "STREETNAME"]}]
```

Next, we loop through the list of indexes and check to see an index by the name of the one we want to create already exists in the geodatabase, and if it does, we remove it:

```
desc = arcpy.Describe(os.path.join(ws, index.get("fc")))
for index in indexes:
    desc = arcpy.Describe(os.path.join(ws, index.get("fc")))
    for g in desc.indexes:
        if g.name == index.get("index_name"):
            arcpy.RemoveIndex_management(os.path.join(cws,
                                         index.get("fc")),")),
                                         index.get("index_name"))
```

Finally, once we know we do not have an index in the geodatabase with the same name as the index we want to create, we will create the index with a call to `AddIndex` tool.

```
    arcpy.AddIndex_management(os.path.join(cws, index.get("fc")),
                              fields, index.get("field"))
```

Summary

A GIS is only as good as the data that powers it. Geodatabase administration is crucial in keeping a well maintained, performant geodatabase that users will be able to utilize efficiently. In this chapter, we discussed how to create a geodatabase, connect to it, create accounts for users to access it, load data into it, manage user privileges, manage user connections, and perform routine maintenance. Next up in `Chapter 3`, *Publishing Content*, we will look at using our data in published content in ArcGIS Server.

3

Publishing Content

Once ArcGIS Server, Portal for ArcGIS, Data Store, and the Web Adaptor are installed and configured, and the geodatabase has data in it, it is time to publish content in the form of services. Services are at the core of ArcGIS Server, ArcGIS Online, and Portal for ArcGIS. Many different service types can be published, with map services, feature services, geoprocessing services, and image services being the most prevalent and popular.

Much like in `Chapter 2`, *Enterprise Geodatabase Administration*, where we dug into what makes up a geodatabase, here, we will first ask ourselves *What exactly is a service?* before we discuss the service types we can publish with ArcGIS Server. Also, before we can publish services, we must register our data sources with ArcGIS Server, so we will discuss what this is and how to do it. Next, we will examine service properties and settings, publish services, and finally, talk about how we can extend services with server-object extensions and server-object interceptors.

After finishing this chapter, you will be familiar with some of the different types of services that can be published with ArcGIS Server, how to get your data ready to be published, and how to publish and configure ArcGIS Server services.

In this chapter, we will cover the following topics:

- What exactly is a service, anyway?
- The different types of services that ArcGIS Enterprise is capable of publishing
- How to publish map, feature, image, and geoprocessing services
- How to properly manage the data behind your services and connections to that data
- How to understand and configure service properties
- How to publish to the ArcGIS Data Store
- Server-object extensions and server-object interceptors

Service types

Many different types of services can be published to ArcGIS Enterprise, each performing a unique task within the ecosystem. All ArcGIS Enterprise services are authored in ArcMap or ArcGIS Pro. Many different service types exist; here, we will focus on several of the most common ones. Consult the ArcGIS Server online documentation for a more extensive list of the available service types.

What is a service?

Before going any further, let's discuss just what exactly a service *is*. At its core, ArcGIS Server operates on a spatially enabled **service-oriented architecture**, or **SOA**. With an SOA, *services* are provided by application components through a protocol over a network. For ArcGIS Server, the services are map services, feature services, and so on; the application component providing the service is ArcGIS Server, the protocol is HTTP, and the network could be an intranet, the internet, or both. With ArcGIS Server, think of services as a representation of that service from where it came. Therefore, a map service is a representation of the ArcMap MXD or ArcGIS Pro Map that it was published from and a geoprocessing service is a representation of the geoprocessing script from which it was published. With a map service, you get the geographic features and attribute data, and with a geoprocessing service, you get the functionality (functions and methods) of the geoprocessing model or script, all exposed over the internet.

ArcGIS Server services are considered RESTful, meaning that they adhere to the **Representational State Transfer** (**REST**) architectural style that powers much of the modern web today. In the REST architectural style, data and functionality are considered resources and are accessed through URLs, such as how we will soon access the ArcGIS Server services using the ArcGIS Server REST API. The URLs define simple, well-defined operations, such as query, identify, or export map, in the case of the ArcGIS Server REST API. The ArcGIS REST API is a window into our ArcGIS Server services, which then lets us list, get, query, find, and perform other standardized, well-defined operations against the services.

Map services

The map service is the most prevalent of the ArcGIS Server service types. It is the way you make maps available over the web using ArcGIS Enterprise. Authored in ArcGIS Desktop or ArcGIS Pro when publishing to a federated ArcGIS Server instance, a map service makes geographic features and their attributes available to client applications.

Map services are typically used to serve dynamic maps, dynamic layers, cached maps (such as basemaps), and features to clients, usually in the form of images that are rendered by the client (a web browser, for example). The most common use of map services is to show business and organizational data on top of a basemap from ArcGIS Online, a local basemap cache, or a basemap from another provider.

Feature services

At a cursory glance, feature services appear to be identical to map services, and on the surface, they essentially are. What sets feature services aside is that a feature service exposes the actual *features* and *symbology* used to display those features as opposed to simply exposing the data as a rendered image. With a feature service, it is also possible to edit features (create, update, delete) over the internet and to do so using feature editing templates for an enhanced editing experience. Editing can also be performed on related tables and non-spatial tables as well.

Geoprocessing services

Geoprocessing services expose the analytical capabilities of ArcGIS Desktop over the internet through a geoprocessing task. A geoprocessing task is the result of a geoprocessing tool that has been published to ArcGIS Server. Geoprocessing services allow you to share complex models and scripts in an abstract way that makes it easy for users to utilize them without having to worry about how the process runs and instead focus on results.

In ArcGIS Pro, a web tool is akin to a geoprocessing tool in ArcGIS Desktop. With a web tool, you share an analysis first run as a custom model or script tool in Pro. A web tool can process data using one or more tools and return output as features, maps, reports, or files.

Image services

While a geoprocessing service exposes a geoprocessing task through a web service over the internet, an image service exposes raster data over the internet. Publishing image services from rasters stored on disk or from a geodatabase is a core capability of ArcGIS Server.

ArcGIS Image Server provides two types of capabilities:

- Dynamic image services allow the publishing of mosaic datasets used to manage large collections of overlapping, often temporal, multiresolution imagery, such as orthophotos, elevation models, and LiDAR, or categorical rasters, such as land use classification
- Raster analysis, the second capability, enables massive distributed processing and analysis of rasters and imagery

ArcGIS Image Server, a separately licensed role from GIS Server, is not required to publish rasters stored on disk or in a geodatabase. It *is* required if those rasters are shared through a mosaic dataset or raster layer containing a mosaic function.

Publishing services

With ArcGIS Enterprise, information can be shared as services published not only to ArcGIS Server, but also to ArcGIS Online and Portal for ArcGIS.

Publishing to ArcGIS Server

Publishing data and processes to ArcGIS Server is a common methodology employed by many organizations to share information. Publishing content to ArcGIS Server allows your services to be consumed internally within your organization and, if you choose, externally, either as public-facing unsecured services or secured services available through authentication.

Creating an ArcGIS Server connection

To publish a service to ArcGIS Server, you must have a publisher or higher (administrator) connection to ArcGIS Server. This connection can be made either prior to service publication or during the service publication process; it is more convenient to have these types of connections made and saved before you need them. With a publisher connection, you can publish GIS resources, configure and publish draft services, and add, delete, start, and stop ArcGIS Server services. With an administrative connection, you can do everything you can with a publisher connection along with the edit server configuration properties and manage the site's data stores. A user connection allows you to only view ArcGIS Server services. The following diagram demonstrates that, as is the case with database permissions, with increased privileges in ArcGIS Server come increased responsibilities:

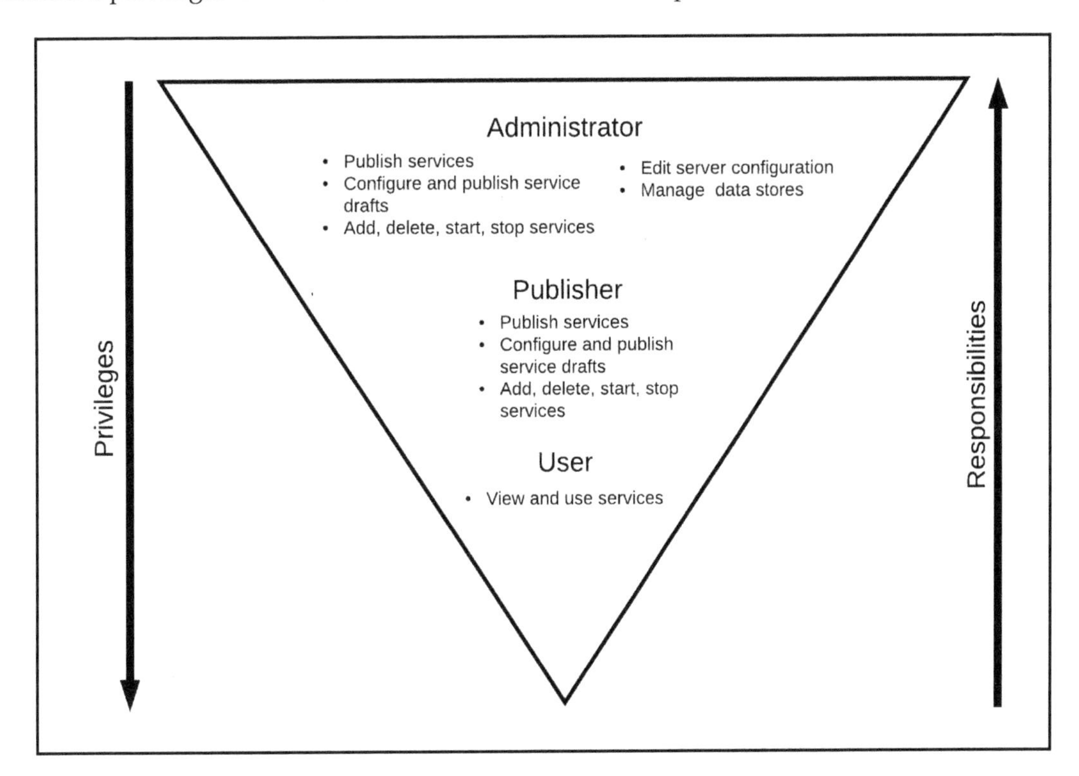

Before publishing a service, you must first decide whether the data behind the service will reside in a registered data source or be copied to the ArcGIS Server instance. See the *Managing service data* section for more information.

To create a connection to ArcGIS Server, do the following:

1. Open ArcCatalog and expand the **GIS Server** section in **Catalog Tree**.
2. Double-click on **Add ArcGIS Server**.
3. You are now at the **Add ArcGIS Server** window. Here, you can create `user`, `publisher`, and `administrator` connections to your ArcGIS Server instance. Select the **Administer GIS Server** radio button. Remember that an administrator is the highest of privileges, one of which is publishing. Click on the **Next** button.

4. In the **General** window, enter your **Server URL**. Depending on what machine you are attempting to access ArcGIS Server from, there are several URL options:
    - If you are on the ArcGIS Server machine, use the localhost form of `http://localhost:6080/arcgis`
    - If you are on another machine on the same internal network and can access the ArcGIS Server machine, use the **fully qualified domain name** (**FQDN**) of `http://gisserver.domain.com:6080/arcgis`
    - If you can access ArcGIS Server through the Web Adaptor, (administrative access is enabled through the Web Adaptor), use this form: `http://FQDN/arcgis/admin`
5. Leave **Server Type** as **ArcGIS Server** and accept the default of **Use ArcGIS Desktop's staging folder**.
6. Enter appropriate administrator credentials. If this is a machine/login where you safely store credentials to admin connections, check the **Save Username/Password** checkbox.
7. Click on **Finish**.
8. Rename the connection appropriately using the username and server URL. For example, `siteadmin@localhost`. Connection naming standards are covered in `Chapter 9`, *ArcGIS Enterprise Standards and Best Practices*.

Service capabilities

ArcGIS Server comes with a standard set of service capabilities that can be enabled for services, some of which do require additional extensions to be installed. These capabilities define the various ways clients can use the services. The following screenshot shows the default service capabilities that are enabled for the standard **SampleWorldCities** service that is published with an ArcGIS Server installation and configuration. Mapping will always be enabled by default, and, typically, KML will also be enabled:

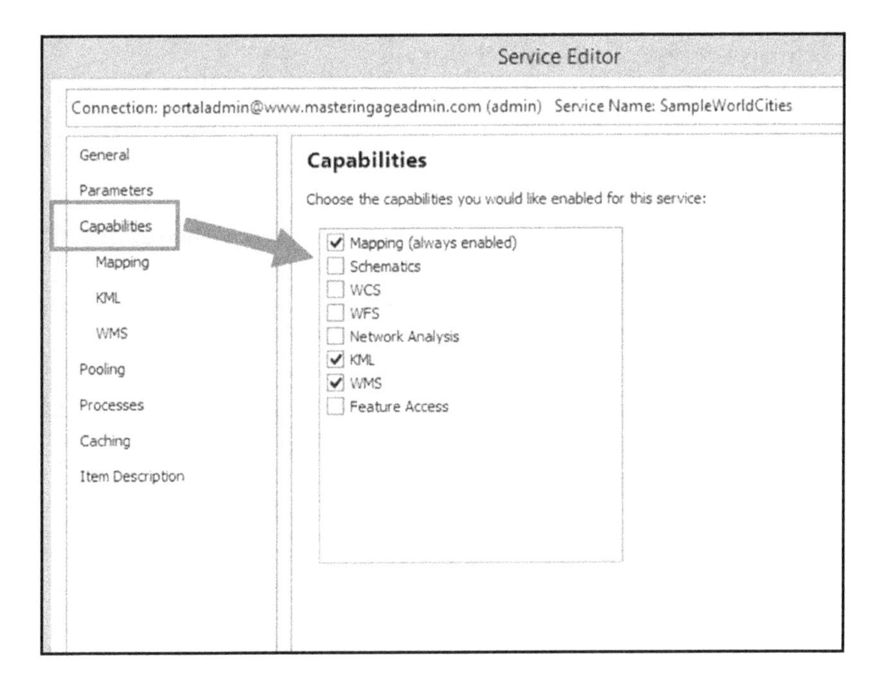

Capabilities available for a given service depend upon the type of service you are publishing; certain service types only expose certain **capabilities**. For example, only a map service exposes the **Mapping** and **Feature Access** capabilities, whereas, only a geoprocessing service exposes the **Geoprocessing** capability.

The list of service capabilities is much too lengthy to discuss here; please search ArcGIS Enterprise online documentation for *What types of services can you publish* for more in-depth information.

Map services

Map services are the most common form of service typically published from ArcGIS Server. Map services make geographic data available to client applications over the internet in the form of maps, features, and attributes. Map services are often used to serve dynamic layers, cached maps, and features. You can also configure map services to have the ability to be queried through the ArcGIS Server REST interface, opening them up to use by many non-Esri applications.

Publishing a map service to ArcGIS Server

To publish a map service to ArcGIS Server, have the map document open in ArcMap and perform the following steps:

1. Go to **File** | **Share As** | **Service...**.
2. If publishing a new service, select **Publish a service**; if the service already exists and you wish to overwrite it, select **Overwrite an existing service**.
3. If publishing a new service, choose a connection to publish to or create a new connection and name your service. If overwriting an existing service, choose a connection to publish to and the service to overwrite. If the current map document has been published as a service previously, ArcMap should select the target service for you, but only if the service is named the same as the source MXD.
4. If publishing a new service, select an existing folder or create a new folder to publish to. Folders can only be one-level deep under the root of the services directory.
5. In the Service Editor, service properties, parameters, and capabilities can be set for a service, determining what end users can do with the service. Set the properties you want. See the *Tuning services* section in `Chapter 9`, *ArcGIS Enterprise Standards and Best Practices*, for more information on this topic.
6. Click on **Analyze**, which will examine your map document for problems (see the next section for more details). Fix any errors in your map document.
7. Click on **Publish**. Once the publishing is completed successfully, it can be accessed on your network at `http://localhost:6080/arcgis/rest/<folder>/<service>`.

Dealing with publishing warnings and errors

During the publication process, ArcMap analyzes the service document for potential problems and errors. During the analysis process, issues such as a layer drawing at all scales, an unregistered data source, or a projection on-the-fly that can impact service performance are detected and reported as warnings and messages in the **Prepare** window. Errors such as unsupported symbology are also reported. Warnings and messages can be ignored; errors, however, cannot and must be dealt with accordingly for publishing to continue. When you publish, ArcMap automatically analyzes. You can also analyze manually anytime during the publishing process by clicking on the **Analyze** button in the **Publishing** window. The following example illustrates some of the more common warnings and errors when attempting to publish a service:

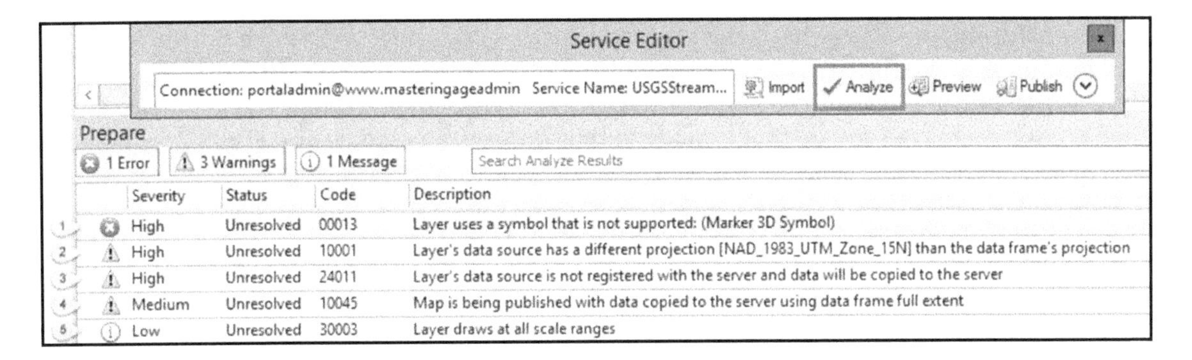

By clicking on the **Analyze** button in **Service Editor**, we are presented with these five issues in the **Prepare** window:

1. **High severity error**: The layer is using a 3D marker symbol for its symbology. ArcGIS Server does not support 3D marker symbols. Solution--choose another non-3D marker symbol for the layer symbology.
2. **High severity warning**: Projection on-the-fly. The source data has a spatial reference different from the data frame and the data will be projected on-the-fly, which could impact performance. Solution--if necessary (the service could be performance-tested to see if on-the-fly projection does indeed impact performance), make the data frame and source data use the same spatial reference.
3. **High severity warning**: Unregistered data source; data will be copied to the server. The data source is not registered with ArcGIS Server; therefore, data will be copied to the server. Solution--register the data source with ArcGIS Server, unless the workflow calls for data to be copied to the server.
4. **Medium severity warning**: Full extent of data layer is being copied to the server. This is related to the issue stated in point 3. Solution--register the data source with ArcGIS Server (if the workflow allows it), and narrow down the extent of the data, if necessary, or do nothing.
5. **Low severity message**: The layer draws at all scales. The layer has no scale dependencies and will draw at all scales. This can impact performance and, oftentimes, it is just bad cartography, often resulting in a *busy*, cluttered map. Solution--set an appropriate scale dependency on the layer, if necessary.

Remember, warnings can be ignored, but keep in mind they usually offer good advice, such as a layer drawing at all scales. Do you want and need to draw all 50,000 of those address points at full extent? Probably not.

While analyzing and reviewing your map document, you can leave **Service Editor** open and minimize it with the up/down arrow button in the upper-right corner, to the right of the **Publish** button. This allows you to **Analyze** for errors, review them, correct them, and then **Analyze** again to ensure that you made the proper adjustments and corrections. Once you are done analyzing, you can **Publish**.

In the **Prepare** window, you can right-click on an issue and be presented with a context menu with options to help resolve the issue. For example, in the first issue, which was mentioned earlier, when right-clicked, the context menu provides options, one of which is to **Change Symbol Properties**. Clicking on this will open the **Symbol Selector**, allowing a different symbol to be selected for the layer.

Feature services

Feature services provide the same capabilities as map services, with the added ability to serve out features and the symbology to use for displaying them along with the ability to edit those features over the internet. Feature editing templates can be utilized to enhance the editing experience. Related data and non-spatial tables can be edited over a feature service.

Feature services can only be published from data stored in an Enterprise geodatabase and can only be edited over the web with a GIS Server Advanced or GIS Server Standard license. Feature services can be created to query only with a GIS Server Basic license.

Publishing a feature service to ArcGIS Server

Publishing a feature service is very similar to publishing a map service, with the added step of enabling feature access in the service. To publish a feature service, do the following:

1. Follow the earlier 1-4 steps to publish a map service.
2. In the **Service Editor** | **Capabilities** tab, select the **Feature Access** checkbox. See the next section for more information on feature service operations and properties.
3. Follow the earlier 6-7 steps mentioned earlier in the *Publishing a map service to ArcGIS Server* section.

Feature service operations and properties

Once **Feature Access** is checked, it becomes a capability in the left property panel of **Service Editor**. Select **Feature Access** in the left panel to view the feature access operations and properties. Search the ArcGIS Enterprise online documentation for *Editor permissions for feature services* for more information. Let's discuss some of these operations and properties that are available, as the following screenshot shows:

1. **Operations allowed**: These control what types of edits users can make to the service including:
    * **Create**: Editors can add geometry features to the service, not just attributes. It is enabled by default, but not required.
    * **Delete**: Editors can delete geometry features from the service, not just attributes. It is enabled by default, but not required.

- **Extract**: Editors can extract a copy of the layer to a file or SQLite database through a custom application. It is disabled by default and not required.
- **Query**: Users can view the data in the feature service; the most basic of all operations. It is required; therefore, it is enabled by default.
- **Sync**: Editors can work with feature service data while they are offline. It is disabled by default and not required.
- **Update**: Editors can update existing features in the feature service. If **Allow geometry updates** is enabled, then geometries can be updated; otherwise, only attributes can be updated. It is enabled by default, but not required.

To control what editors can and cannot do while editing, creative combinations of the earlier operations can be utilized. Publishing a feature service for a Collector application where you want users to be able to add new features but not make changes to attributes of existing features or delete features? Publish the feature service with only the **Create** and **Query** operations allowed.

To allow a feature service to be edited in ArcMap, you must enable the **Create**, **Delete**, and **Update** capabilities; otherwise, ArcMap editors will encounter errors.

2. **Allow geometry updates**: Checking this will allow editors to edit the geometry of a feature in the feature service. It is enabled by default. Disabling this will allow editors to only update attributes of features. A use case for this is for a Collector inspections application where field crews are to inspect only existing features and update attributes. In this case, you would want to allow the **Query** and **Update** operations and disable **Allow geometry updates**.
3. **Apply default z-value**: This applies a default *z*-value (if *z*-values are enabled on the underlying feature class) in the event that the user either does not enter one or the application they are using to edit the feature service does not support specifying a *z*-value.
4. **Allow geometry updates without m-value**: Similar to **Applying default z-value**, this applies default *m*-values if the underlying feature class is *m*-enabled. The difference is that if your underlying data is i-enabled and either the user doesn't enter an *m*-value or the client application doesn't support specifying an *m*-value, they will be blocked from editing the layer.

Enabling this property on a feature service with existing *m*-values could result in those *m*-values being overwritten with a `NaN` if the feature service is edited with a client that does not support specifying an *m*-value.

5. **Enable ownership-based access control on features**: This property is only available with Enterprise geodatabases and allows you to not only record information about who created each feature but restrict users from accessing features they do not own. It requires **Editor Tracking** and a field designated to hold the feature creator name. Search the ArcGIS Enterprise online documentation for *Ownership-based access control for feature services* for more information on this property.
6. **Advanced options**: An interesting property here is **Add realm to user name when applying edits**, which allows you to append a realm in the form of `@realm` to the name recorded as the feature editor. This requires that **Editor Tracking** is enabled on the underlying feature class. For example, you could add `@server` as your realm in the feature service to differentiate edits that are performed over the web from those performed directly to the source data in ArcGIS Desktop. Search the ArcGIS Enterprise online documentation for *Editor tracking for feature services* for more information on this subject.

Geoprocessing services

Geoprocessing services expose the analytical and data processing capabilities of ArcGIS Server through the publishing of the ArcGIS Model Builder models, Python Toolboxes, and custom Python scripts. A geoprocessing service is comprised of one or more geoprocessing tasks, each of which represents a geoprocessing tool. In a geoprocessing service, a task takes input from an application (typically, as input from a user) as input, processes it, and performs any calculations, and returns meaningful output in the form of maps, features, reports, files, or, in the case of a custom geoprocessing service, some sort of custom output.

If a geoprocessing service requires a great deal of server resources (memory or CPU power), it may be best to publish that service on a separate instance of ArcGIS Server to prevent impacting published map and feature service performance on your primary ArcGIS Server instance.

Publishing a geoprocessing service to ArcGIS Server

Publishing a geoprocessing service is somewhat different than publishing other types of services. With a geoprocessing service, you publish the result of the geoprocessing model, tool, or script. To publish a geoprocessing service, do the following:

1. In ArcCatalog or ArcMap, execute a successful run of your geoprocessing model, tool, or script.
2. Go to the **Geoprocessing** menu, and select **Results**.

Results can be managed under **Geoprocessing** | **Geoprocessing Options**. Here, you can set the length of time to keep results under the **Keep results younger than** setting. The default is 2 weeks.

3. Find the result of the run you just successfully completed. Right-click on the main task and select **Share As** | **Geoprocessing Service**, as follows:

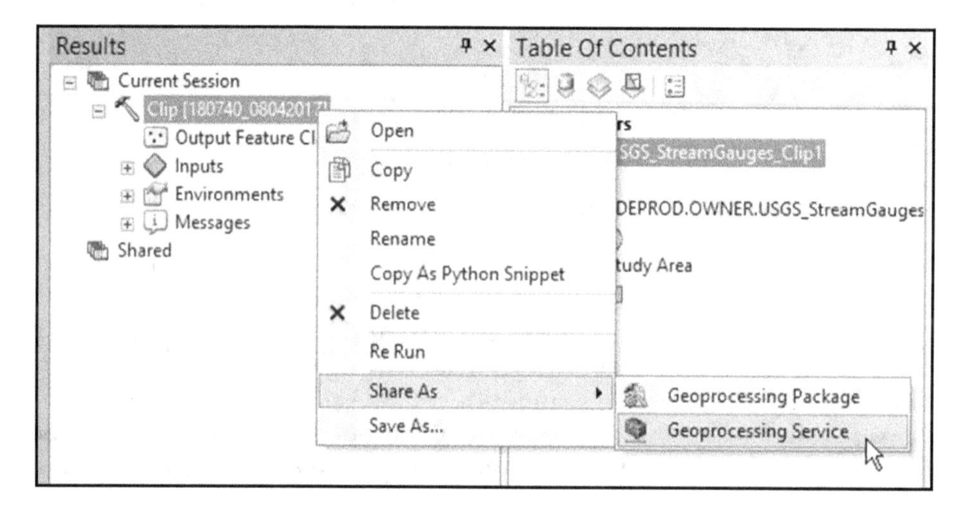

4. Next, just as with publishing a map or feature service, either create a new service or overwrite an existing one.
5. The **Service Editor** will now be presented, with many of the same settings we have seen for map and feature services. Note that there is only one capability in the left panel, **Geoprocessing**.

6. After configuring your service properties and settings (see the following section for more information on select geoprocessing service settings), click on **Publish**. Note that prior to publishing, the service is analyzed just as other services are and the **Prepare** window will show you any warnings or errors. Publishing will halt on errors.

Geoprocessing service settings and properties

Geoprocessing tasks have many settings and properties; here, we will highlight some of the more important ones:

- **Parameters**:
    - **Execution mode**: **Synchronous** means the client application waits for the task to finish before anything else can be done. This is typically reserved for quick tasks taking 5 seconds or less. For longer running tasks, choose **Asynchronous**, where the client application does not wait, but must have logic in place to poll the task periodically (every second, for example) to check for and get the results.
    - **Properties | Message Level**: Whether or not to return messages of varying levels to the client upon task execution. The default is **None** and is recommended for production environments, as messages can contain sensitive information. Set to **Error** or higher while developing tools to help with debugging, but switch to **None** once the task is migrated to production.
    - **Properties | Maximum number of records returned by the server**: This is the maximum number of records that can be returned to a client application. The default value is 1,000 but can be adjusted higher to return more features to the map or in a query result. Be careful with this number, as setting it too high can cause a heavy load on your server as it tries to send the features to the client.

- **Geoprocessing task settings**: The geoprocessing task represents the result of the geoprocessing tool, model, or script that is being published. The task is shown by its name in the left panel of **Service Editor**. Following is a simple geoprocessing task created from the result of the **Clip** tool:

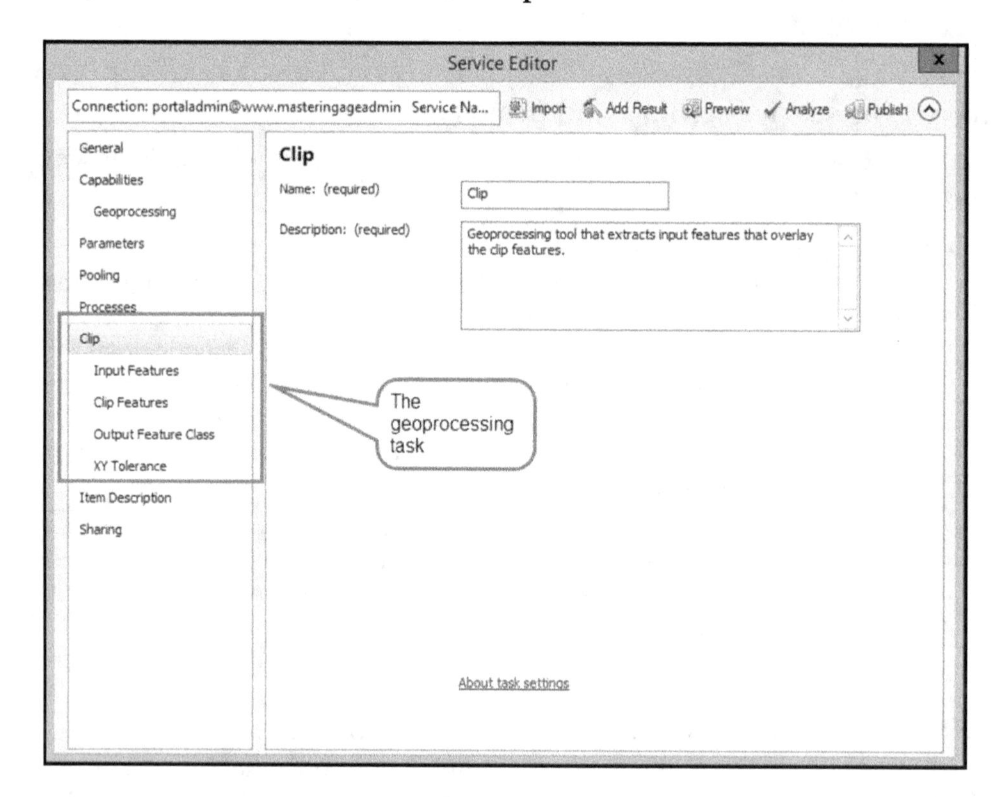

- **Geoprocessing inputs and outputs**: Tasks have inputs and outputs just like the tools, models, or scripts they are derived from--you put something (inputs) in, the task processes it, you get results (outputs) back. Depending on your task, you will define parameters such as the input mode (whether the user is presented with a choice list, do they enter the input, and so on) and schema of inputs and outputs. Search the ArcGIS Enterprise online documentation for *Overview of geoprocessing task settings* for more information on this lengthy topic.

Publishing to ArcGIS Online

In addition to publishing to ArcGIS Server, feature services and tiled map services can be published to ArcGIS Online directly from ArcMap with no additional software installation necessary. Publishing content to ArcGIS Online is a straightforward process, similar to publishing to ArcGIS Server. To publish a feature service to ArcGIS Online, sign in to your ArcGIS Online account with publisher or higher access. In ArcMap, go to **File | SignIn** and enter your credentials. Next, just as you would do to publish to ArcGIS Server, go to **File | Share as Service**, select **Publish a service**, and then hit **Next**. The next step of selecting the publishing connection is where this process differs from publishing to ArcGIS Server. For the connection, choose **My Hosted Services (<My Org>)** from the drop-down and name your service accordingly.

Publishing to Portal for ArcGIS

There are several dozen types of items that can be added to your Portal, including geoJSON files, layer files, Excel workbooks, PDFs, and many files as zips such as shapefiles and file geodatabases. In addition to the items that can be added to your portal, services published to ArcGIS Server sites that are federated with your portal are automatically added as items in your portal (see `Chapter 5`, *Portal for ArcGIS Administration* for more information on federation). All items added by either method appear in the **My Content** section of the account that was used to add or publish them.

To add content to your portal you must have Publisher or higher privileges.

There are four ways to add items to your portal:

- **Add a file from your computer**: This method allows you to add supported files from your computer to your portal. Some of these items can only be downloaded and used by others in your organization in desktop applications. Other items, such as image files, can be referenced in Portal applications through their URL. To add an item to your portal, do the following:
    1. Sign in to Portal for ArcGIS with privileges to create content.
    2. Go to **My Content** and click on **Add Item | From my computer**.
    3. Choose the file, adjust the title if necessary (it defaults to the filename the minus extension), and add tags. Click on **Add Item**.
    4. The item will be added to your portal and you will be redirected to its item details page where you can then share the item as needed.

- **Add an item from the web**: This lets you add a reference to an item's REST endpoint. You can, of course, reference ArcGIS Server web services, along with other types of web services, documents, and images. To add an item from the web, do the following:
    1. Sign in to Portal for ArcGIS with privileges to create content.
    2. Go to **My Content** and click on **Add Item** | **From the web**.
        1. Choose the type, enter the URL to the item, and enter a title and tags. If the service is secured, you will be prompted to enter credentials to access the service and select whether to store those credentials. If you choose not to store credentials, users will be prompted for credentials when they attempt to access the item.
    3. Click on **Add Item**. The reference to the item will be added to your portal and you will be redirected to its item details page where you can then share the item as needed.

- **Add an application**: Much like adding an item from the web, adding an application simply shares the URL of the application with your organization and users. To add an application, do the following:
    1. Sign in to Portal for ArcGIS with privileges to create content.
    2. Go to **My Content** and click on **Add Item** | **An application**.
    3. Set the type of the application, the application purpose, and other options based on the type of application.
    4. Add a title and tags, and then click on **Add Item**. The reference to the item will be added to your portal and you will be redirected to its item details page where you can then share the item as needed.

- Publish a feature service or tiled map service. See the *Publishing a feature service from ArcMap* section for more information.

Managing service data

For a service to display data, it must be able to access that data. This applies to ArcGIS Server, Portal for ArcGIS, and ArcGIS Online services. Vector data can be stored in an enterprise or file geodatabase; raster data can be stored in an enterprise or file geodatabase or on disk.

Making data accessible to ArcGIS Server

Your ArcGIS Server machines need to be able to access your data. To do this, follow these three rules:

1. Store your data where your ArcGIS Server machines can see it and access it. Here there are several options:
    - **Store data locally on each ArcGIS Server machine**: This is not an optimal solution, especially if you have multiple machines in your server site. This can be beneficial for performance, but not for maintenance.
    - **Store data in a shared directory**: Many organizations do this already; share data on a network drive using a **Universal Naming Convention** (**UNC**) path, such as `\\server\folder\data`. This has its merits in that if used properly, UNC paths provide a uniform, consistent method to reference data from anywhere on your network. The biggest drawback to this method is network latency; traffic on the network can introduce bottlenecks and hinder performance.

If sharing data over your network, **never** use mapped drives. **Always** use UNC paths. Drive letters can change, and mapped drives are usually set based on permissions of the logged on user, so access can vary widely. UNC paths are a much more stable and safer method.

- **Store data in an Enterprise geodatabase**: The geodatabase provides a standardized, powerful, robust method of storing and sharing data. Esri recommends storing your source data for services in an Enterprise geodatabase. However, this is often not optimal for some organizations due to licensing costs and the administrative overhead (need for a database administrator) associated with using an Enterprise geodatabase.

2. Grant the ArcGIS Server account at least read access to your data--if your data is stored on a shared network UNC path, that folder needs to be accessible to the ArcGIS Server account. This need for data accessibility is one of the driving forces behind using a domain account for the ArcGIS Server account (see `Chapter 1`, *ArcGIS Enterprise Introduction and Installation* for more on the ArcGIS Server account). If you are using a local account for the ArcGIS Server account, ArcGIS Server will not be able to access network resources such as UNC shares.

3. **Register your data with ArcGIS Server**: This makes the data source a well-known location that is approved by the ArcGIS Publisher and/or administrator. See the following section for more information on data registration.

Enterprise geodatabase or file geodatabase?

A common question is whether to house the data behind map services in a file geodatabase or in the Enterprise geodatabase. Rarely is there a quick and easy answer to this question, as there are many factors that can come into play and scenarios can vary widely from organization to organization.

You are advised to give serious consideration to this topic and consult the ArcGIS Enterprise online documentation for in-depth details on this topic.

As stated previously, Esri recommends storing source data for services in an enterprise geodatabase. Unfortunately, this is not an option for many smaller organizations due to lack of funding for licensing and/or staffing of a geodatabase administrator. Typically, using a file geodatabase on the ArcGIS Server machine to store data for services is a more than satisfactory approach for smaller organizations, especially those with a single-server deployment. Even for medium to larger-sized organizations, it is not uncommon to house publication data in a file geodatabase separate from the enterprise geodatabase and have that file geodatabase regularly updated via custom Python scripts or with geodatabase replication. There are considerations for using file geodatabases:

- If you have a multiple server configuration, each GIS server needs its own copy of the file geodatabase for performance reasons. This could be accomplished easily with a Python script.
- File geodatabases as the data source behind ArcGIS Server services that are intended to be used as read-only.
- Enterprise geodatabase functionality is, of course, absent. This includes versioning, replication, failover, logging, and backup and recovery.

Regardless of which method you employ, you will either be required to register your data sources with ArcGIS Server or copy data to the server.

Registering data sources

Registering data sources can only be done by those with Publisher or Administrative access to ArcGIS Server, thus making it a way to force users to reference data locations known and approved by Publishers and Admins. Both enterprise and file geodatabases are required to be registered with ArcGIS Server. If not registered, the data will be copied to the GIS server (see the next section). Data sources can be registered from ArcMap, ArcCatalog, and ArcGIS Server Manager.

One of the simplest methods of registering a data source is doing so only when you must, at the time of service publishing. This works well on a fresh installation of ArcGIS Server, when no data sources have yet been registered. While publishing, the **Prepare** window will inform you (as a warning) of any unregistered data sources. When the warning is right-clicked, the context menu provides options, one of which is to **Register Data Source With Server**, as shown in the following screenshot:

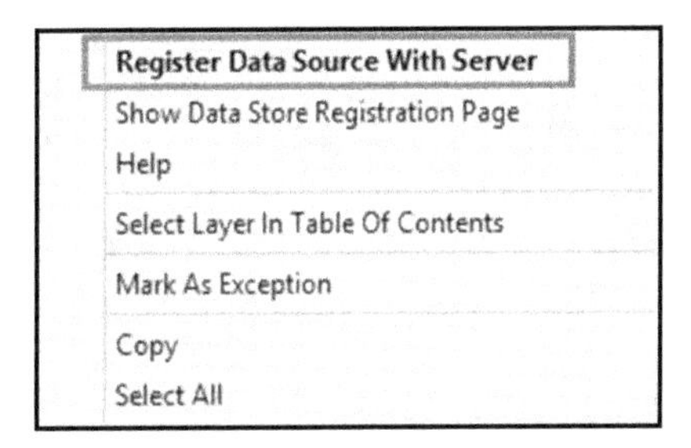

Selecting this option reads the connection information from the layer source and creates a connection string to the database. Give the registration a **Name** and click on **OK**:

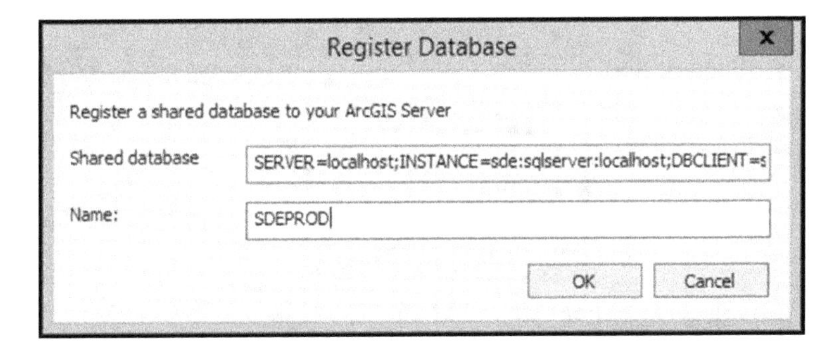

Run **Analyze** again, and your data registration warning should no longer appear, as the data source is now registered. All subsequent connections using that same connection will also be registered.

A data source, either an enterprise geodatabase connection or a folder location, must only be registered once with ArcGIS Server. All subsequent connections to that data source will then use the same registration.

To register a data source through ArcCatalog, connect to your server instance as a publisher or admin. Right-click on the connection in the Catalog tree and select **Server Properties**. You are now presented with the **ArcGIS Server Properties** window. Go to the **Data Store** tab. To register a database, follow these steps:

1. Click on the + button to the right of the **Registered Databases** section.
2. In the **Register Database** window, click on **Import**, as you can see in the following screenshot. Browse to the enterprise geodatabase connection you want to register. Click on **Select**. Give the connection an appropriate meaningful name, such as `webreader@sdeprod`. Click on **OK** in the **Register Database** window:

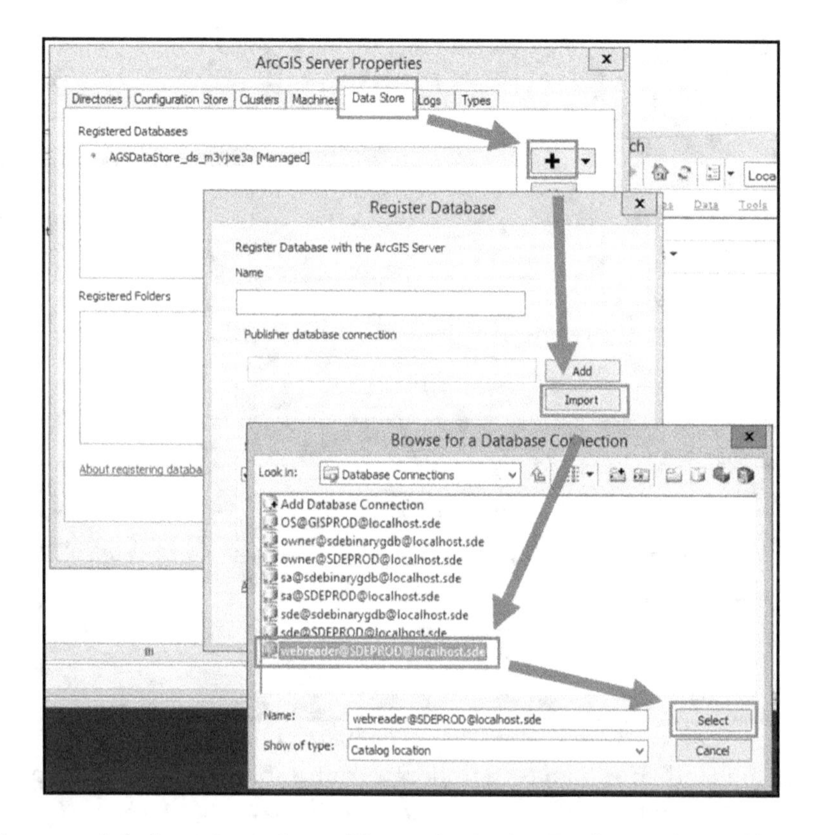

3. If successful, the connection will now be in the **Registered Databases** section with a green check mark to the left of it.

To register a folder, do the following:

1. Click on the + button to the right of the **Registered Folders** section.
2. Give the folder an appropriately descriptive name and click on the **Add** button. Navigate to the folder you wish to register and select it.
3. Click on **OK** in the **Register Folder** window.
4. If successful, the connection will now be in the **Registered Folders** section with a green check mark to the left of it.

Copying data to the server

Copying data to the GIS server is a method ArcGIS Server can use to make your data accessible to the server. Typically, most organizations do not want to copy data to the server, as it, of course, makes a copy of the data that is then disconnected from the source. However, there may be scenarios and workflows where you would want or need to copy data to the server when publishing. Some examples are as follows:

- **Small organization with a handful of services**: If your data does not require an enterprise geodatabase or you simply cannot have one due to licensing constraints, and you only have a small number of services, it is best and easiest for you to just let ArcGIS Server automatically copy the data to the server. Be careful of data duplication with this method and design your services carefully to avoid duplication.
- **Your ArcGIS Server instance is in the cloud**: If your server is in the cloud, it will more than likely need its own copy of the data if one is not already there. If you cannot log on to the cloud server, you could still publish to it directly and copy the data automatically. Be careful of data duplication with this method and design your services carefully to avoid duplication.

Publishing to the ArcGIS Data Store

Data can be added and published to the Data Store in several ways. We will discuss a few of those methods here. Data can be added to Portal and then published to a feature service--feature services can be published from ArcMap, and feature services can be published from ArcGIS Pro.

Publishing a CSV file

A CSV file with either an address field or latitude/longitude in decimal degrees can be added to Portal and subsequently published as a feature service. Features are published in the Web Mercator coordinate system.

Portal reads the first 10 rows of the CSV file to determine the data type and maximum length to use for each field in the hosted feature layer. Once these field types and lengths are set, any subsequent rows that do not match the types and lengths will not be included in the hosted feature layer. Pre-process your CSV to make sure records with the longest fields are in the first 10 rows. If your file has fields that contain mixed numeric and text fields, move one of those to the top as well so Portal sets it as a text field (numbers can be written to text fields, but text cannot be written to numeric fields).

To add and publish a CSV file, follow these steps:

1. Sign in to Portal for ArcGIS with privileges to create content.
2. Go to **My Content** and click on **Add Item** | **From my computer**.
3. Choose the file and add an appropriate title and tags.
4. Check **Publish this file as a hosted layer**.
5. Choose how you want to locate features. You can use latitude/longitude (decimal degrees only), address fields, or you can publish the CSV as a non-spatial table. To set up how to locate features, do the following:
    - For latitude/longitude, locate the fields containing the latitude and longitude coordinates and assign them in the **Location Fields** column. Also, review and change any field types.
    - For an address, choose the appropriate **Country** from the drop-down list. Locate the address fields and assign them in the **Location Fields** column. Also, review and change any field types.
6. Click on **Add Item**. Portal will now add the item and create a hosted feature service from it. This could take some time, depending on the size of the input CSV file.
7. Once the publishing is successful, you will be taken to the item details page where you can share the feature service and test it out by adding to a web map.

Publishing a feature service from ArcMap

Hosted feature services can be published to your Portal if ArcMap is connected to your Portal. To connect ArcMap to your Portal, do the following:

1. Close all the ArcGIS Desktop applications and launch **ArcGIS Administrator**. Click on the **Advanced** button.
2. Under the **ArcGIS Online** section, click on **Manage Portal Connections...**.
3. In **Manage Portal for ArcGIS Connections**, click on the **Add** button.
4. Add the Web Adaptor URL of your Portal in the **Add Portal for ArcGIS** window. Back in the **Manage Portal for ArcGIS Connections** window, select your new connection and click on the **Connect** button.

To publish a feature layer to be hosted on your portal, follow these steps:

1. Open the MXD you wish to publish in ArcMap.
2. Go to **File** | **Sign In**. Sign in to your portal with publisher or higher credentials.
3. Go to **File** | **Share As** | **Service**.
4. Select **Publish a service** and click on **Next**.
5. In the **Connections** drop-down of **Publish a Service windows**, select **My Hosted Services (Portal for ArcGIS)** and name your service. Click on **Continue**.
6. Set the **Feature Access** properties for your feature service in **Service Editor**.
7. Add in the required information under **Item Description**.
8. Click on **Publish**.
9. Once publishing is successful, your feature service will be available in the **My Content** section of the account that was used for publishing.

Publishing a feature service from ArcGIS Pro

As with ArcMap, Pro must also be connected to your portal. To connect to your Portal from Pro, do the following:

1. Launch Pro. Go to the **Projects** tab. Click on **Portals**, then the **Add Portal** button. Add your Portal **Web Adaptor** URL.
2. Right-click on the newly added portal in the **Portals** list and choose **Sign In**. Once signed in, right-click on the portal connection again and select **Set as Active Portal**. This sets this portal as the one we will publish to.

To publish a hosted feature layer, do the following:

1. Start ArcGIS Pro and open your project.
2. Either right-click on a single layer you wish to publish and select **Share As Web Layer** or to publish all the layers in the map, go to the **Share** tab and click on **Web Layer | Publish Web Layer**.
3. Give the feature layer a name; choose to **Copy all data**.
4. Choose **Feature** for **Layer Type**, as we are publishing a hosted feature layer.
5. Set the **Item Description** properties.
6. Share the item accordingly. This can also be done in Portal later.
7. Under **Configuration**, select the **Configure layers** button, and then click on the **Configure Web Layer Properties** button. Set the allowed operations and properties as needed (similar to when we published a feature service to ArcGIS Server earlier in the chapter) in the *Feature service operations and properties* section.
8. Once publishing is successful, you can select **Manage the web layer** to be taken to the item in your portal.

Extending services

Using **Server object extensions** (**SOEs**) and **Server object interceptors** (**SOIs**), it is possible to extend ArcGIS Server map and image services with custom Java or .NET code that is executed on the GIS server through a client application.

Server object extensions

An SOE creates new service operations to *extend* the base functionality of a map or image service. SOEs are appropriate if you need to expose custom functionality that is not available in any other manner or that needs to be executed quickly. An example use-case for an SOE would be a situation where an ArcObjects code must be used to accomplish a task that cannot be done with a geoprocessing task. Before embarking on developing an SOE, look closely to see if out-of-the-box tools can accomplish the task at hand. Also, remember that custom Python scripts can accomplish many tasks that only a few years ago were attainable only by using through ArcObjects.

Server object interceptors

SOIs enable you to change the behavior of existing map or image service operations, such as the behavior of a query or map image request. This is done by *intercepting* the requests of the built-in operations and executing custom code that overrides those operations. Examples of SOIs include adding a watermark to all images created by ArcGIS Server and layer-level security in map services.

Summary

This was an important chapter, as we discussed the publication of services, a key component in ArcGIS Enterprise. A wide range of service types can be published with various capabilities to ArcGIS Server, Portal for ArcGIS, and ArcGIS Online, allowing organizational data and information to be easily shared and managed. These capabilities each have their own array of settings and properties, all of which can be changed and tuned for your needs. An important step in the process is determining where your data needs to reside and managing service data sources. We also touched lightly on several ways to add and publish data to Portal for ArcGIS. Next, we will dive into Portal for ArcGIS and discuss how to leverage Portal to manage your organization's content, groups, and users and how those users access information.

4

ArcGIS Server Administration

ArcGIS Server is literally and figuratively a core component of ArcGIS Enterprise; literally, in the sense that it is one of the four components of Enterprise; figuratively, in the sense that the functionalities it provides are central to the principles behind web GIS. Knowing this, it is easy to see the importance and grasp the gravity of ArcGIS Server administration.

Just as with every other component of ArcGIS Enterprise, a well-maintained and clean ArcGIS Server environment lends to an efficient and smooth-running site. There are many interfaces in ArcGIS Server administration, and we will discuss them all here. From **ArcCatalog** to **ArcGIS Server Manager** to the **REST Administrator**, these all provide different windows in ArcGIS Server from which you can perform varying levels of tasks. Sometimes, for quick tasks, these can be accomplished through ArcCatalog (many tasks that can be completed in ArcGIS Server Manager can also be completed through an administrative connection in ArcCatalog) and other fine-grained settings can only be accessed through the REST Administrator.

We can by no means cover all aspects of ArcGIS Server administration in a single chapter. That said, once you have completed this chapter, you will be comfortable with exploring, navigating, and changing server settings.

In this chapter we will cover the following topics:

- Connecting to an ArcGIS Server site
- What the ArcGIS Server config store is and how it is structured
- Working with the ArcGIS Server logs
- Backup and restore of an ArcGIS Server site
- Retrieving primary site administrator credentials
- Working with the ArcGIS Server REST Administrative directory to manage tokens, services, and system settings
- Out-of-the-box command line utilities useful for administrative tasks

Connecting to an ArcGIS Server site

Before any administrative tasks can be completed, we need to first know where and how to connect to our administrative interfaces.

Accessing ArcGIS Server Manager

Since the earliest days of ArcGIS Server, ArcGIS Server Manager has been the central, web-based management tool for ArcGIS Server. When we installed ArcGIS Server in `Chapter 1`, *ArcGIS Enterprise Introduction and Installation*, we were presented with ArcGIS Server Manager after the initial site setup. Depending on your Web Adaptor configuration, there are several ways to access ArcGIS Server Manager in a web browser, such as the following:

- From the ArcGIS Server machine, ArcGIS Server Manager can be accessed at `https://localhost:6443/arcgis/manager`. A shortcut to ArcGIS Server Manager at this URL is also installed with ArcGIS Server and can be found under the Windows Start menu as **ArcGIS Server Manager**.
- From another machine on the internal network, ArcGIS Server Manager can be accessed using the server name instead of `localhost`, such as `https://servername:6443/arcgis/manager`.
- If, during your Web Adaptor configuration, you choose to **Enable administrative access to your site through the Web Adaptor**, you also will be able to access ArcGIS Server Manager through your **fully qualified domain name** (**FQDN**), such as `https://www.masteringageadmin.com/arcgis/manager`.

Regardless of what URL you use, you will log in as the **primary site administrator** (**PSA**) designated during installation or with other administrator credentials.

Note that all the preceding URLs use HTTPS. ArcGIS Server automatically redirects access attempts to `Manager` over HTTP to HTTPS unless your site is set to use **HTTP Only for communication**, which is typically not recommended.

Accessing the ArcGIS Server REST Administrator directory

Recall from earlier discussions that ArcGIS Server exposes its functionality through web services using REST. With this architecture comes the ArcGIS Server REST **Application Programming Interface**, or **API**, that, in addition to exposing ArcGIS Server services, exposes *every* administrative task that ArcGIS Server supports. In the API, ArcGIS Server administrative tasks are considered resources and are accessed through URLs (which are Uniform *Resource* Locators, after all). Operations act on these resources and update their information or state. Resources and their operations are hierarchical and standardized and have unique URLs. Like the web, the REST API is stateless, meaning that it does not retain information from one request to another by either the sender or receiver. Each request that is sent is expected to contain all the necessary information to process that request. If it does, the server processes the request and sends back a well-defined response. As it is accessed over the web, the ArcGIS Server REST API can also be invoked from any programming language that can make a web service call, such as Python.

As with accessing ArcGIS Server Manager, accessing the ArcGIS **Server Administrator Directory** can be done in several ways, depending upon your Web Adaptor configuration:

1. From the ArcGIS Server machine, the Server Administrator Directory can be accessed at `https://localhost:6443/arcgis/admin`. There is no shortcut to this URL in the Windows Start menu.
2. From another machine on the internal network, the Server Administrator Directory can be accessed using the server name instead of `localhost`, such as `https://servername:6443/arcgis/admin`.
3. If, during your Web Adaptor configuration, you chose to **Enable administrative access to your site through the Web Adaptor**, you also will be able to access the Server Administrator Directory through your FQDN, such as `https://www.masteringageadmin.com/arcgis/admin`.

As with Server Manager, you will log in as the PSA designated during installation or with other administrator credentials.

Prior to ArcGIS 10.1, server configuration was held in plain text configuration files in the configuration store. These files are no longer a part of ArcGIS Server architecture. The ArcGIS Server REST Administrator Directory now exposes these settings.

Accessing server settings through ArcCatalog

Once you have an administrator connection to ArcGIS Server setup in ArcCatalog, which we did in `Chapter 3`, *Publishing Content*, accessing ArcGIS Server to perform administrative tasks is simple. To access administrative settings of ArcGIS Server through ArcCatalog, do the following:

1. Launch ArcCatalog and go to the **GIS Servers** section of **Catalog Tree**.
2. Right-click on your administrative connection to ArcGIS Server and select **Connect**.
3. Right-click on your administrative connection again in the **Catalog Tree** and select **Server Properties...**:

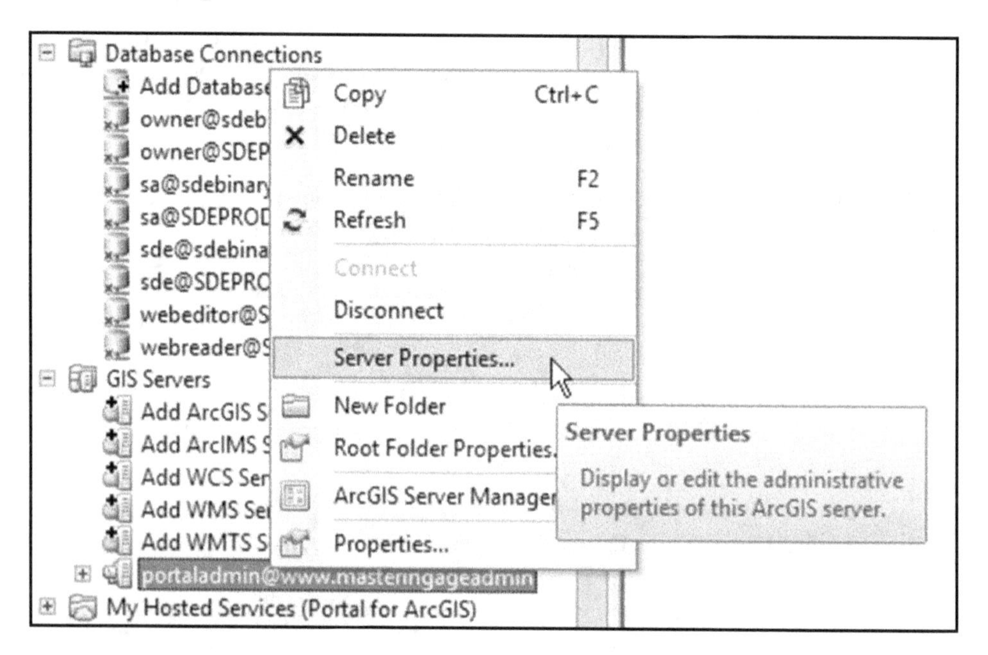

4. You are now at the **ArcGIS Server Properties** window. Here, you can access items such as the ArcGIS Server directories settings and configuration store settings. You can also work with Registered Data Stores under the **Data Stores** tab (the biggest reason I typically access the ArcGIS Server settings from ArcCatalog).

A quick tour of the configuration store and ArcGIS Server directories

The ArcGIS Server configuration store, commonly referred to as the **config store**, is a directory of files (many of which are JSON formatted) that contain all the essential properties of the server site, including clusters, machines, services, web adaptors, and security to name a few--in short, the configuration. For this reason, the config store is a crucial piece of ArcGIS Server and should be treated with great care and respect. If possible, the configuration store should reside in a redundant location. In a single server site, the config store typically resides on the same machine that is running ArcGIS Server. In a multiple machine site setup, the config store is required to be stored in a location accessible by all ArcGIS Server machines, such as a network share. If desired, and the hardware resources are available, the config store can even be isolated on its own file server.

The location of the config store can be changed at any time to accommodate growth of your site and changes in your architecture. When starting out with ArcGIS Server, you may only have one server, for example. As usage increases, so do your hardware requirements. If you move to a multiple machine setup, you will need to move your config store. One way to do this is through an admin connection to ArcGIS Server in ArcCatalog, as discussed in the previous section. On the **Configuration Store** tab, set the **Configuration file path** parameter and click on **Apply**. When you change the configuration store location, ArcGIS Server moves the existing configuration store for you and then restarts. The config store location can also be changed in ArcGIS Server Manager.

Keep file permissions in mind when changing the configuration store location. Remember that the ArcGIS Server account will need full access to the new config store location.

Like the config store, the ArcGIS Server directories, or **server directories**, represent physical directories that are specifically designated to store server site information and data. The server directories are created during ArcGIS Server site setup (see `Chapter 1`, *ArcGIS Enterprise Introduction and Installation*, for more information), and like the config store, can be changed at any time, either in the server properties of an administrative connection in ArcCatalog or in ArcGIS Server Manager. There are four server directories, each with its own purpose:

- **arcgiscache**: This holds cache tiles for cached map services. Caches can get large in file size, so ensure that your server directories reside in a location that is adequately provisioned. Remember you can always change the location of your directories to a new location later.

- **arcgisjobs**: This stores files used by geoprocessing services such as temporary files and information about ongoing jobs and results.
- **arcgisoutput**: This holds temporary files needed by the server. It is required by geoprocessing services as this is where geoprocessing results are written to.
- **arcgissystem**: This is used to manage information required for maintaining ArcGIS Servers, services, and database connections.

Neither the config store nor the server directories are intended to be used as storage locations. Place data and other site-specific folders in a separate location.

The **arcgisjobs**, **arcgisoutput**, and **arcgissystem** directories are periodically cleaned to get rid of old, stale content. These directories are set to be cleaned using the criteria as follows. These parameters can be changed in both the ArcGIS Server Administrator and the ArcGIS Server REST Administrator Directory. The following table shows the default cleanup settings for the ArcGIS Server directories:

Directory	Cleanup mode	Maximum file age	Notes
`jobs`	Time since last modified	360 minutes (6 hours)	A job is only deleted if it has completed, been canceled, or failed
`output`	Time since last modified	10 minutes	Only files or folders that begin with `_ags` are deleted
`system`	Time since last modified	1,440 minutes (24 hours)	Only items that have been completely uploaded to the server are deleted

Now that we have discussed the config store and server directories so that you are familiar with their structure and purpose, let me advise you to **never** directly edit, change, or modify the files or folders within either of these locations. One exception to this may be that if you are experiencing an issue, Esri Technical Support may instruct you to make changes directly to the files in your config store or server directories.

Changes that you may need to make to any of the files in the config store or server directories can be made either in ArcGIS Server Manager or the ArcGIS REST Administrator Directory.

Carrying out administrative tasks

Now that we have discussed the different methods to connect the admin interfaces to ArcGIS Server, let's cover some administrative tasks and scenarios, and how we can tackle them.

Adding and removing machines from an ArcGIS Server site

ArcGIS Server machines are the work horses of your server site. The beauty of ArcGIS Server is that it is fully scalable, meaning that, if in the future you need to add more processing power to your site, you can simply add another ArcGIS Server machine to help distribute the workload. To add a machine to your existing site, it is recommended to provision the new machine to have the exact specs of existing servers in the site. Some of the requirements of a new machine that must be the same as the other ArcGIS Server machines in the site include the following:

- Same operating system
- Same hardware specs (RAM, processor, and more)
- Same version number of ArcGIS Server
- Same licence
- Same ArcGIS Server account (domain is best)

Other requirements include the following:

- It must be able to read/write to the site's config store and server directories
- It must be able to communicate with other ArcGIS Server machines in the site through the proper ports
- It must have read access to data referenced by any services in the site

You can add or remove a machine in ArcGIS Server Manager by going to **Site** | **GISServer** | **Machines** | **Add Machine**. Enter the **fully qualified domain name** (**FQDN**) of the ArcGIS Server machine you want to add to the site and its URL, including the appropriate protocol and port (HTTP:`6080`/HTTPS:`6443`):

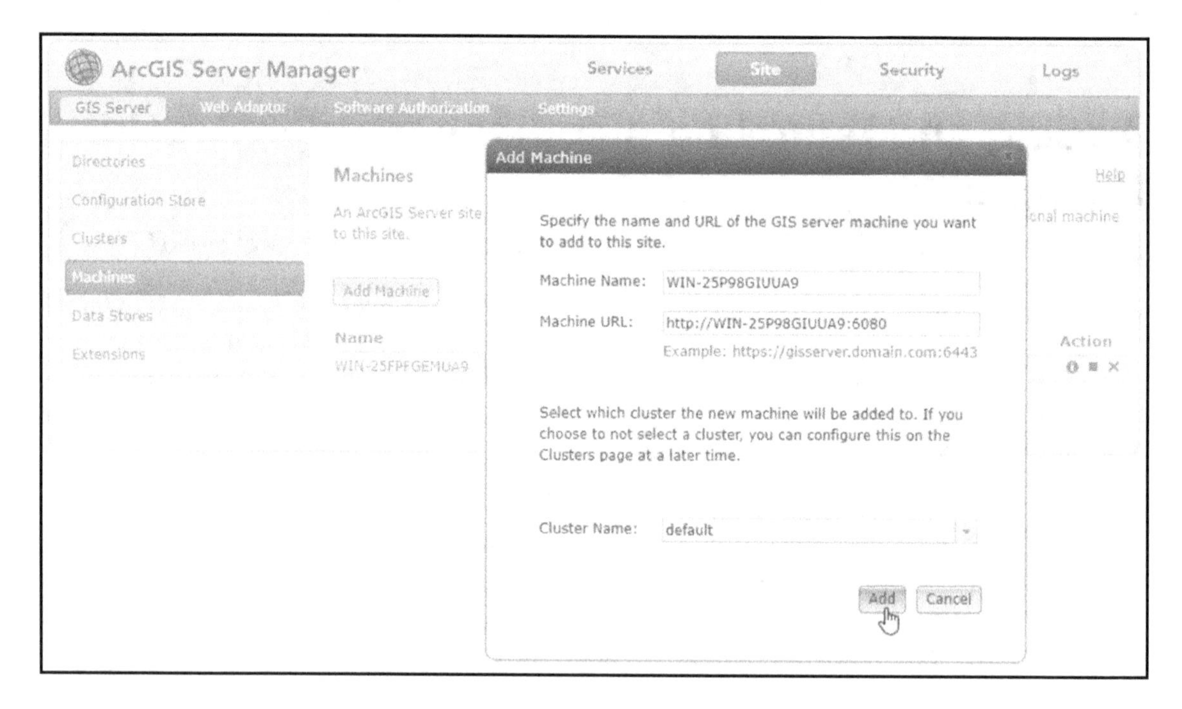

Not sure what the FQDN is of the machine you are adding? If you can, log on to the server, open Windows Explorer, right-click on **This PC** in the catalog tree and go to **Properties**. Under **Computer name, domain, and workgroup settings**, you will find the **Full computer name**:

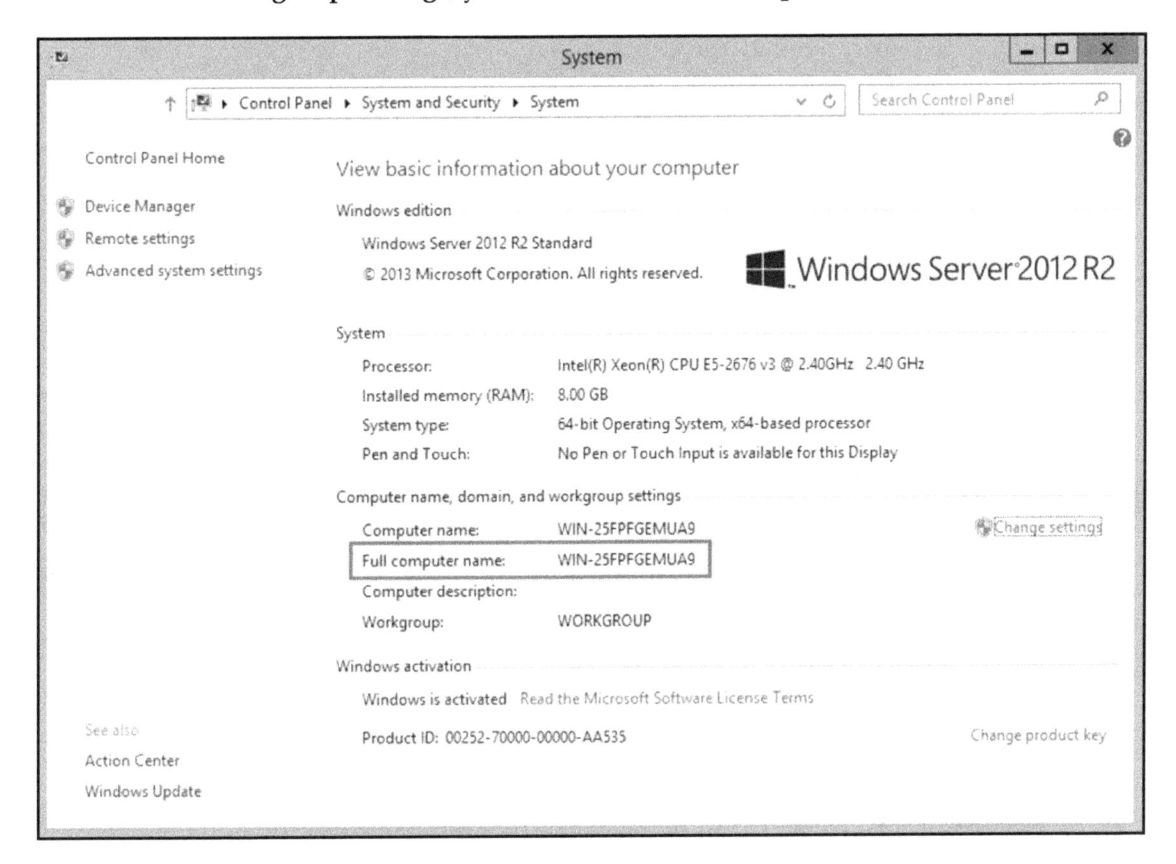

A server machine can be added or removed from your site through the **ArcGIS Server Properties** windows in ArcCatalog as well. Right-click on your administrative connection to ArcGIS Server under **GIS Servers** in the **Catalog Tree,** and go to **Server Properties**. On the **Machines** tab, add a machine using the same parameters discussed earlier, when using ArcGIS Server Manager.

Using and managing ArcGIS Server logs

ArcGIS Server is constantly recording events that occur, along with any errors associated with those events, to log files. Logs are crucial to troubleshooting errors with your site and provide a history of events that occur over time. A wide range of events are recorded in the logs, including, but not limited to, the following:

- Installation and upgrade events
- Service publication
- Data store registration
- Site management events such as adding or removing servers or modifying the config store
- Security events
- Service events such as starting and stopping and layer draw times

To access the logs in ArcGIS Server Manager, go to **Logs** | **View Logs**.

Log settings

There are not many settings related to logging, but they are all important. You will eventually need to change some of these to help you either capture event information or manage the lifespan or location of your server logs. To access log settings in ArcGIS Server Manager, go to **Logs** | **View Logs** | **Settings**.

Log level

Logging level is an important tool for diagnosing issues and debugging errors in ArcGIS Server. The default log level is **Warning**, which will log moderate to severe problems that require attention. Each level includes all levels above it in its messaging. The following table describes the different logging levels available in ArcGIS Server and gives examples of the items logged by each:

Log Level	Description
Severe	Serious problems that require immediate attention. Includes only severe messages.
Warning	Moderate problems that require attention. Includes severe messages.
Info	Common administrative messages of the server, such as service creation and startup. Includes severe and warning messages.

Fine	Common messages from use of the server. Includes severe, warning, and info messages.
Verbose	Messages about how fast the server draws layers, if each layer in a service was drawn successfully, or how long it took for the server to access a layer's source data. Includes severe, warning, info, and fine messages.
Debug	Highly verbose messages intended to be used for troubleshooting issues. Not recommended for use in production environments as it may cause a decrease in performance. If you enable **Debug** logging, you **must** remember to turn it back off; if you do not, your logs can grow substantially in size and could possibly fill up your disk.
Off	Logging is turned off and no events are logged. Not recommended for production environments.

Changing the logging level of ArcGIS Server is easy through ArcGIS Server Manager. In Server Manager, go to **Logs** | **View Logs** | **Settings**. Select your logging level from the dropdown, and click on **Save**:

In `Chapter 7`, *Scripting Administrative Tasks*, we will look at how to access and query the ArcGIS Server logs programmatically with Python.

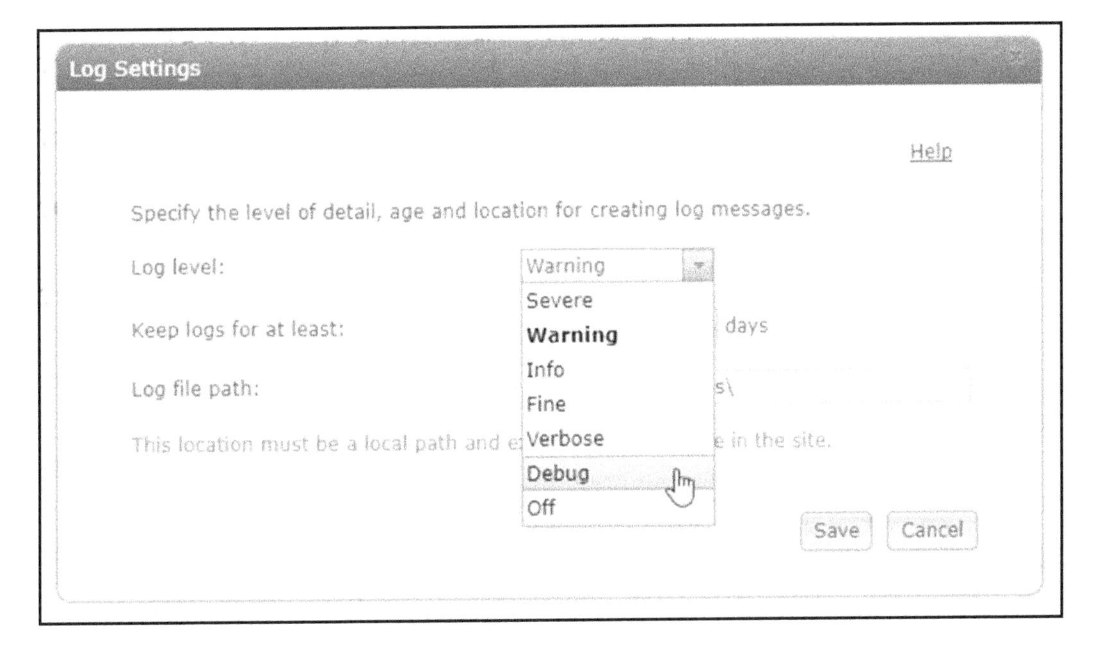

If you ever set your log level to **Debug**, do so only for as long as you absolutely must, and **always** remember to change it back to your default setting afterwards.

Log retention time

Log retention time, or the number of days that logs are kept on disk, is another important log setting in ArcGIS Server. By default, logs are retained on disk for 90 days. When a log file becomes older than 90 days, it is deleted by ArcGIS Server. Your organization may have **standard operating procedures** (**SOPs**) in place that determine the lifespan of your logs; consult with your systems administrator if you are unsure of how long your organization may want to keep the ArcGIS Server logs for auditing purposes.

Log retention time can be used in conjunction with log level to keep your log directory from growing too large. If your organization requires verbose logging, perhaps you can lower the log retention time to 30 days from 90.

To change the log retention time in ArcGIS Server Manager, go to **Logs** | **View Logs** | **Settings** and change the **Keep logs for at least** parameter accordingly.

Logs directory

The log directory location is important as it pertains to available disk space. The default directory where ArcGIS Server writes logs is `C:\arcgisserver\logs`. This is created on initial site setup and is based on the location chosen for the config store and server directories. It is not uncommon, especially with the use of virtual machines, to have C drives that are small and dedicated to the operating system only, and an additional attached named drive for ancillary software installs. If you have the option to use an additional attached named drive to store your config store and server directories, it is best to put your logs on this drive as well. As was discussed with log retention times, it is important to find a balance between log level and log retention; the log storage location also plays into this equation.

Esri recommends storing the ArcGIS Server logs directory on a local disk and not on a shared network drive. It recommends keeping the logs directory on a local disk resource.

To change the log directory location in ArcGIS Server Manager, go to **Logs** | **View Logs** | **Settings**, change the **Log file path**, and click on **Save**.

All the preceding log settings can also be changed easily through an administrative connection to ArcGIS Server in the ArcCatalog under the **Logs** tab.

Backup and restore of an ArcGIS Server site

Just as you back up your enterprise geodatabase (you *do* have backups scheduled, don't you?), you should routinely back up your ArcGIS Server site configuration so that it could be restored in the event of catastrophic hardware failure, corruption, or even something as innocent as human error.

Tucked away in your ArcGIS Server installation directory (`D:\Program Files\ArcGIS\Server\tools\admin`, in my case) are the ArcGIS Server command-line utilities, a set of scriptable Python utilities that can perform some very handy administrative tasks, two of which are server site backup and restore. Take a minute to find these scripts in your ArcGIS Server installation and examine some of them--`manageservice.py` is a good one that lets you programmatically stop and start services.

To back up your server site, follow these sites:

1. On the ArcGIS Server machine, open a command prompt. Navigate to the command-line utilities folder:

   ```
   C:\Users\Administrator>d:
   D:\>cd "Program Files\ArcGIS\Server\tools\admin"
   ```

2. Call the script and pass in the necessary parameters:

   ```
   "D:\Python27\ArcGIS10.5\python.exe" backup.py -u siteadmin
   -p pass -s http://localhost:6080 -f "D:\backups\ags"
   ```

   Here, `-u` is the administrative username, `-p` is the administrative password, `-s` is the server URL, and `-f` is the path to the folder to store the back up. If successful, you will get a message like this:

   ```
   Backing up the site running at "localhost"
   Site has been successfully backed up and is available at this
   location: D:\backups\ags\August-5-2017-10-43-32-PM-UTC.agssite
   ```

To restore your server site using the restore utility, follow these steps:

1. On the ArcGIS Server machine, open a command prompt. Navigate to the command-line utilities folder:

```
C:\Users\Administrator>d:
D:\>cd "Program Files\ArcGIS\Server\tools\admin"
```

2. Call the script and pass in the necessary parameters:

```
"D:\Python27\ArcGIS10.5\python.exe" restore.py -u siteadmin
-p pass -s http://localhost:6080
-f "D:\backups\ags\August-5-2017-10-43-32-PM-UTC.agssite"
-r D:\backups\ags
```

Where `-u` is administrative username, `-p` is administrative password, `-s` is server URL, `-f` is the path to the backup to restore, and `-r` is the path to where the utility will generate its report. If successful, you will get a message like the following:

```
Beginning to restore the site running on "localhost" using the
site backup available at:  D:\backups\ags\August-5-2017-10-43-
32-PM-UTC.agssite
This operation can take some time. You will not receive any
status messages and will not be able to access the site until
the operation is complete...
Site has been successfully restored. Import operation completed
in '00hr:02min:12sec'.
Below are the messages returned from the restore operation.
You should review these messages and update your site
configuration as needed:
-------------------------------------------------------
1. Configuration options pertaining to HTTPS were changed
   during the import operation and the Server is restarting.
   The site should become available shortly.
 A file with the report from the restore utility has been
 saved at:  D:\backups\ags\report.txt
```

Knowing that these scripts are callable from the command line, we can easily schedule a weekly backup of our ArcGIS Server configuration and move the backup off the ArcGIS Server machine onto a redundant file server for safe keeping. We could call `backup.py` from a Windows batch file, and then use another Python script to move the latest backup to another location. Our batch file would look like the following:

```
"D:\Python27\ArcGIS10.5\python.exe" "D:\Program
 Files\ArcGIS\Server\tools\admin\backup.py" -u siteadmin
 -p pass -s http://localhost:6080 -f "D:\backups\ags"
"D:\Python27\ArcGIS10.5\python.exe"
```

```
"D:\Scripts\ags\backup\move_ags_backup.py"
```

Our script to move the backup, `move_ags_backup.py`, would simply get the file from its source directory and move it to the target directory:

```
import os
import shutil

src_dir = r"D:\backups\ags"
tgt_dir = r"\\server\share\folder"
files = os.listdir(src_dir)
for f in files:
    shutil.move(os.path.join(src_dir, f),
        os.path.join(tgt_dir, f))
```

The batch file could be scheduled to run at a frequency best suited for your organization's workflows and backup policies.

When running `backup.py` from a batch file, if your password contains a special character, it may be necessary to enclose it in double quotes; `-p "*iuj7^%L"`, for example.

Resetting or changing the ArcGIS Server service account password

I hate to tell you this, but, sooner or later, you will have to change the password on an ArcGIS Server account. Either the password will get lost, expire, or you'll have some sort of permissions issue with ArcGIS Server, and the easiest thing to do is reconfigure the ArcGIS Server account to reset permissions on all the appropriate directories. I know, it sounds scary, but it's not. Trust me. Esri makes this pretty painless with the **Configure ArcGIS Server Account** wizard. Get to it by going to the Windows Start menu and then **Apps** | **Configure ArcGIS Server Account**.

Note that the heading for this section mentions resetting or changing. I bring up resetting of the ArcGIS Server account because you can do just that; reset the password to the same password you've been using all along--this is something you would do to take care of the permissions issues.

To run the **Configure ArcGIS Server Account** wizard, follow these steps:

1. Enter in the **ArcGIS Server Account**, **Password**, and **Confirm password**. You may optionally load a configuration file containing all the ArcGIS Server account setup information if you saved one previously. Do not select this option if you want to change the ArcGIS Server account or the account password. Click on **Next**:

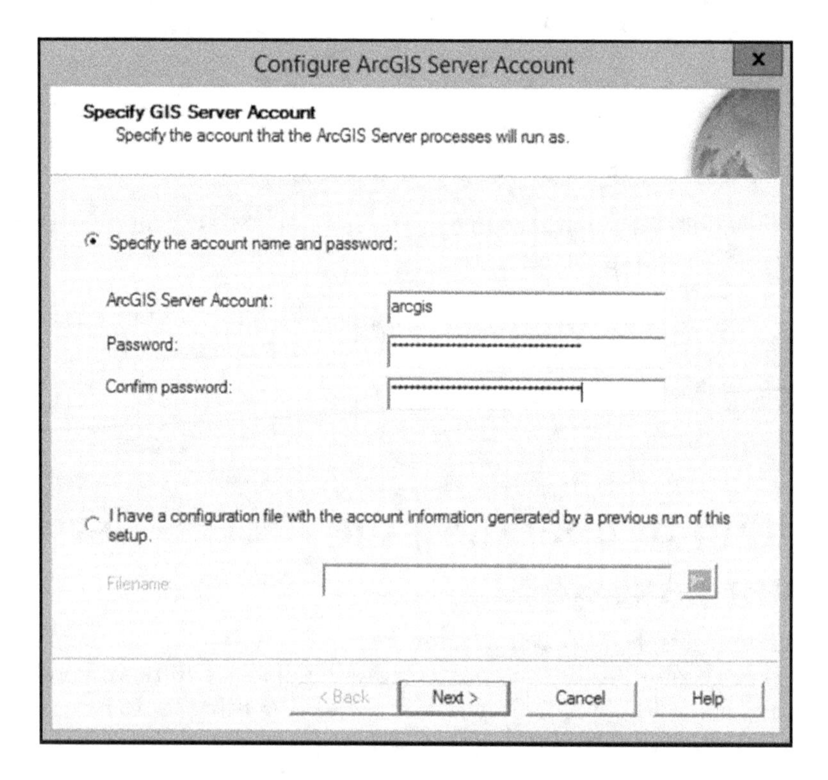

2. Enter in your **Root Server Directory**, **Configuration Store**, and **Logs Directory** paths. It is easiest and safest to open Windows Explorer, navigate to each of these, and copy/paste the full paths into the wizard. Click on **Next**.
3. Optionally, export a server configuration file. Click on **Next**.
4. You are now at the summary window. Click on **Configure**. The configuration process should take less than one minute.

5. Upon successful completion, you will be presented with a summary of the configuration process, which will look like the following:

```
              ArcGIS Server Account Summary
---------------------------------------------------------
Server configuration begins.
Adding user arcgis to computer WIN-25FPFGEMUA9.
The account already exists.
Setting ArcGIS Server user account rights.
Granting read permissions on folder d:\program files\arcgis\server\
for arcgis
Granting full control permissions on folder d:\program
files\arcgis\server\framework\ for arcgis
Granting full control permissions on folder d:\program
files\arcgis\server\geronimo\ for arcgis
Granting full control permissions on folder d:\program
files\arcgis\server\usr\ for arcgis
Granting full control permissions on folder d:\program
files\arcgis\server\bin\ for arcgis
Granting full control permissions on folder d:\program
files\arcgis\server\XMLSchema\ for arcgis
Granting full control permissions on folder d:\program
files\arcgis\server\DatabaseSupport\ for arcgis
Granting full control permissions on folder
D:\arcgisserver\directories for arcgis
Granting full control permissions on folder D:\arcgisserver\config-
store for arcgis
Granting full control permissions on folder D:\arcgisserver\logs
for arcgis
Setting ArcGIS Server service to run as account arcgis
Setting Log On As property for ArcGIS Server has succeeded.
Server Configuration completed
```

As you can see, the wizard simply adds the account, grants permissions to the appropriate directories, sets the ArcGIS Server service to run as the ArcGIS Server account, and grants the ArcGIS Server account log on as a service right. You have now successfully reset or changed the ArcGIS Server account password.

Don't let the ease of the **Configure ArcGIS Server Account** wizard fool you, there may be instances where you will need to manually go in and set permissions on some or all these directories. Permissions issues can be the worst. See `Chapter 10`, *Troubleshooting ArcGIS Enterprise Issues and Errors,* for more information on troubleshooting permissions issues.

Retrieve, reset, or change the ArcGIS Server PSA account credentials

Yes, you guessed it. Just like with the ArcGIS Server account, you will eventually have to retrieve, reset, or change an ArcGIS Server PSA account. You may inherit a system where no one knows the password (been there) or someone might have changed it, forgot to put it in the password manager (you *are* using a password manager, aren't you?), and has no idea what they changed it to (been there too). No worries here, Esri provides several methods to take care of these matters.

Retrieving a forgotten PSA account name

Esri provides a batch file utility to retrieve a forgotten PSA account name. The utility, `PasswordReset.bat`, is located in the `\tools\passwordreset` directory of your ArcGIS Server installation (`D:\Program Files\ArcGIS\Server\tools\passwordreset`, in my case). To get your lost account name, simply run the utility with the `-l` flag in a command prompt:

```
D:\Program Files\ArcGIS\Server\tools\passwordreset>passwordreset -l
```

This will give you the following output:

```
Primary site administrator account: siteadmin
```

Changing a forgotten PSA account password

The same utility, `PasswordReset.bat`, is used to change a forgotten PSA account password. To change a password, use the `-p` flag along with your new password:

```
D:\Program Files\ArcGIS\Server\tools\passwordreset>passwordreset
-p newpassword
```

When using the password reset utility, if your new password contains special characters, it may be necessary to enclose it in double quotes, `-p "$$ju&jJHY&"`, for example.

Changing a PSA account credentials when you know the current password

When you know the current PSA credentials and need to change the account credentials, there are two ways to accomplish the task.

The first and more traditional way to change the PSA credentials is through ArcGIS Server Manager. Log in to Manager as an administrator (you may log in as the current PSA, but know that as soon as you save the changes to the account, you will be logged out of the Manager). Go to **Security** | **Settings** and click on the pencil icon under the **Primary Site Administrator Account** section. In the **Edit Primary Site Administrator Account** window, enter in a new username (optional) and password and click on **Save**.

The second method utilizes the ArcGIS Server REST Administrator Directory. Log in to the REST Admin Directory (see earlier in this chapter for information on logging in) and go to **security** | **psa** | **update**. Enter a **Primary administrator account name** and **Password**. Select your response format (personally, I prefer **JSON**) from the dropdown and click on **Update**. You will get a response back, and, if successful, it will simply be as follows:

```
{"status": "success"}
```

Passwords with special characters do not need to be enclosed in double quotes when using ArcGIS Server Manager or the REST Admin Directory to change passwords.

Utilizing the ArcGIS Server REST Administrator Directory

The ArcGIS Server REST Administrator Directory, or **REST Admin**, as it will be herein referred to, is a powerful way to manage all aspects of ArcGIS Server administration, as it exposes *every* administrative task that ArcGIS Server supports. Remember from earlier that the API is organized into resources and operations. Resources are settings within ArcGIS Server and operations act those resources to update their information or change their well-defined state usually through a HTTP `GET` or `POST` method.

HTTP `GET` requests data from a resource while HTTP `POST` submits data to be processed to a resource. In other words, `GET` retrieves data, `POST` inserts/updates data.

An example of a resource is a service. An existing service can have a well-defined state of stopped or started; it must be one or the other. Operations available on the service resource in the REST API include **Start Service**, **Stop Service**, **Edit Service**, and **Delete Service**. The start, stop, and delete operations change the state of the service, from stopped to started and started to stopped, and either stopped or started to deleted (technically, if the service is started, it is first stopped before it is deleted) respectively. The **Edit Service** operation changes the information in the resource.

Resources can also have child resources that can, in turn, have their own set of operations and child resources. Remember that the API is hierarchical, so, for example, in the case of a service resource, it has the child resource **Item Information**, which has the **Edit Item Information** operation. To get to this operation in the REST Admin, we would log in to the REST Admin and go to **services** | **<servicename>** | **iteminfo** | **edit**, which would resemble the following in a URL form:

```
https://www.masteringageadmin.com/arcgis/admin/services/SampleWorldCiti
es.MapServer/iteminfo/edit.
```

In the REST Admin, we could now edit the service `description`, `summary`, `tags`, and `thumbnail`:

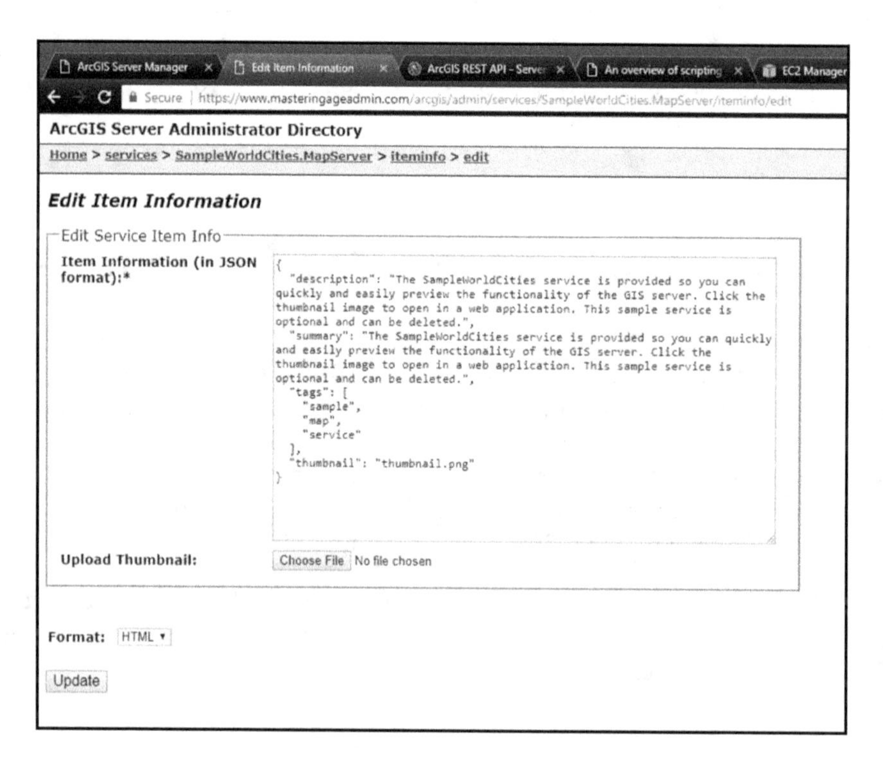

By updating the **Item Information** in the preceding example and clicking on the **Update** button, we would be sending an edit HTTP `POST` operation to the `https://www.masteringageadmin.com/arcgis/admin/services/SampleWorldCities.MapServer/iteminfo` resource. ArcGIS Server Manager equivalent for this process would be to go to **Services** | **Manage Services** | **Edit Service** (the pencil button to the right of the service name) | **Item Description**.

Hopefully, this gives you a better understanding of how the REST API works and how actions carried out in Server Manager and Server are executed by the API on the backend. Now that we have looked closely at the inner workings of the API, let's look at some examples of how you can use it to manage and administer ArcGIS Server.

For more information on the REST API and full reference on all resources and operations available, there is a link in the upper right of your ArcGIS Server Administrator Directory site, called **API Reference**. Here, you can find full documentation on all aspects of the REST API.

Navigating the REST Admin

At first glance, the REST Admin seems very plain in design, and it is. Remember that it is meant to be primarily accessed programmatically, not through a user interface. For this reason, the interface is simply pages of links to resources and operations with plain web forms for making configuration edits and changes. As previously stated, the API structure is hierarchical (the API Reference will help you navigate this), and you are safe just navigating around following links up and down resources and operations. Typically, to make an actual change, you must get into an operation page and click on a button, such as **Edit**, **Update**, or **Delete**. **Supported Operations** will always be listed at the bottom of a page. The following examples are to familiarize you with the structure and capabilities of the API, which we will continue to build on in later chapters.

Working with tokens

As an ArcGIS Server administrator, it is your responsibility to maintain and guard access to your ArcGIS Server site and its credentials. However, you will find that others in your organization will eventually need administrative access to your server to perform maintenance tasks. For example, you may have a developer writing custom scripts that need administrative access to ArcGIS Server. To avoid providing them with administrative credentials, you could instead generate a long-lived token that they could use in their script to authenticate to ArcGIS Server.

Token basics

Tokens are a way for users to authenticate themselves without having to provide a username and password. A username and password are required to generate the token. The token is an encrypted string that contains the username, password, the token expiration time, and other proprietary information. To access a secured resource, the user presents the token in place of typical credentials, such as a username and password. For security reasons, tokens have lifespans, and, therefore, they expire. If a user attempts to use an expired token, they will not be authenticated and allowed access.

Token lifespans

There are two lifespans available for the ArcGIS token, short-lived and long-lived. The default lifespan for a short-lived token is 60 minutes, and for a long-lived token, the default lifespan is 1,440 minutes (24 hours). These are configurable properties, and we will change them momentarily. Which type of token gets returned for a request depends on how it is requested from the client. The following table summarizes how ArcGIS Server issues tokens:

What client requests	What client receives
A token with no time-out value	A short-lived token with a time-out that matches the short-lived token lifespan property
A token with a time-out value less than the short-lived token lifespan setting	A short-lived token with the requested time-out value
A token with a time-out value less than the long-lived token lifespan setting	A long-lived token with the requested time-out value
A token with a time-out value that exceeds the long-lived token lifespan setting	A long-lived token with a time-out that matches the long-lived token lifespan property

In other words, if no time-out value is provided in the request, a short-lived token is returned with the default lifespan. If a time-out value less than the short-lived lifespan setting is provided, a short-lived token is returned with the requested lifespan. If a time-out value less than the long-lived lifespan but greater than the short-lived lifespan is provided, a long-lived token with the requested time span is returned. Finally, if a timeout value that exceeds the long-lived lifespan is requested, a long-lived token with the default long-lived lifespan is returned. A token will never be provided with a lifespan longer than the long-lived token default lifespan.

Changing token settings

That said, before you start giving out tokens, you're probably going to need to make some token configuration changes, seeing that the default long-lived lifespan is only one day. You want the token you provide to last for more than a day, but, at the same time, you don't want it to last for five years either. The shorter the life of a token, the more secure it is. The shorter the lifespan of the token, the more often an administrator has to reissue it.

Token settings can be changed in two ways:

- To change token settings using ArcGIS Server Manager, follow these steps:
    1. In ArcGIS Server Manager, go to **Security** | **Settings** | **Token Settings** | **Edit Token Settings** (pencil icon).
    2. Change **Lifespan of Short-lived Tokens** and/or **Lifespan of Long-lived Tokens** to the appropriate values. Note that in the Server Manager, a short-lived lifespan is in minutes, and a long-lived lifespan is in days.
    3. Click on **Save**.

- To change token settings in the REST Admin, follow these steps:
    1. In the REST Admin, go to **security** | **tokens** | **update**.
    2. Change the `longTimeout` and/or `shortTimeout` values in the JSON Token configuration. Note that in the REST Admin, both short-lived and long-lived lifespans are in minutes. Here, we are changing the `longTimeout` to 10 days from 1:

```
{
  "type": "BUILTIN",
  "properties": {
    "sharedKey": "lnvyyi56vVg/UPGS64iUkyjHyKBY=",
    "longTimeout": "14400",
    "shortTimeout": "60"
   }
}
```

    3. Choose your response **Format** from the drop-down.
    4. Click on **Update**.

Generating a token

Tokens are generated within the REST Admin at `http://server:port/arcgis/admin/generateToken`, or, if your Web Adaptor has administrative access, use the FQDN, such as `https://www.masteringageadmin.com/arcgis/admin/generateToken`. To generate a token, you will need administrative credentials, possibly the IP address of the target user of the token or a referrer URL, and an idea of how long you want the token to live for. The following screenshot shows the **Generate Token** interface:

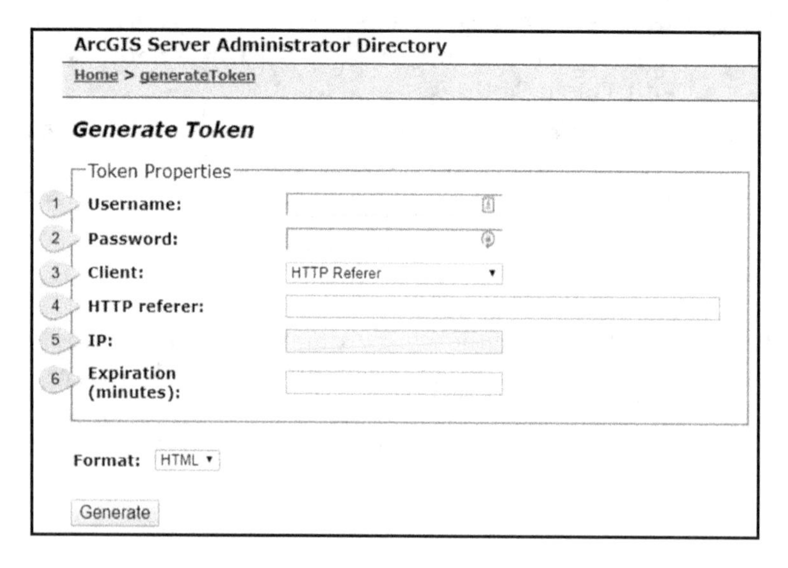

Let's break down the required inputs:

1. **Username**: This is the administrative username.
2. **Password**: This is the administrative password for the preceding username.
3. **Client**: This forces HTTP and IP-based restrictions on how the token may be used:
    - `HTTP Referrer`: This is the default value. With this option, the issued token can only be used in requests from the specified URL. This option is good when you have a web application with a set URL from which users will be making requests.
    - `IP Address`: With this option, the token can only be used in requests issued from the specified IP address. This option would be good to use in the case we discussed earlier of needing to provide a developer with a token to use for development purposes--you would use the IP of their development machine.

- `IP address of this request's origin`: With this option, the token can only be used in requests made from the machine from which the token was generated.

4. **HTTP referrer**: If **HTTP referrer** is selected as the **Client**, enter the URL from which requests will be made.
5. **IP**: If **IP Address** is selected as the **Client**, enter in the IP address from which requests will be made.
6. **Expiration (minutes)**: Lifespan of the token. This is the amount of time, in minutes, for which the token will be valid from the time of issue.

Select your response **Format** and click on the **Generate** button. You will get your token and an expiration Unix timestamp returned to you, for example, in **JSON** format:

```
{
  "token": "rspUZTL93FZG94Y2r6I_g5TDFmS7ME2p0fDC4pDs-mQJu",
  "expires": "1502247606590"
}
```

You can now copy your token to distribute it appropriately.

To translate your expiration timestamp from Unix time, go to a site such as `https://www.epochconverter.com/`, which will convert the Unix timestamp to UTC and your local time zone. Note that the expiration timestamp is in milliseconds, so divide it by 1,000 before entering it into a timestamp converter.

Managing services

Service management is a key aspect of ArcGIS Server administration. To access services in the REST Admin, go to the **services** link. When you get there, you'll notice that it looks very similar to the ArcGIS Server REST services page, listing the **Folders**, **Services**, **Resources**, **Supported Operations**, and **Supported Interfaces**. Click on a service though, and things look much different.

What you see here is every property currently set for that service, and the list can be quite long:

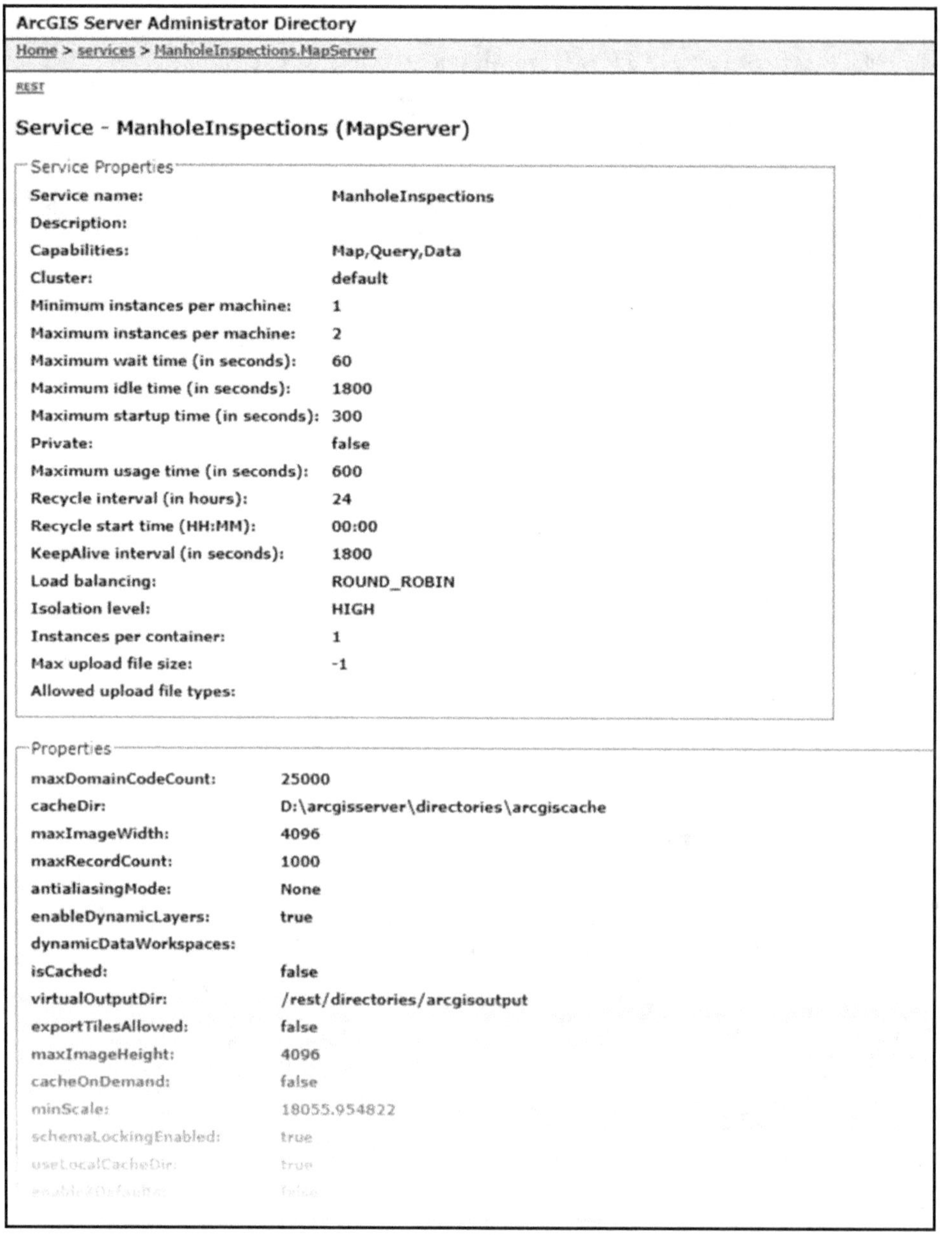

ArcGIS Server Administrator Directory

Home > services > ManholeInspections.MapServer

REST

Service - ManholeInspections (MapServer)

Service Properties

Service name:	ManholeInspections
Description:	
Capabilities:	Map,Query,Data
Cluster:	default
Minimum instances per machine:	1
Maximum instances per machine:	2
Maximum wait time (in seconds):	60
Maximum idle time (in seconds):	1800
Maximum startup time (in seconds):	300
Private:	false
Maximum usage time (in seconds):	600
Recycle interval (in hours):	24
Recycle start time (HH:MM):	00:00
KeepAlive interval (in seconds):	1800
Load balancing:	ROUND_ROBIN
Isolation level:	HIGH
Instances per container:	1
Max upload file size:	-1
Allowed upload file types:	

Properties

maxDomainCodeCount:	25000
cacheDir:	D:\arcgisserver\directories\arcgiscache
maxImageWidth:	4096
maxRecordCount:	1000
antialiasingMode:	None
enableDynamicLayers:	true
dynamicDataWorkspaces:	
isCached:	false
virtualOutputDir:	/rest/directories/arcgisoutput
exportTilesAllowed:	false
maxImageHeight:	4096
cacheOnDemand:	false
minScale:	18055.954822
schemaLockingEnabled:	true
useLocalCacheDir:	true
enableZDefaults:	false

Scroll to the bottom of the services properties, to the **Supported Operations**, and click on **edit**. Here, we can edit our **Service Properties**. Look through these properties; many should look or, at least, sound familiar, as they are properties you set in **Service Editor** when you initially created a service.

Hiding a service

When viewing a service's JSON **Service Properties**, the first section you will see is **Service Framework Properties**, which looks similar to the following snippet of JSON:

```
"serviceName": "ManholeInspections",
"type": "MapServer",
"description": "",
"capabilities": "Map,Query,Data",
"provider": "ArcObjects",
"clusterName": "default",
"minInstancesPerNode": 1,
"maxInstancesPerNode": 2,
"instancesPerContainer": 1,
"maxWaitTime": 60,
"maxStartupTime": 300,
"maxIdleTime": 1800,
"maxUsageTime": 600,
"loadBalancing": "ROUND_ROBIN",
"isolationLevel": "HIGH",
"configuredState": "STARTED",
"recycleInterval": 24,
"recycleStartTime": "00:00",
"keepAliveInterval": 1800,
"private": false,
"isDefault": false,
"maxUploadFileSize": 0,
"allowedUploadFileTypes": "",
```

Here, for example, you could edit the `minInstancesPerNode` or `maxInstancesPerNode` to increase or decrease the minimum/maximum number of instances of services that get spun up on the node within your cluster. There is an interesting property that you won't find in your default service framework properties, but it is available in the API. The `deprecated` property is hidden by default and must be added manually (if you set `"deprecated: false"` in your service properties, save it, and then go back in to view your service properties JSON again; you won't see `"deprecated: false"`, as, by default, its value is `false` and `hidden`). If set to `true`, the service will not display in the ArcGIS Server REST Services Directory, even if it is a fully unsecured service.

This property is useful if you need to retire a service, but still need it to function in existing applications but not allow others to add it to new applications (it's still there, they just cannot see it). Another use case could be to hide a fully public unsecured service from users unless they know about it and explicitly know the URL to its REST endpoint. In this case, users that know of the service can add it to applications, but the public is no longer able to see the service and, therefore, it will not consume it.

To set a service as deprecated, add the `"deprecated: true"` property at the very top of the **Service Properties**, and then click on the **Save Edits** button:

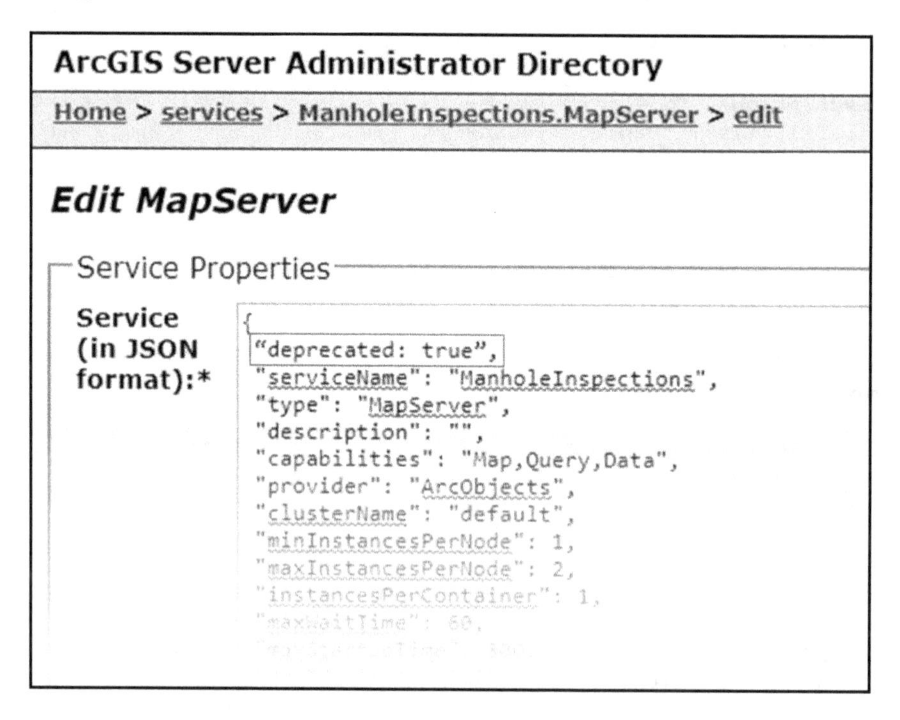

It will take a few seconds to make the changes to the service, after which you will be returned to the service properties page. Note that the deprecated property does not show up here. Go to your ArcGIS Server REST endpoint and see if the service shows up (remember to log in if it is a secured service). Your service should not show up. Go back to the REST Admin **Service Properties**, go to **edit**, and near the bottom of the **Service Framework Properties**, you should see your `"deprecated": true,` property.

System settings

The system resources covers a wide range of server-wide resources, such as server properties, the config store, and directories and web adaptors. In the ArcGIS Server REST API, there are several resources to note. Most of these can be accessed to some degree in ArcGIS Server Manager, but I'm sure by now you have realized that it is just as easy (if not easier) to make administrative changes in the REST Admin. To get to the system settings in the REST Admin, log in to the REST Admin and go to **system**.

Web Adaptors

Once you are at the **System** page in the REST Admin, go to **webadaptors**, where you will be presented with a list of **Web Adaptors** that you have configured (you're more than likely only going to have one Web Adaptor registered with ArcGIS Server) for ArcGIS Server. The long alphanumeric name of the Web Adaptor is meaningless, it's just an internal ID; click on it. At the **Web Adaptor Properties**, you can view all the properties of the Web Adaptor and **Supported Operations**. Note that the only operation listed is **unregister**, which, if carried out, unregisters that particular Web Adaptor and removes it from the server's trusted list. However, unregister isn't really the *only* operation available. While at your **Web Adaptor Properties** page, go to the browser address bar window and append `/update` to the end of the URL, so it will look something like `https://www.masteringageadmin.com/arcgis/admin/system/webadaptors/9b58t09d-c645-4ty2-9634-7487eb7ca589/update`, and hit the *Enter* key. You are now at the **Update Web Adaptor** page, where you can really only update the **Description** and **HTTP/HTTPS port** properties of your Web Adaptor. The **update** operation is documented in the API, but the operator is not present in the API interface of the REST Admin. The point here is that, sometimes, by reading the documentation carefully, you can find ways of doing things that might not be so readily apparent. We will see something similar in the next chapter when we are discussing Portal for ArcGIS administration.

Properties

Properties are a container for some of the more intricate properties and are typically empty by default. One property that can be set here is to change the geoprocessing service and service extension publishing privileges. Starting at ArcGIS Server 10.4, only administrators can publish geoprocessing services and deploy **server object extensions** (**SOEs**) and **server object interceptors** (**SOIs**). In the **System Properties** of the REST Admin, this property can be changed so users with publisher privileges can publish these resources as well. To change this setting, do the following:

1. From the REST Admin, go to **system** | **properties** | **update**.

2. Enter the property in the following format:

```
{"allowGPAndExtensionPublishingToPublishers": true}
```

3. Click on **Update.**

Esri recommends leaving this property set to `false` in production environments.

Directories

When we configured our instance of ArcGIS Server after installation, we set the location of the ArcGIS Server directories. In the REST Admin, we can view (`GET`) and update (`POST`) the ArcGIS Server directory locations. The `arcgiscache`, `arcgisjobs`, `arcgisoutput`, and `arcgissystem` directory locations can be changed. To change the **arcgiscache** folder location, for example, follow these steps:

1. In the REST Admin, go to **system** | **directories** | **arcgiscache** | **edit**.
2. Set the **Physical Path** property to the new folder location.
3. Click on the **Update** button.

If the `arcgiscache` directory path is edited, any service caches will have to be manually copied from the old `arcgiscache` directory to the new one. For the `arcgisjobs`, `arcgisoutput`, and `arcgissystem` paths, the content is copied from the old location to the new one by ArcGIS Server. For all four folders, if the path is edited, the ArcGIS Server site will be restarted.

Other parameters, such as the **Max file age** and **Cleanup mode** , can be set for directories as well. Directories can also be purged of their content with the **clean** operation.

Config store

As with the ArcGIS Server directories, the ArcGIS config store location is accessible in the REST Admin. To edit the configuration store properties, do the following:

1. In the REST Admin, go to **system** | **configstore** | **edit**.
2. Change the **Connection string** to the new location.
3. Select **Move current configuration** if you want to move the current configuration store to the new location. The default is `true`.
4. Click on the **Save Edits** button.

The config store (and ArcGIS Server directories) can be located on a network share and referenced by a UNC path.

Logs

The logs directory, log level, and log retention properties can all be changed in the REST Admin under **logs** | **settings** | **edit**. Another parameter, **Usage metering**, can also be enabled here. When enabled, the usage metering logs the `INFO` level usage parameters, such as the number of bytes returned from a call and the type of task that was utilized (`find`, `identify`, `export`, and more), and this information can be used along with a username from the logs to track usage.

Data

By default, ArcGIS Server allows publishers to copy service source data to the ArcGIS Server machine when they publish. In some organizations or workflows, this is an acceptable and standard operating procedure. However, if you do not want your users to be able to copy data to the server during publishing, or if you want to force them to only reference data in registered databases and folders, you can set the `blockDataCopy` property. To change this property in the REST Admin, do the following:

1. In the REST Admin, go to **data** | **config** | **update**.
2. In **Configuration**, enter the property in the following format:

   ```
   {"blockDataCopy": true}
   ```

3. Click on the **Update** button. Users will now not be able to copy data to the server.

After this setting is changed, if a user attempts to publish a service with an unregistered data source, they will get a high severity error during the Analysis process telling them that the layer's source data must be registered with the server. They will not receive a warning that the source data will be copied to the server, as that is not an option anymore.

The ArcGIS Server command-line utilities

Tucked away in the ArcGIS Server installation directory is a set of Python scripts, collectively referred to as the *ArcGIS Server command-line utilities*. These can be found at <ArcGIS Server installation location>\tools\admin:

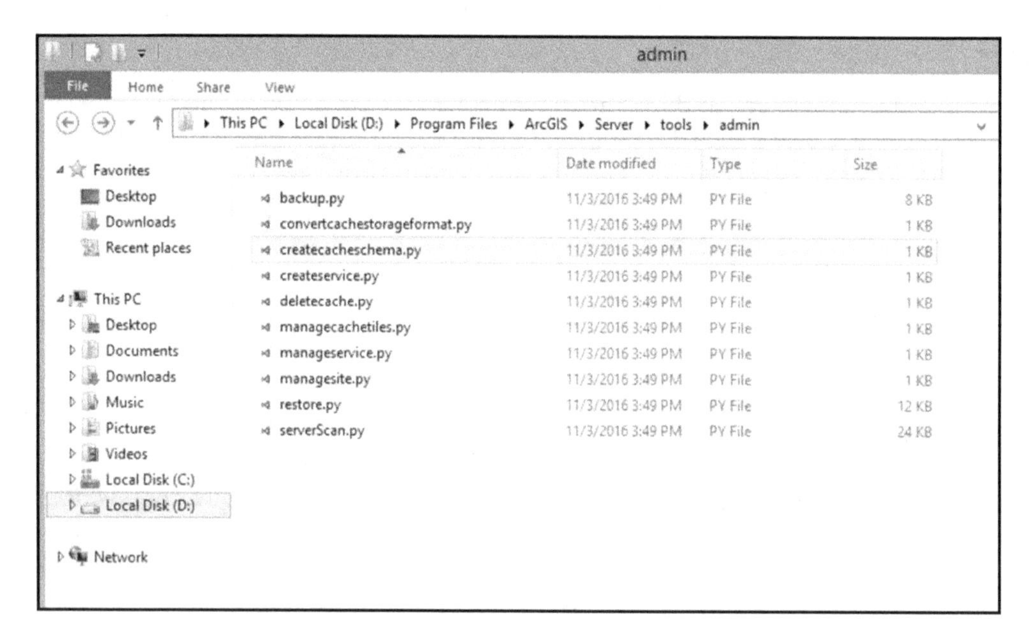

There are scripts for backing up and restoring an ArcGIS Server site (previously discussed in this chapter), creating services, managing services, and managing caches, among others. To run any of these scripts, call them from the command line and pass in any required arguments.

The `manageservice.py` script, for example, can be used to list all services in a site:

```
python.exe manageservice.py -u siteadmin -p pass -s https://www.m
asteringageadmin.com -l
```

This gives the following result, printed to standard output:

```
ManholeInspections.MapServer               | STARTED
SampleWorldCities.MapServer                | STARTED
System/CachingControllers.GPServer         | STARTED
System/CachingTools.GPServer               | STARTED
System/CachingToolsEx.GPServer             | STARTED
System/DistributedWorker.GPServer          | STOPPED
System/DynamicMappingHost.MapServer        | STARTED
```

```
System/GeoAnalyticsManagement.GPServer     | STOPPED
System/GeoAnalyticsTools.GPServer          | STOPPED
System/PublishingTools.GPServer            | STARTED
System/PublishingToolsEx.GPServer          | STARTED
System/RasterAnalysisTools.GPServer        | STOPPED
System/RasterProcessing.ImageServer        | STOPPED
System/RasterRendering.ImageServer         | STOPPED
System/ReportingTools.GPServer             | STARTED
System/SceneCachingControllers.GPServer    | STARTED
System/SceneCachingTools.GPServer          | STARTED
System/SpatialAnalysisTools.GPServer       | STARTED
System/SyncTools.GPServer                  | STARTED
System/SyncToolsEx.GPServer                | STARTED
Utilities/GeocodingTools.GPServer          | STARTED
Utilities/Geometry.GeometryServer          | STOPPED
Utilities/PrintingTools.GPServer           | STARTED
Utilities/Search.SearchServer              | STOPPED
```

Summary

Administration is a crucial task that must be diligently carried out. A properly maintained and administered ArcGIS Server site will run efficiently, smoothly, and dependably. In this chapter, we started at the ground floor and learned about how to access administrative functions through ArcGIS Server Manager, ArcCatalog, and the REST Admin. We looked at a wide array of administrative tasks from log settings, backup and restore, and resetting and changing the ArcGIS Server primary site account. We then dove into the ArcGIS Server REST Admin and learned how to navigate around to work with tokens, manage services, system settings, logs, and data settings. We will soon use our familiarity with the REST Admin in `Chapter 7`, *Scripting Administrative Tasks*, where we will interact with the REST Admin through Python scripts, allowing us to automate administrative tasks. However, first, in the next chapter, we will look at Portal for ArcGIS administration.

5

Portal for ArcGIS Administration

With the release of ArcGIS Enterprise 10.5, Portal for ArcGIS, referred to as **Portal** here, became a first-class citizen, joining the ranks of ArcGIS Server and becoming a fully integrated, standard component of the Esri platform. Using Portal, your users can create, edit, and share web maps and web mapping applications. GIS content can be searched, shared, and accessed in ArcGIS Desktop and ArcGIS Pro. When ArcGIS Server is federated with your Portal, Portal provides an interface and a window into your ArcGIS Server content. Even without federation, Portal still provides an intuitive ArcGIS Online-like experience and central access point into your enterprise GIS. In many organizations, Portal is *the* method used to interact with the enterprise GIS. Also, administrative tasks for Portal determine and set which base content and templates are available to your users and how they interact with these resources. For these reasons, it is easy to see how important a task Portal administration is.

In this chapter, we will look at the different interfaces and methods that can be used to administer Portal. We will also look at a wide array of Portal settings and how to change them, how to customize content for your users, and how to use currently available (and free!) management tools to make Portal administration easier.

Topics to be covered include:

- Administering Portal through the web interface
- Administering Portal through the REST Administrative interface
- Backup and restore of Portal
- Management tools for Portal

Connecting to Portal

Just as with ArcGIS Server in the last chapter, there are multiple methods available to connect to and access Portal administrative tasks. We will first discuss the two standard ways Esri provides. Later in this chapter, in *Administering through the Portal REST Administrative Directory* section, we will discuss other windows into Portal administrative functions.

Accessing Portal through the standard web interface

The standard Portal web interface (https://www.yourdomain.com/portal) is how most users will interact with Portal; it is also how you can access it for a portion of your administrative duties. How you access the Portal web interface depends on your configuration and whether you are accessing Portal externally through the Web Adaptor or on the internal network, bypassing the Web Adaptor. If the former, then the URL you use will look like the following:

```
https://<FQDN>/<webadaptor/home/
```

In our case, it would be like this:

https://www.masteringageadmin.com/portal/home/.

If the latter, the URL will look like the following:

```
https://<FQDN>:7443/arcgis/home/.
```

In our case, it would be this:

https://win-25fpfgemua9:7443/arcgis/home/.

Now, before going any further, let's discuss some differences here that can be confusing. In the first instance, going through the Web Adaptor, the URL ends with /portal/home/, but in the second instance, where we bypass the Web Adaptor, the URL ends with /arcgis/home/. Why is that? When we configured our Web Adaptor for Portal, we named it portal, so that is why, when going through the Web Adaptor, we use /portal/home/.

When bypassing the Web Adaptor and explicitly going in over port 7443, we use /arcgis/home/ because, for some reason, Esri named the default Portal site on port 7443 **arcgis**, which, confusingly, is the name typically given to ArcGIS Server's default site arcgis on port 6443. If you try to go to https://<FQDN>:7443/portal/home/, you will get a 404 error code returned, as /portal/home/ only works with the Web Adaptor.

To log in to Portal as an administrator, click on **Sign In** in the upper-right of the Portal for ArcGIS window with an account with administrator privileges.

Accessing Portal through the Portal Admin

The Portal Administrative Directory, referred to as **Portal Admin** here, is what the ArcGIS Server REST Admin is to ArcGIS Server--an interface to all administrative settings for Portal for ArcGIS. As with the Portal web interface, Portal Admin can be accessed from within the internal network (bypassing the Web Adaptor) at a URL such as this:

```
https://<FQDN>:7443/arcgis/portaladmin/
```

In our case, it would be as follows:

```
https://win-25fpfgemua9:7443/arcgis/portaladmin/
```

If administrative access is enabled on the Portal Web Adaptor, then we can access Portal Admin outside of our internal network at a URL such as this:

```
https://<FQDN>/webadaptor/portaladmin/
```

In our case, it would be as follows:

```
https://www.masteringageadmin.com/portal/portaladmin/.
```

To access Portal Admin through the Web Adaptor, administrative access must be enabled in the Web Adaptor (see Chapter 1, *ArcGIS Enterprise Introduction and Installation*).

To log in to Portal Admin as an administrator, enter the **Username** and **Password** of an account with administrator privileges on the **Portal Administrative Directory Login** page and click on the **Login** button.

Administering through the web interface

As an administrator in Portal, you have access to administrative functions under **My Organization**, where there are links to **EDIT SETTINGS**, **ADD MEMBERS**, and **VIEW STATUS**. Under **My Organization**, you can also search for users and view/manage users, as well as viewing information regarding your named user allotments:

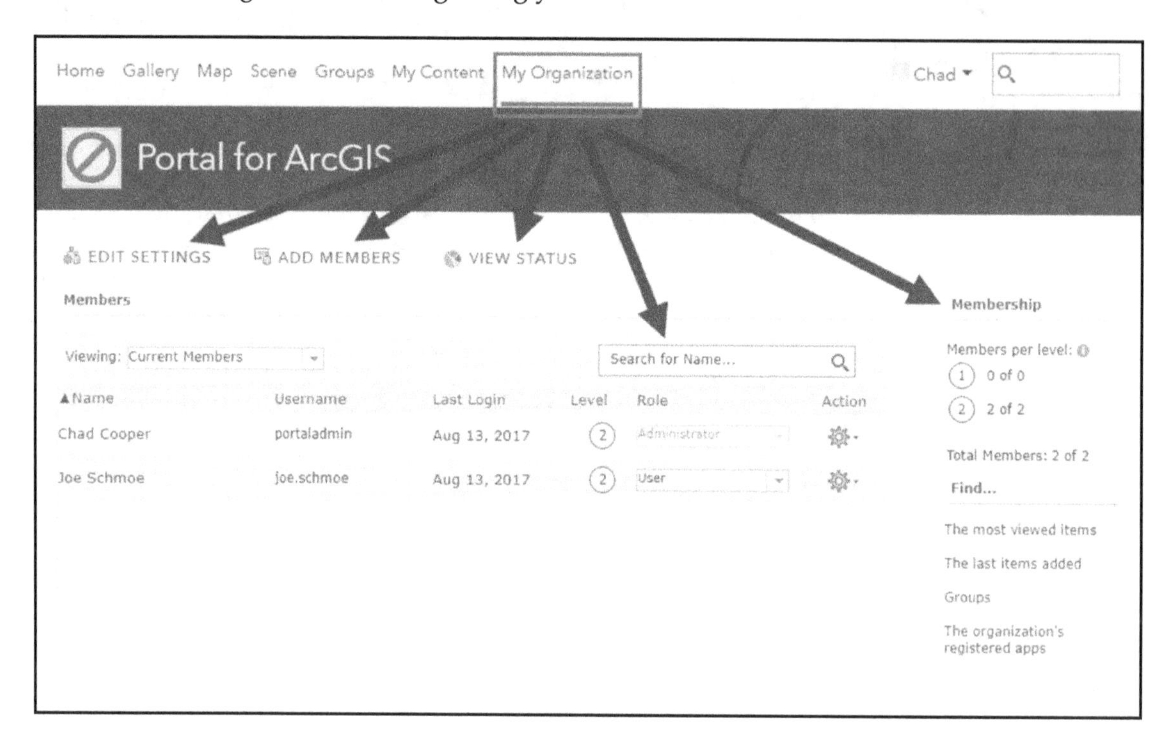

Changing the look and feel of your Portal

When your Portal is first configured and launched, it has a generic appearance, as illustrated in the following screenshot:

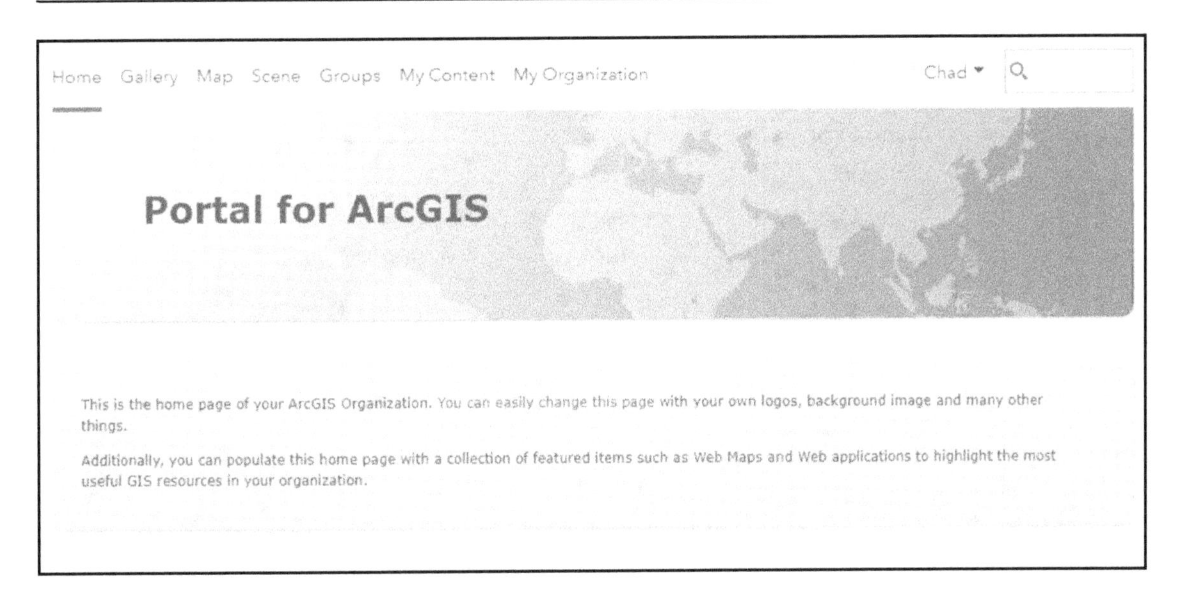

Fortunately, this look can be upgraded easily in the Portal settings. To edit the home page of your Portal, log in as an administrative user and perform the following steps:

1. Go to **My Organization** | **General**.
2. If you want a description (the second gray box in the preceding figure), add one in the **Description** section and check the **Show description toward bottom of Home Page** checkbox. You can add text, images, and links. You also have full control over the HTML.
3. In the right menu, go to **Home Page**. To fully customize your Portal home page, change the following items:

- **Background Image**: You can remove this, go with the provided one, or add an image of your own. Try to use an image 1920 pixels wide and no larger than 1 MB in file size. The image will be positioned at the top and center of the page and will repeat horizontally if smaller than the browser or device window.
- **Banner**: You can use an image or a custom design. With **Custom design**, you have full control over the HTML and can modify font sizes and faces, images, links, and more.
- **Featured Content**: You can optionally place a carousel of featured content on your home page under the banner. You can use any group's content in the carousel and can change the sort attribute and number of items to display from the group.

In around 45 minutes, I edited the preceding settings and transformed my Portal from the vanilla screenshot seen earlier to the following:

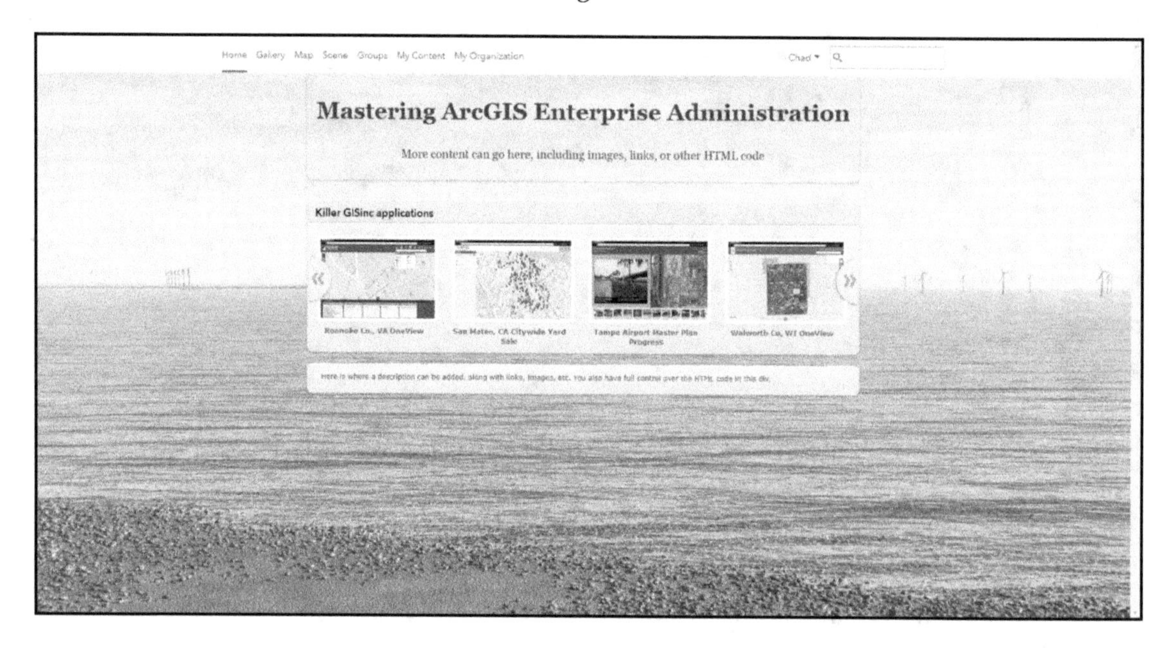

I had to get a little creative with the `div` tags to get the font and alignments I wanted, but nothing fancy at all; the following is the HTML for the title block:

```
<div style="position: absolute; width: 960px;">
<div style="font-size:32px;"><br /></div>
<div style="font-size:40px; text-align:center; font-family:
 georgia"><b><font color="#214072">Mastering ArcGIS Enterprise
 Administration</font></b></div>
<div style="font-size:32px;"><br /></div>
<div style="font-size:20px; text-align:center; font-family:
 georgia"><font color="#214072">More content can go here,
 including images, links, or other HTML code</font></div>
</div>
```

Managing content

There are many aspects to content management in Portal; some impact your content creators, others impact your content consumers and viewers.

Featured content

We discussed this briefly earlier, but let's expand on it. **Featured Content** is a group containing the items that appear in the carousel on your home page (**Killer GISinc applications** in the preceding screenshot). These are your top items, the items you want anyone entering your Portal to see and take notice of. As an administrator, you have complete control over which group is designated as the Featured Content group and which items are in that group and, therefore, show up in the home page carousel. The name of the group is what displays in the carousel (again, **Killer GISinc applications** in the preceding screenshot), so if you want the carousel header to display something other than **Featured Content**, you will have to create a group, share the items you want in the **Featured Content** carousel with that group, then set that group as the **Featured Content** group in **My Organization** | **EDIT SETTINGS** | **Home Page**:

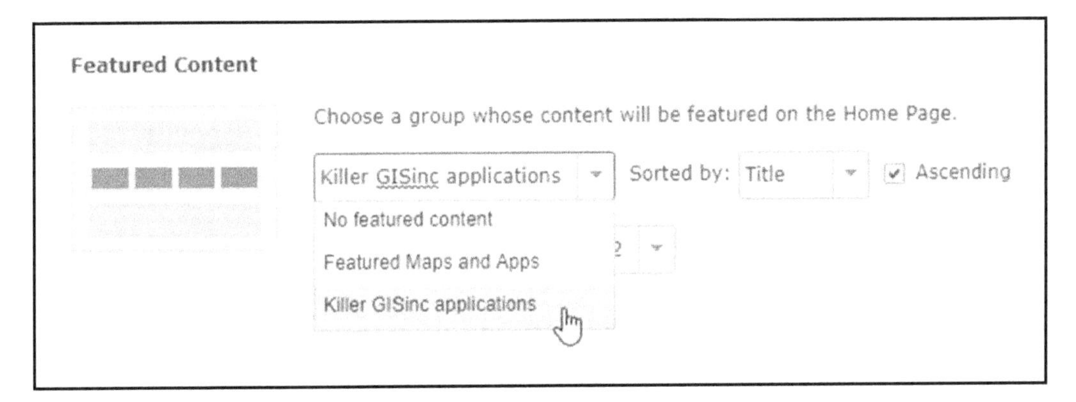

Customizing basemaps

Just as in ArcGIS Online, the Portal web map viewer offers a basemap gallery that content creators and editors can use to change the web map basemap. As was the case with **Featured Content**, Portal administrators have full control over which group is utilized as the basemap group and which basemaps are in that group. To customize the basemap group to use a custom group, follow these steps:

1. While logged in as an administrative user, go to **Groups** | **CREATE A GROUP**.
2. Enter **Name**, **Summary**, **Description**, and **Tags** for the group.
3. Keep **Status** as **Public**, and if you want users to be able to request membership to the group so they can contribute basemaps, check whether Users can apply to join the group.

4. Decide whether you want to have contributors or only allow the group owner to edit group content.
5. Click on the **Save** button.

Next, add the maps you wish to have as basemaps in the basemap picker in your map viewer. To do this, share the maps with your newly created basemap group. For example, you can go to **Gallery** | **Esri Featured Content** and search or browse for maps. In the following screenshot, I have found the **USGS National Map** basemap. If I hover over the thumbnail, I am presented with the **DETAILS** button, which, when clicked, will take me to the item details page:

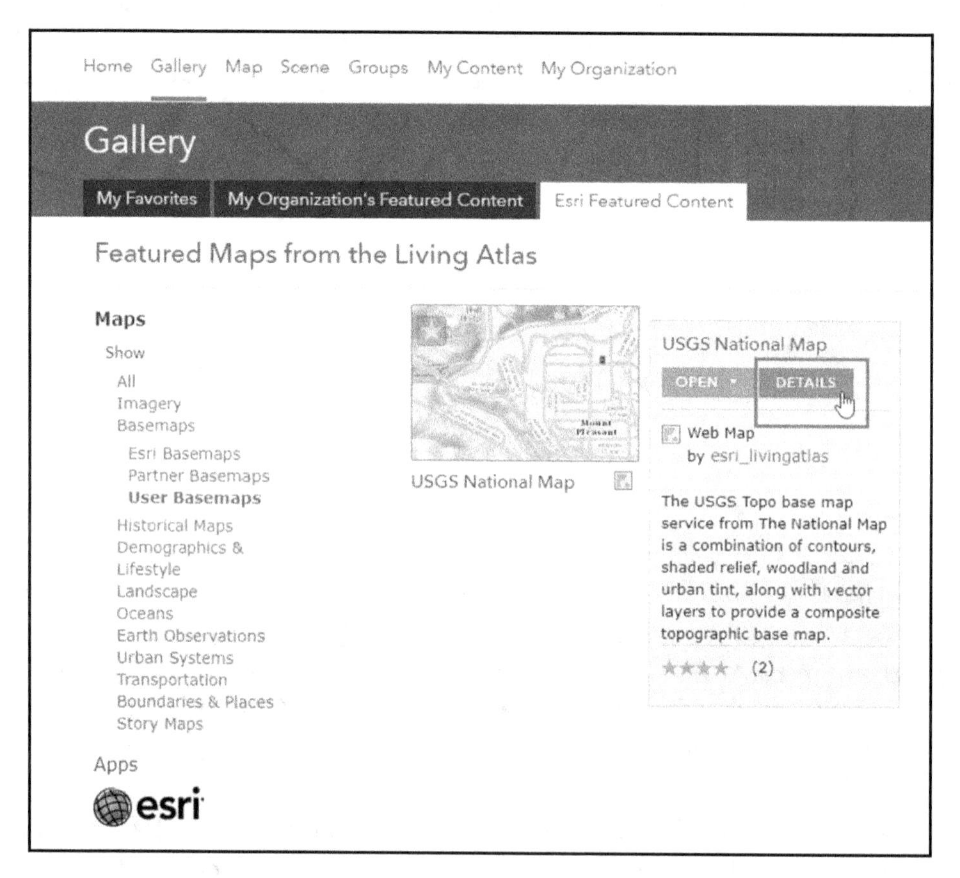

Earlier, I created a group named `Our Org Basemaps` to house the basemaps that I, as a Portal administrator, want to use as the basemaps group in my Portal.

Once at the **USGS National Map** item details, I can click on **Share** and choose to share this item with my group, **Our Org Basemaps**:

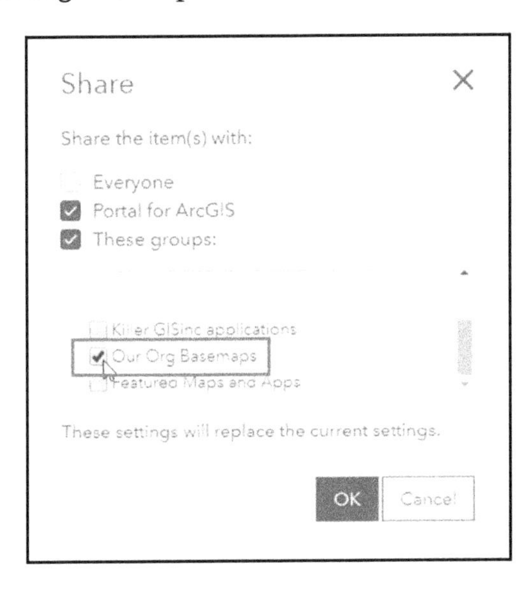

The item is now a member of the **Our Org Basemaps** group. If we go to our Portal Map or create a new web map, then select the **Basemap** chooser; we will see that the **USGS National Map** basemap is now available for use:

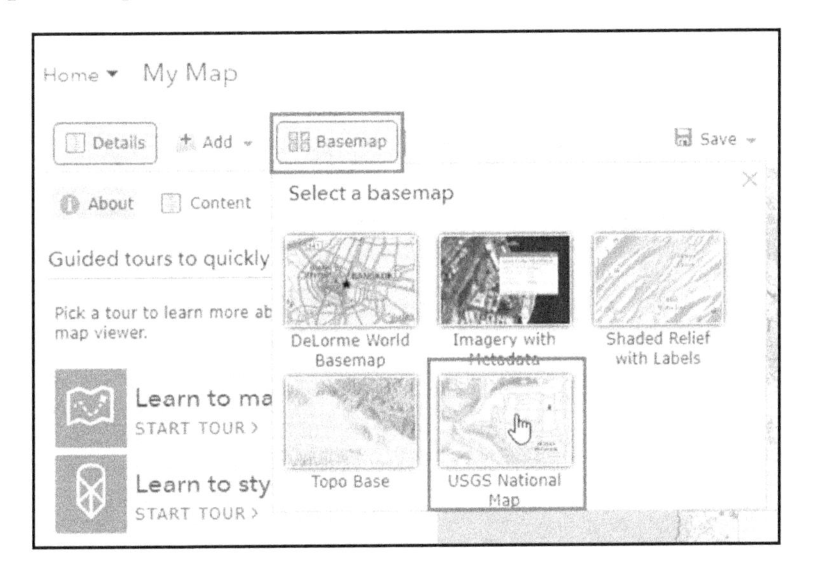

Configuring the map viewer

In addition to configuring basemaps for the map viewer, several other parameters of the map viewer can be controlled under the map settings:

- **Default Basemap**: Here, you can change the default basemap that is active when a user creates a new web map. The basemap must come from the default **Basemap** gallery (discussed earlier).
- **Default Extent**: This sets the extent of any new web map that is created. The options are to use either the extent of the default basemap or to choose a custom extent. When choosing a custom extent, you can search for an address or draw an extent. To remove a custom extent and go back to using the default extent of your default basemap, select **USE MAP EXTENT AS DEFAULT** from the **Default Basemap** section.
- **Bing maps**: To use **Bing maps** in your applications, you must provide a Bing maps key, available at `https://www.bingmapsportal.com/`. If you enable this option, users can add Bing maps layers to a web map by selecting **Add** | **Add Layer from Web** and then selecting **A Bing basemap** from the data type drop-down in the **Add Layer from Web** dialog. **Bing Roads**, **Aerial**, and **Hybrid** can be added to the map.
- **Units**: This sets the default units for the map scale bar, measure tool, directions, and analysis. The choices are **US Standard** or **Metric**. Also note that users can set the units they see on their profile page.

Configuring utility services

There are several utility services that can be configured to work with your Portal. These services include printing, geocoding, geometry, and elevation, to name a few. To get to the utility service settings, log in to Portal as an administrator and go to **My Organization** | **EDIT SETTINGS** | **Utility Services**.

Search the ArcGIS Enterprise online help for *configure utility services* for more information on the utility services available.

Printing

If you've worked with ArcGIS Server since the `9.x` days, it's hard to argue that printing has come a long way. That said, however, the out-of-the-box **print** functionality in Portal is quite lackluster. Never fear though, as this is easy to change. We can quickly configure a custom print service in using our own templates and consume that service in Portal as our default print service. Before we do that, let's take a look at the default print service that comes with Portal.

Go to **Utility Services** in Portal settings and look for **Printing**, which is at the top of the list. If you have not set this to use another print service, it will be set to **Default**. This is the default Portal print service that will print the `MAP_ONLY` view of the web map to a PNG image at a resolution of 675x500 pixels (see what I meant earlier by **lackluster**?). This default print service cannot be used in web app templates or in Web AppBuilder applications.

Using the default ArcGIS Server print service

Besides configuring a custom print service, ArcGIS Server comes with a preconfigured printing service in the `Utilities` folder named **PrintingTools**. The service includes basic landscape and portrait layouts in several common paper sizes. This service is stopped by default, but can be easily started by navigating to it in the `Utilities` folder of your ArcGIS Server instance:

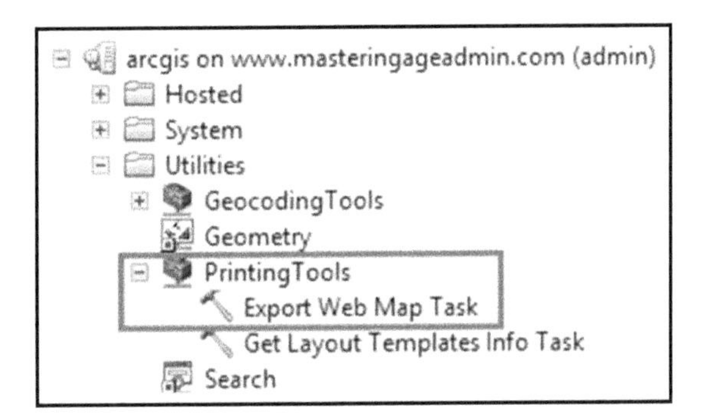

Once the service is started, double-click on the **Export Web Map Task** to open it. Examine the contents of the **Format** and **Layout Template** drop-downs to see what this print task offers:

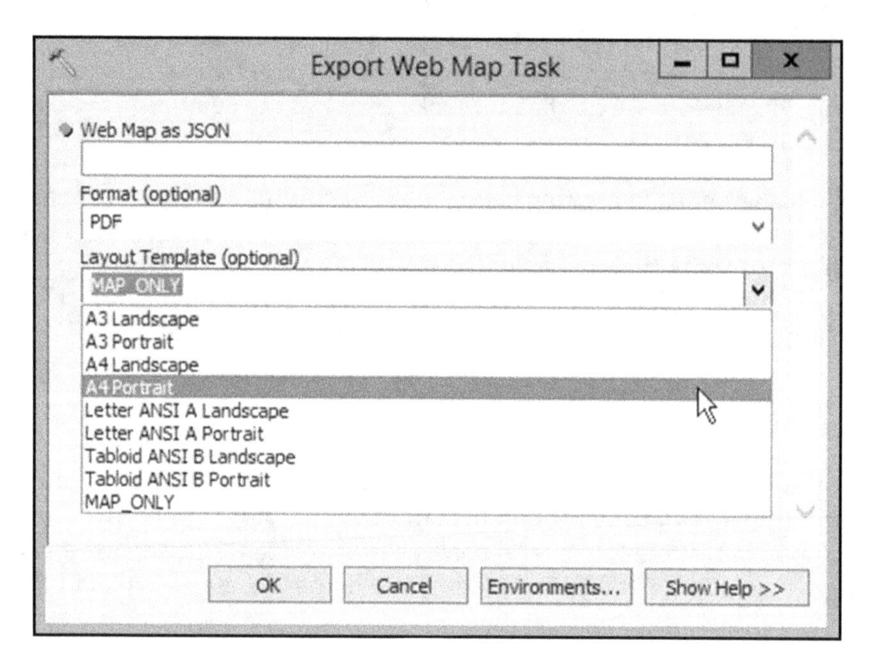

To set this as the default print service for Portal, go back to the Portal settings and, under **Printing** on the **Utility Services** tab, enter in the full URL to the REST endpoint of the print task for the print service, which is as follows:

```
https://www.masteringageadmin.com/arcgis/rest/services/Utilities/Printi
ngTools/GPServer/Export%20Web%20Map%20Task.
```

Portal will read the print task parameters and list the templates available. Here, you can reorder, delete, and edit the properties of the templates available in the task. The properties set here are honored by the print functionality in the web map viewer and standard web applications. For printing in these applications, by default, the **Print Template Format** is **Image**, which creates a PNG file. The other option is PDF.

For both, all templates in the task are available:

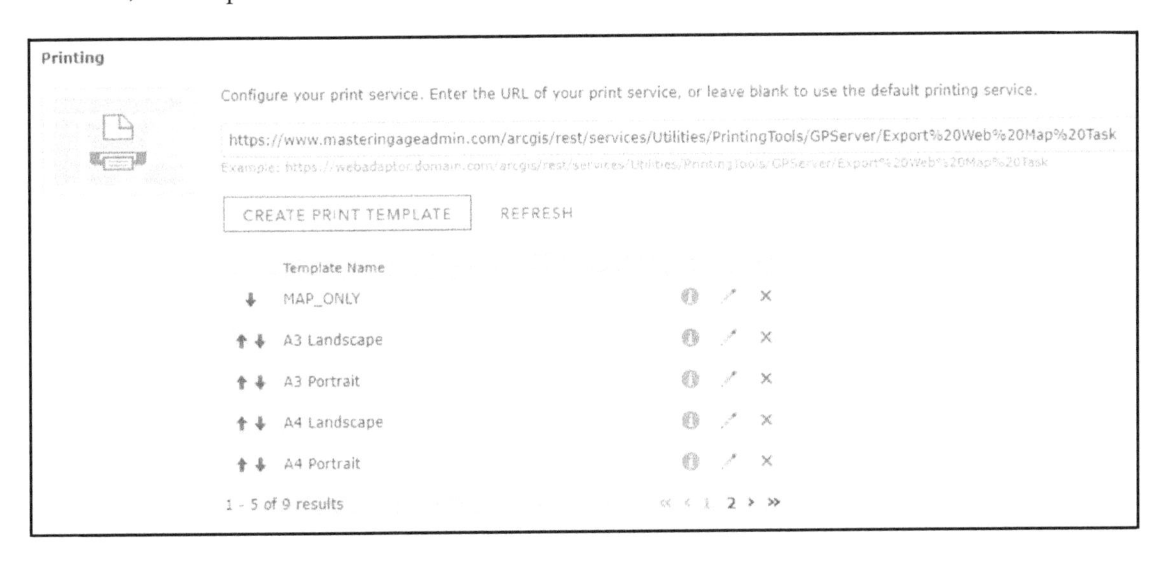

Click on the **Save** button to the upper-right above the left menu. When this print task is set as the default print service in Portal, it will be set as the default print service in any standard web application or Web AppBuilder web app. In the following screenshot, in a Web AppBuilder app, when the **Print** widget was added to the application, the print service URL was automatically set to the default Portal print service:

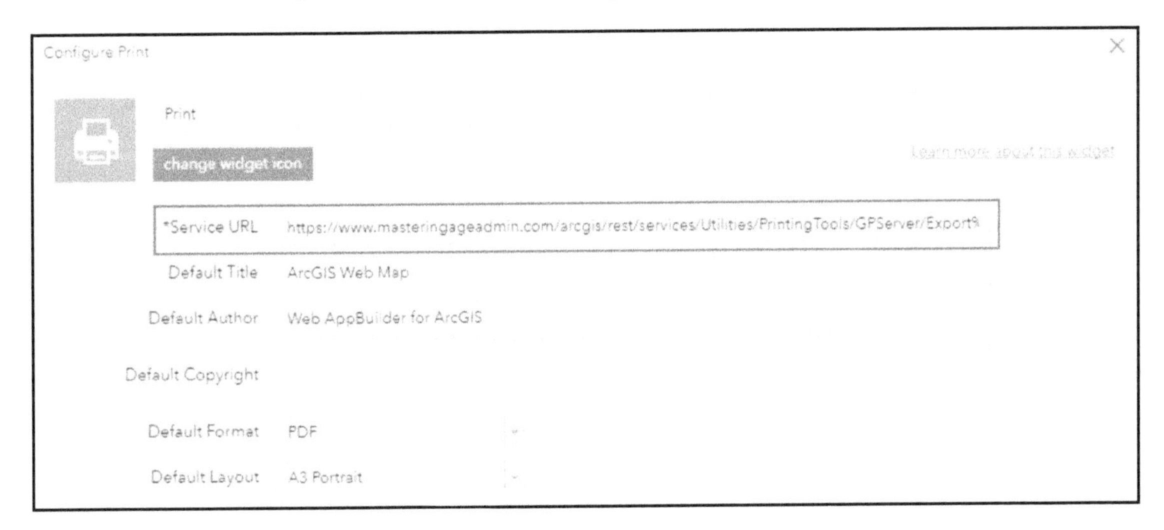

When added to a Web AppBuilder Print Widget, all options of the ArcGIS Server print task are available, including multiple formats and layouts. In a web map or standard web app, all layouts are available, but Image (PNG) and PDF formats are the only available options.

Resetting any of the utility services back to their default value is easy; simply remove the custom URL value, leave it blank, and click on **Save**. This resets the parameters back to their default values.

Using a custom ArcGIS Server print service

Sure, you can use the ArcGIS Server **PrintingTools** print service out-of-the-box, but what if I told you that making your own custom print service using *your own* print templates was actually very easy? Well, it is. A custom print service consists of a geoprocessing task that takes a map input, output format, and template as inputs. It returns an exported map in the selected output format.

What we cover in this section goes for any ArcGIS Server print service. If you are creating a print service to be consumed in a custom web application utilizing the Esri JavaScript API, these same principles apply.

Print templates

At the heart of any respectable custom print service are print templates tailored to your organization, content, and users. A print template is simply an MXD with your custom layout but no map layers; the geoprocessing service that calls the print service and utilizes the template provides the map input. When considering how you want your custom print templates to look, think about what your organization may already be using. Does your organization have standard print layouts that are used to provide maps internally or externally? If so, you may be able to use those, or at least base your templates on those existing ones, keeping that standard look and feel in place. Remember, consistency is something you are trying to provide with these templates.

See `Chapter 9`, *ArcGIS Enterprise Standards and Best Practices*, for a section on best practices for print templates.

Like many things in life, there are two ways to create custom print templates, the easy way and the hard way. Let's say you want to have two sizes of exports available in your print service: letter 8.5 x 11 inches and tabloid 11 x 17 inches with landscape and portrait orientation in each for a total of four different templates. The hard way of creating these would be to start each one as its own MXD from scratch and add all the layout elements required to each one. You will probably have to get the templates approved, so if you need to make any changes (and you probably will), you'll have to make the changes to all four templates. The easy way to create these would be to create one, let's say the portrait 8.5 x 11-inch size, with all the required layout elements. Next, get it approved through any necessary channels within your organization and get everyone on your team to agree on the design. Once everyone likes the design and signs off on it, make a copy of the portrait 8.5 x 11 MXD, save it as the landscape 8.5 x 11 template, and make the necessary modifications to the layout to accommodate the orientation change (with a small paper size like this, the changes should be minimal). When saving the print templates, save them in a central location. If you have a multiple GIS machine configuration, this will need to be a network location accessible by the ArcGIS Server account. I like to store print templates in a folder named `PrintTemplates` within my map services folder (which is usually named `Services`). Run this landscape template through the same approval process you did earlier. Finally, once everyone likes the landscape template, make copies of both the portrait and landscape 8.5 x 11 MXDs, save them as the 11 x 17-inch templates, and make the appropriate changes to them to account for the page size change (again, these will be minor at such small page sizes). You now have your four print templates.

Register the print templates with ArcGIS Server

When you created your print templates earlier, you stored them all in one central folder, accessible to ArcGIS Server and the ArcGIS Server account. This is an important step in the creation of your print service, as we need to register this folder with ArcGIS Server. Registering your `PrintTemplates` folder with ArcGIS Server will allow you to make changes to the existing templates in the published print service and to add new print templates to the service *without even having to republish the print service*. That's right; need to move or update one little text element in just one template (or all of them)? Simply make the change(s), save the MXD(s), and since the folder is registered with ArcGIS Server, it sees that a change has been made and propagates the change to the print service. Need to add a new template? Simply place the template MXD in the `PrintTemplates` folder and, again, ArcGIS Server sees the change and propagates it out to the print service.

Publish the print service

The final step in building a custom print service is to publish the print service using your custom print templates. To do this, complete the following:

1. Launch the **Export Web Map** geoprocessing tool from either ArcMap or ArcCatalog. It is in the **Server Tools** toolbox, but remember, you can always search for it as well.
2. Fill out the appropriate parameters:
    - **Web Map as JSON**: This parameter can be left blank, as the calling application will provide the map data to the service; publishing the service initially requires no map data.
    - **Output File**: The default value can be accepted and used for this parameter. This is the location that the output file will be written to; when published to ArcGIS Server, the output location will be within the `arcgisoutput` folder in the ArcGIS Server directories.
    - **Format**: This is the default value of the export format. **PDF** is usually a good choice here.
    - **Layout Templates Folder**: This is the most important parameter in the tool. Enter the path to your **PrintTemplates** directory here. This tells the print task to reference and use all MXDs in the folder.
    - **Layout Template**: This sets the default template for the print service, from the dropdown, select the layout you wish to serve as the default:

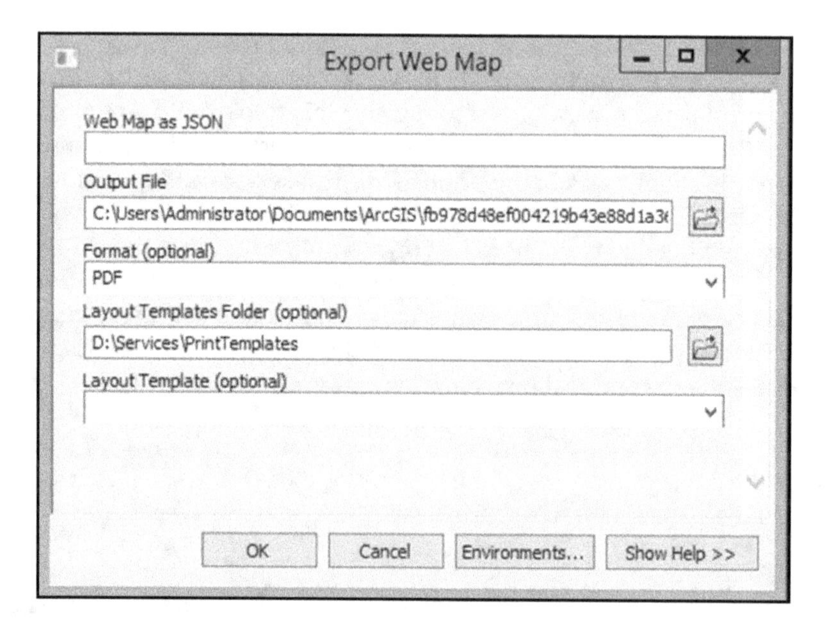

3. Click on the **OK** button. The tool should complete quickly and notify you in the lower-left window when it is complete.
4. Go to **Results** on the **Geoprocessing** menu. Find the result of your run of the **Export Web Map** tool, right-click on it, and go to **Share As** | **Geoprocessing Service**, as shown in the following screenshot:

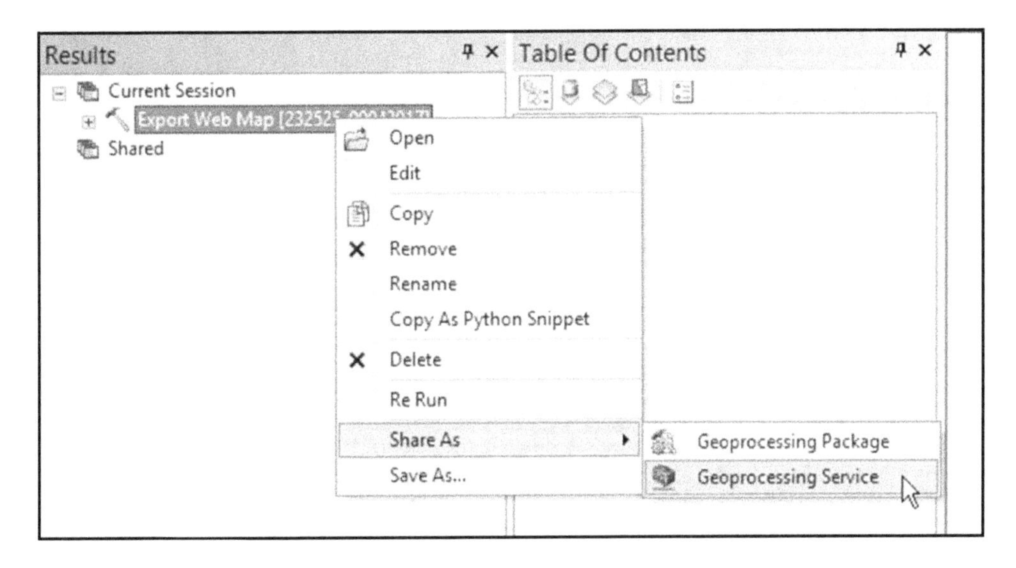

5. In the **Share a Service** window, select **Publish a service** and click on **Next**.
6. Next, in the **Publish a Service** window, **Choose a connection** if one already exists or create a new one, **Select a Service** name, and click on **Next**.
7. Next, choose an existing folder to publish the service to or create a new folder.

8. In the **Service Editor**, there are several parameters to note:

- Under **Export Web Map** | **Format**, you can give users the option to have a drop-down of export format choices or to simply export to the default format. With an **Input mode** of **Choice list**, the default export format is the format set earlier in the **Export Web Map** task. Here, you can also set the other export formats you want to make available to users:

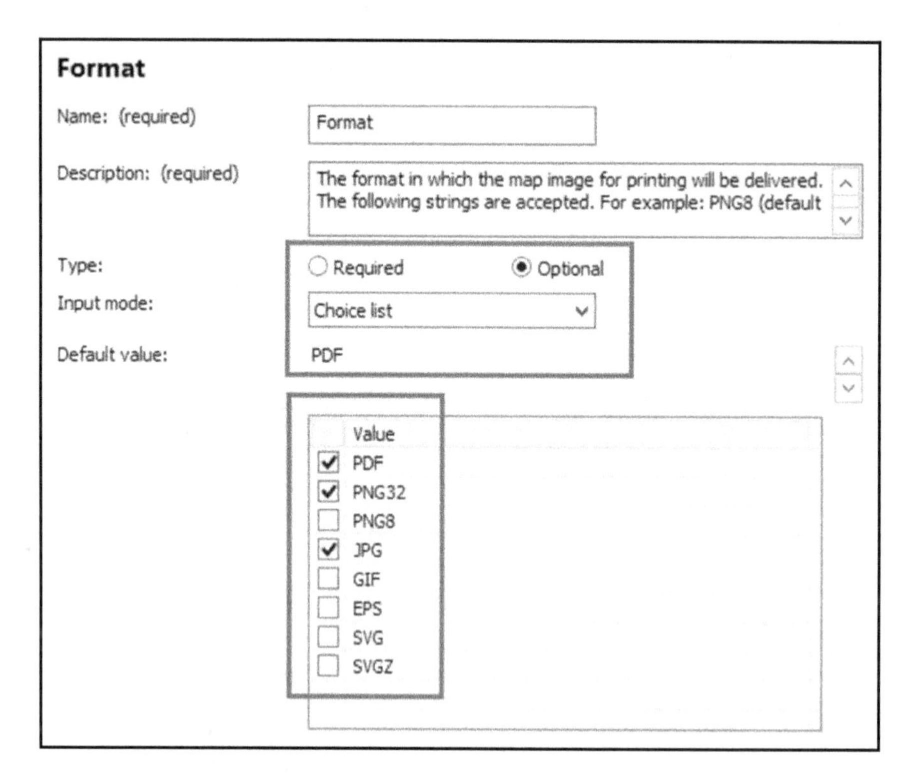

- The checklist of formats allows you to remove some of the lesser-used formats, such as **GIF**, **EPS**, **SVG**, and **SVGZ**. Setting the **Input mode** to the **Constant** value removes the option list and exports only to the **Default** value.
- Under **Export Web Map** | **Layout Template**, you can turn off templates you don't want to be made available to your users, such as the **MAP_ONLY** option.

9. Once the print service is successfully published, it can be accessed through your ArcGIS Server REST endpoint at `http://<webadaptor>/arcgis/rest/services/ExportWebMap/GPServer/Export%20Web%20Map`.

Using the custom print service

Now that you've created a custom print service, you can consume it just as you would the standard Esri print service.

Portal to Portal collaboration

Introduced in ArcGIS Enterprise 10.5, distributed Portal collaboration allows content to be shared between individuals as well as across organizations and communities. Portal collaborations are based on trust, where one organization initiates the collaboration and one to many other organizations accept the invitations and are participants. Portal collaborations can serve as hubs for larger organizations, simple shares between portals, or shares between portals and ArcGIS Online (available with the 10.5.1 release of ArcGIS Enterprise).

Setting up a collaboration

At a high level, to set up a collaboration, a host must first create a collaboration and invite guests to it. Invited guest administrators must then accept the invitation, in turn providing a response for the host administrator to import. Finally, the guest administrator joins the collaboration workspace. Detailed requirements, instructions, and workflows for creating, joins, and maintaining collaborations can be found in the ArcGIS Enterprise online documentation by searching for *About distributed collaboration*.

A key component to making collaboration work is ensuring users have equal rights in each organization. The host and guest Portals must use the same type of identity store as well.

Administering through the Portal REST Administrative Directory

Just like ArcGIS Server, Portal has a REST backend from which all administrative tasks can be performed.

We previously covered how the web interface for ArcGIS Server is a frontend to the ArcGIS Server REST API, and Portal is no different. We also covered services and how REST calls are made to the API. With all that covered, let's dive in and look at some of the administrative actions that can be performed in the Portal REST Admin.

System properties

Portal system properties include items such as Web Adaptors and licensing. Information about these items can be viewed, and each has operations to make configuration changes.

Web Adaptor

To get to your Portal Web Adaptor in Portal Admin, go to **Home** | **System** | **Web Adaptors** | **<web adaptor ID>**. If you've examined the Web Adaptor settings in the ArcGIS Server REST Administrator, then this should look familiar to you. Here, you can see **Machine Name**, **Machine IP**, **URL**, and the ports associated with your Web Adaptor:

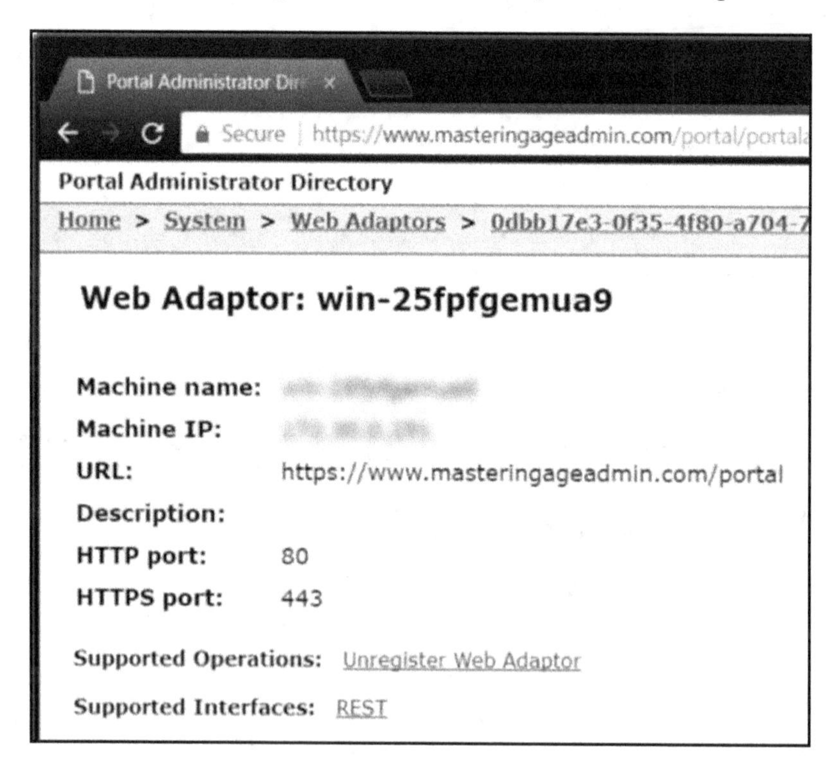

Here, there is a link to the **Unregister Web Adaptor** operation, but there is also a hidden operation here, just as there is in the ArcGIS Server REST Admin. Append `edit` to the end of your Web Adaptor properties URL, so it looks like the following:

```
https://www.masteringageadmin.com/portal/portaladmin/system/webadaptors
/0dbb17e3-0f35-4f80-a704-7c28uidr97eba0a/edit.
```

You are now at the **Edit Web Adaptor** endpoint, where you can edit several properties of your Web Adaptor:

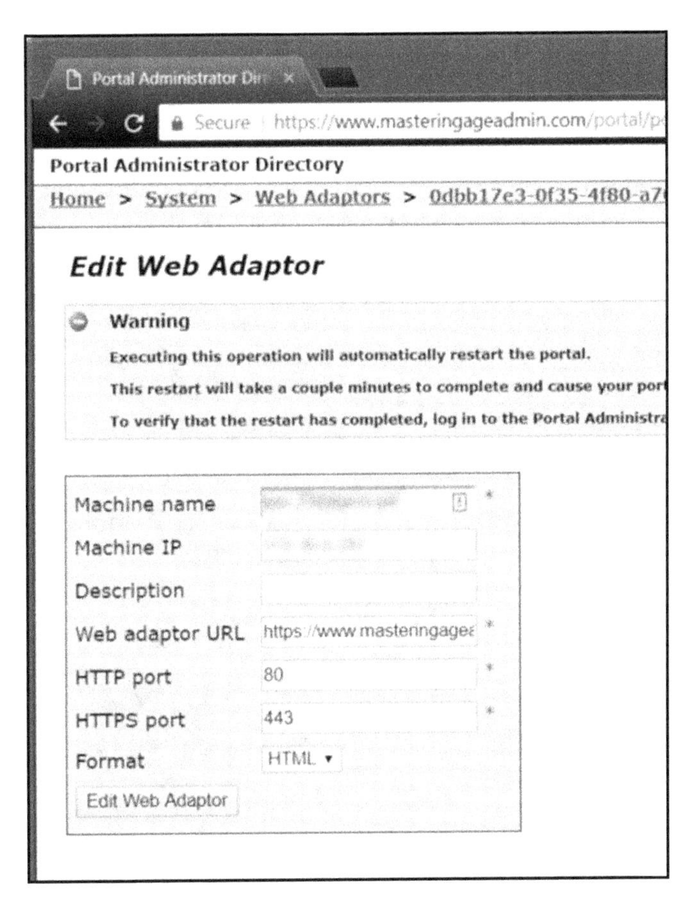

A common usecase here is, sometimes, during installations, the **Web adaptor URL** will not get registered properly and you will need to come in to this configuration endpoint and add the proper FQDN. Now, just because you can come in here and easily change properties of your Web Adaptor, doesn't mean that you should. Realize that unregistering, uninstalling, reinstalling, and reconfiguring a Web Adaptor is about a fifteen-minute task altogether, and if you are having issues that you think could be resolved by a Web Adaptor modification, making those modifications on the REST Admin may not be sufficient, and a full uninstall/reinstall of your Web Adaptor may be in order.

Licensing

Information on current Portal licensing can be viewed by going to **Home** | **System** | **Licenses**. Here, information on the validity and expiration of licensing and registered members can be viewed. The **Import Entitlements** operation allows the import of entitlements for ArcGIS Pro and additional products, such as Business Analyst or Insights. For ArcGIS Pro, the operation requires the entitlements file to be exported out of My Esri. Once the entitlements have been imported, licenses can be assigned to users within Portal. Entitlements can have effective parts and parts that become effective on a certain date. These all get imported, with the effective parts available immediately and the non-effective parts placed into a queue that Portal will automatically apply once they become effective. To import entitlements for ArcGIS Pro, follow these steps:

1. Have your entitlements file ready.
2. In Portal Admin, go to **Home** | **System** | **Licenses** | **Import Entitlements**.
3. Choose your entitlements file under **Choose File**.
4. For Application, choose **ArcGISPro**.
5. For Format, choose **JSON** or **HTML** (this is only the response format).
6. Click on **Import**.

Once the entitlements are imported, the licenses can be assigned to users in Portal under **My Organization** | **Manage Licenses**.

Logs

Just like ArcGIS Server, Portal has a complete logging system that runs during installation, normal everyday operation, and upgrades. The most important thing to remember about the Portal logs is to *use them*. It is very easy to get caught up in troubleshooting an issue and completely forget about looking at the logs. Always remember to step back and look at the logs.

Installation and upgrade logging

During installation and upgrading, Portal's log level is set to Verbose for detailed message collection. If an error is encountered during installation or upgrade, follow these steps to examine the logs for information on the error:

- First, determine whether the Portal Admin directory can be accessed (`https://<webadaptor>/portal/portaladmin`). If it can be accessed, then query the logs there (refer to the next section, *Working with Portal logs*).
- If the Portal Admin cannot be accessed, browse to the logs on disk at `<Portal for ArcGIS installation directory>/arcgisportal/logs/<machine name>/portal/`, find the most recent log file, and open it in a text editor to review it for errors.

Everyday logging

- Once installation or an upgrade has successfully completed, logging for the Portal is set to Warning by default. Now, logs can be accessed from Portal Admin (refer to the next section, *Working with Portal logs*). Depending on the level of logging configured for Portal, there are a multitude of events that can be recorded in the logs, such as the following.
- Publishing events such as publishing web maps, applications, and hosted services
- Security events such as updating the HTTP/HTTPS protocol settings, updating of the portal's identity store, and users logging in to Portal
- Organizational management events such as the creation of groups, group user management, and the configuration of Portal items such as basemaps, gallery, logs, and federation

Working with Portal logs

With Portal, there is no web-based frontend to view logs like is available in ArcGIS Server; all log querying and display is done from Portal Admin. To query and view Portal logs, follow these steps:

1. Open Portal Admin (`https://<webadaptor>/portal/portaladmin`) and log in as an Administrator.
2. Go to **Logs** | **Query**.

3. Leave all default settings and click on **Query**. You will be presented with a list of warning and severe messages:

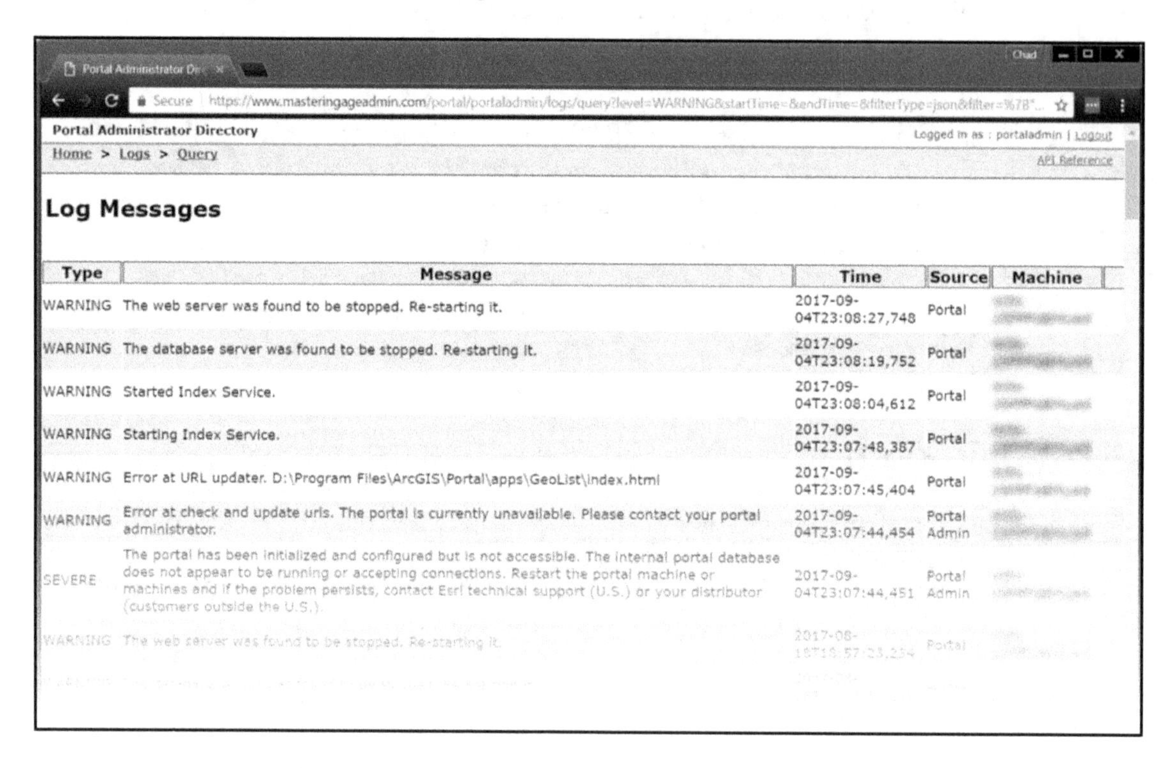

Portal for ArcGIS log levels are very similar to those for ArcGIS Server, as is illustrated in the following table:

Log Level	Description
Severe	These are serious problems that require immediate attention. It includes only severe messages.
Warning	These are moderate problems that require attention. It includes severe messages.
Info	These are common administrative messages for the server, such as service creation and startup. It includes severe and warning messages.
Fine	These are common messages from use of the server. It includes severe, warning, and info messages.

Verbose	These are messages about how fast the server draws layers, whether each layer in a service was drawn successfully, or how long it took for the server to access a layer's source data. It includes severe, warning, info, and fine messages.
Debug	These are highly verbose messages intended to be used for troubleshooting issues. Not recommended for use in production environments, as it may cause a decrease in performance. If you enable **Debug** logging, you must remember to turn it back off; if you do not, your logs can grow substantially in size and could possibly fill up your disk.
Off	Logging is turned off and no events are logged. Not recommended for production environments.

Log parameters will be covered in detail under troubleshooting in `Chapter 10`, *Troubleshooting ArcGIS Enterprise Issues and Errors*.

Backing up Portal

ArcGIS Enterprise ships with a Java utility called `webgisdr` that can be used to back up and restore portal items and settings, GIS services and settings, and the ArcGIS Data Store managed database (relational data store) and hosted scene layer cache (tile cache data store). When using the `webgisdr` utility, there are requirements and considerations:

- The utility requires Java 1.7+ to run. The `JAVA_HOME` operating system environmental variable must be set to the Java installation directory on the machine that `webgisdr` will run from.
- The utility is version dependent, meaning that the deployment you restore *must* be the same version from which the backup was created.
- The utility can create both full and incremental backups, where an incremental contains all changes made since the last full backup.
- The utility does not back up the following (backup of these items is discussed shortly):
    - Map service cache tiles and hosted tile layer caches
    - File-based and enterprise data sources used by web services, such as file geodatabases or file-based imagery
    - Spatiotemporal data stores

- Backups require *substantial* disk space. A deployment backup devoid of any data or services is approximately 400 MB in size alone. Storage locations are required on each component server and should be sized accordingly, where the more of the following there are in the Portal configuration, the more storage space will be required for backups:
    - How many items are in your portal?
    - The number and type of hosted web layers
    - The number of federated servers
    - The number of ArcGIS Server machines
- The backup process writes to a temporary location on each component machine before being moved to the shared directory:
    - **Portal**: `<Portal content directory>\temp`, as in `D:\arcgisportal\temp`
    - **ArcGIS Server**: `C:\Users\<user>\AppData\Local\Temp`, where `user` is the account that `webgisdr` runs as
    - **Data Store**: `<Data Store directory>\temp`, as in `D:\arcgisdatastore\temp`
- The domain accounts used to run the components of ArcGIS Enterprise, along with the account used to run `webgisdr` must all have write access to the shared directory specified for the main backup file.

Running the webgisdr utility

The webgisdr utility is a command-line utility that requires Java 1.7+. Before running the utility, ensure that Java is installed on the machine running `webgisdr` and that the `JAVA_HOME` system environmental variable has been set to the Java Runtime Environment directory, such as `C:\Program Files (x86)\Java\jre1.8.0_144`, where 1.8.0 is the Java version and 144 is the update number.

When setting the `JAVA_HOME` environmental variable, make sure you don't include the `/bin` directory, as this will result in a **The system cannot find the path specified error** when running `webgisdr`.

Configuration

The `webgisdr` utility uses a simple text configuration file, `webgisdr.properties`, which is located in the `webgisdr` directory. Make a copy of this file (it can reside anywhere the executing account has access to) and open it in a text editor.

There are several dozen properties for `webgisdr`, some of which are documented in the `webgisdr.properties` file, some of which are not. All are covered in the ArcGIS Enterprise online documentation. Search the documentation for *Create an ArcGIS Enterprise backup* for a full list of properties and their usage.

To configure a basic on-premise backup, set the following parameters:

1. `SHARED_LOCATION`: This is where backup temp files are copied to before they are moved to the location specified by `BACKUP_LOCATION`. This location needs to be writable by the ArcGIS Enterprise component domain service accounts and the account executing the script. Be sure to escape all backslashes, that is, `\\\\server\\share\\folder`.
2. `BACKUP_STORE_PROVIDER`: Set this to `FileSystem`.
3. `BACKUP_LOCATION`: This is the file share where the final site backup file will be stored. Be sure to escape all backslashes, that is, `\\\\server\\share\\folder`.
4. `PORTAL_ADMIN_URL`: This is set using the FQDN of your Portal instance. For example, `https://portalmachine.domain.com:7443/arcgis`. Note that since we will not go through the Web Adaptor, we will use the `7443` port and `arcgis` for the instance name instead of `portal`.
5. `PORTAL_ADMIN_USERNAME`: This is an account with administrative access to your Portal.
6. `PORTAL_ADMIN_PASSWORD`: This is a plain-text password for the preceding account.
7. `PORTAL_ADMIN_PASSWORD_ENCRYPTED`: Set this to `true`. Once the utility runs, it will encrypt the password in the `PORTAL_ADMIN_PASSWORD` parameter and change this flag to `false`.
8. `BACKUP_RESTORE_MODE`: The options here are `full` or `incremental`. The default is `full`. `Full` creates a full backup, whereas `incremental` will generate an incremental backup between full backups. This flag would be good to use in a scheduled task to fully back up the system.

Backup

To execute a backup with the utility, call it from a command prompt, passing in the `--export` flag followed by the `--file` flag and the path to the configuration file:

```
C:\Users\Administrator>"D:\ProgramFiles\ArcGIS\Portal\tools\webgisdr\webgis
dr.bat" --export -file
"D:\ProgramFiles\ArcGIS\Portal\tools\webgisdr\webgisdr.properties"
```

Upon a successful run, you will receive the following messages:

```
===========================================

Starting the webgisdr utility.

===========================================


The configuration and base backup time in the current Web GIS
--------------------------------------------------------------
  Portal: https://www.masteringageadmin.com/portal

Starting the backup process with the webgisdr utility.

Starting the backup for Portal for ArcGIS:
Admin Url: https://www.masteringageadmin.com/portal.

The following Portal for ArcGIS has been backed up successfully:
Admin Url: https://www.masteringageadmin.com/portal.

The backup of Portal for ArcGIS has completed in 00hr:01min:33sec.

Packaging web GIS backup ...
The backup file has been packaged successfully in 00hr:00min:52sec.

The backup file for the current web GIS site is located at D:\backups\age-
backup
\September-5-2017-3-57-25-PM-UTC-FULL.webgissite.

The backup of Web GIS components has completed in 00hr:02min:29sec.

Stopping the webgisdr utility.
C:\Users\Administrator>
```

The backup for the preceding, very small ArcGIS Enterprise development instance took approximately 2.5 minutes and the backup file `September-5-2017-3-57-25-PM-UTC-FULL.webgissite` was almost 600 MB in size.

Restore

Once you have backed up your ArcGIS Enterprise deployment, you can use the backup file and `webgisdr` to restore it. When restoring, note the following:

- Any items or services created since the last backup export (full or incremental) will be lost
- Anything not backed up by the utility will not be restored, for example, file geodatabases
- If restoring to a new machine, installation and content directories on the new machine must match those of the old machine

To restore a deployment, follow these steps:

1. If changes are necessary, make a copy of the `webgisdr.properties` file used for the backup and name it something like `webgisdr-import.properties`. By default, the most recent backup file is restored. To restore to a specific file, specify the full path to the file in the `SHARED_LOCATION` parameter.
2. At the command prompt, call the utility and pass in the `--import` flag, followed by the `--file` flag and the full path to import the configuration file, as shown here:

```
C:\Users\Administrator>"D:\Program
Files\ArcGIS\Portal\tools\webgisdr\webgisdr.bat"
--import --file "D:\Program Files\ArcGIS\Portal\tools\webgisdr\webgisdr-
import.properties"
```

3. If you created incremental backups, the latest of those will need to be restored after your latest full backup has been restored. Run the utility again using a configuration file that references the full path of the latest incremental backup in the `SHARED_LOCATION` parameter.
4. Upon successful completion, the following messages will be displayed:

```
==========================================

Starting the webgisdr utility.
==========================================

The configuration and base backup time in the current Web GIS
-------------------------------------------------------------
  Portal: https://www.masteringageadmin.com/portal

Unzipping the backup file:
D:\backups\age-backup\September-5-2017-3-57-25-PM-UTC-FULL.webgissite
```

```
The backup file has been unzipped in 00hr:00min:06sec.

The backup file was created at September 5, 2017 3:57:25 PM UTC.

The configuration and base backup time in the incoming Web GIS
----------------------------------------------------------------
  Portal: https://www.masteringageadmin.com/portal at 9/5/17 3:55 PM


Starting the restore process with the webgisdr utility.

Starting the restore of Portal for ArcGIS:
Admin Url: https://www.masteringageadmin.com/portal.

The following Portal for ArcGIS has been restored successfully:
Admin Url: https://www.masteringageadmin.com/portal.

The restore of Portal for ArcGIS has completed in 00hr:11min:00sec.

The Portal for ArcGIS has been restarted successfully in 00hr:01min:06sec.

The restore of Web GIS components has completed in 00hr:12min:18sec.

Stopping the webgisdr utility.
C:\Users\Administrator>
```

The preceding restore of a small ArcGIS Enterprise development deployment, where the backup file was around 600 MB, took 12 minutes to complete.

Backup of other items

As mentioned earlier, map service cache tiles, file geodatabases and file base data, and spatiotemporal data stores are not backed up by the `webgisdr` utility. A variety of methods can be utilized to back up these types of items.

File-based data

Backing up file-based data can be done manually with any number of software packages, such as TeraCopy (`http://www.codesector.com/teracopy`), or with virtually any self-respecting programming language. That said, in its most basic form, what we need to do here can probably, in most cases, be handled with a simple Windows batch file utilizing Robocopy, which has been a standard feature since Windows Vista and Windows Server 2008. Let's take a look at what that might look like.

First, we set ECHO OFF to suppress messaging. Next, through a series of FOR loops, we repeatedly parse the DATE system variable to get the month, day, and year parts we need to construct a date string for today's date in the form of yyyy-mm-dd or 2017-09-05. We set that date string into the _date_today variable for use later:

```
@ECHO OFF

FOR /F "TOKENS=1* DELIMS= " %%A IN ('DATE/T') DO SET CDATE=%%B
FOR /F "TOKENS=1,2 eol=/ DELIMS=/ " %%A IN ('DATE/T') DO SET mm=%%B
FOR /F "TOKENS=1,2 DELIMS=/ eol=/" %%A IN ('echo %CDATE%') DO SET dd=%%B
FOR /F "TOKENS=2,3 DELIMS=/ " %%A IN ('echo %CDATE%') DO SET yyyy=%%B
SET _date_today=%yyyy%-%mm%-%dd%
```

Next, we SET several variables. The _src is the source folder under which our data resides and the _dest is the target where we want our data copied to. Here, we use the date string variable as the target folder. With this configuration, we can do a backup every day and have each target folder timestamped in the yyyy-mm-dd format. Under the _what variable, we set the /COPYALL flag, which copies all file attributes along with the data, and the /MIR flag, which mirrors directory trees. Finally, set _options, where the /LOG logs the Robocopy output to a log file with today's date in the filename and /XF excludes files matching *.lock (those pesky .lock files in file geodatabases):

```
SET _src=D:\data
SET _dest=D:\backup\%_date_today%
SET _what=/COPYALL /MIR
SET _options=/LOG:file_copy_%_date_today%.txt /XF *.lock
```

Lastly, we call Robocopy, passing in all our variables we set in the preceding code:

```
ROBOCOPY %_src% %_dest% %_what% %_options%
```

This script could easily be modified to include multiple source directories and simple error handling.

Conversely, a Python script utilizing arcpy methods could also perform these simple data backup tasks.

Spatiotemporal data stores

Data Store ships with its own set of command-line administration utilities. The backupdatastore utility can be used with relational, tile cache, and spatiotemporal data stores. These utilities are Windows batch files and are located at <Data Store installation directory\datastore\tools.

Prior to backing up any data store with `backupdatastore`, the backup location must be set using the `configurebackuplocation` utility.

To run the Data Store utilities, your login must be a member of the Windows Administrator group and you must launch Command Prompt under **Run as Administrator**.

The configurebackuplocation utility

This utility can be used to configure (and change existing) the backup location for relational, tile cache, and spatiotemporal data stores. A backup location must be registered with Data Store, and the backup location must be a network share; local drives cannot be used for spatiotemporal data store backup files. The `configurebackuplocation` utility uses the following syntax:

```
configurebackuplocation --location <backup_location> [operations]
```

The command to register a spatiotemporal data store backup location looks like this:

```
D:\Program Files\
ArcGIS\DataStore\tools\configurebackuplocation.bat --operation -register --
store spatiotemporal --location \\server\share\folder
```

The backupdatastore utility

The first time a spatiotemporal data store is backed up, a full backup is created. Subsequent runs create incremental backups. The `backupdatastore` utility can be run from any machine that is a member of the spatiotemporal data store. It uses the following syntax:

```
backupdatastore [<backup name>] [--store
{relational|tileCache|spatiotemporal}] [--prompt <yes | no>]
```

The command to backup a spatiotemporal data store looks like this:

```
D:\Program Files\ArcGIS\DataStore\tools\ backupdatastore.bat
spatiotemporal-backup --store spatiotemporal --prompt no
```

We have only covered two of the 20+ Data Store utilities. Search the ArcGIS Enterprise online documentation for *ArcGIS Data Store command utility reference* for a full list of all Data Store utilities and their usage.

Changing the Portal for ArcGIS account

Just as with ArcGIS Server, you can (and trust me, one day you *will* need to) reset or change the Portal for an ArcGIS service account. One difference is that the tool to do this for Portal is not located on the Windows Start menu as it is for ArcGIS Server. Instead, it is an executable file utility located at `<Portal installation directory>\tools\ConfigUtility` on the Portal machine. Like the Configure ArcGIS Server Account tool, `ConfigurationUtilityCL.exe` sets the account (domain or local) to run the Portal service and grants the account privileges on Portal system folders and files. `ConfigurationUtilityCL.exe` uses the following syntax:

```
configureserviceaccount.bat --username mydomain\username --password
password -writeconfig c:\temp\config.xml
```

The available parameters for the utility are the following:

- `username`: This is the username of the Portal service account.
- `password`: This is the password for the Portal service account.
- `writeconfig`: This is optional. It is a path to the configuration file to be saved so the same configuration can be applied in future runs of the utility.
- `readconfig`: This is optional. It is a path to the configuration file saved from the previous run of the utility.

To run the utility, enter a command such as the following:

```
D:\Program
Files\ArcGIS\DataStore\tools\ConfigUtility\ConfigurationUtilityCL.exe --
username mydomain\username --password trytoguess --writeconfig
D:\backup\portal-account-recovery.xml
```

Management tools

In addition to using the administrative settings in Portal and the Portal Admin, there are solutions available to assist in managing your Portal content.

AGO Assistant

ArcGIS Online Assistant, or AGO Assistant for short, is a web application created and hosted by Esri that can be found at `https://ago-assistant.esri.com/`. Esri calls it *A swiss army knife for your ArcGIS Online and Portal for ArcGIS accounts,* which sums it up quite well.

AGO Assistant uses the ArcGIS REST API to work with content in ArcGIS Online and Portal for ArcGIS through a simple interface. Some of the tools available include viewing the underlying JSON for any item in your Portal or ArcGIS Online (a personal favorite of mine), copying content from one account to another (Portal to Portal, AGO to AGO, Portal to AGO, or vice versa!), and updating service URLs in web maps and applications (another huge timesaver). Let's take a look at some ways to use AGO Assistant for administrative purposes.

Accessing AGO Assistant

To access the AGO Assistant for your Portal, go to `https://ago-assistant.esri.com/` and select **Log in to Portal for ArcGIS**. You can log in directly using a Portal for ArcGIS account (shown as follows), a SAML-based identity (OAuth Login), or a **single sign-on** (**SSO**) through **Integrated Windows Authentication** (**IWA**):

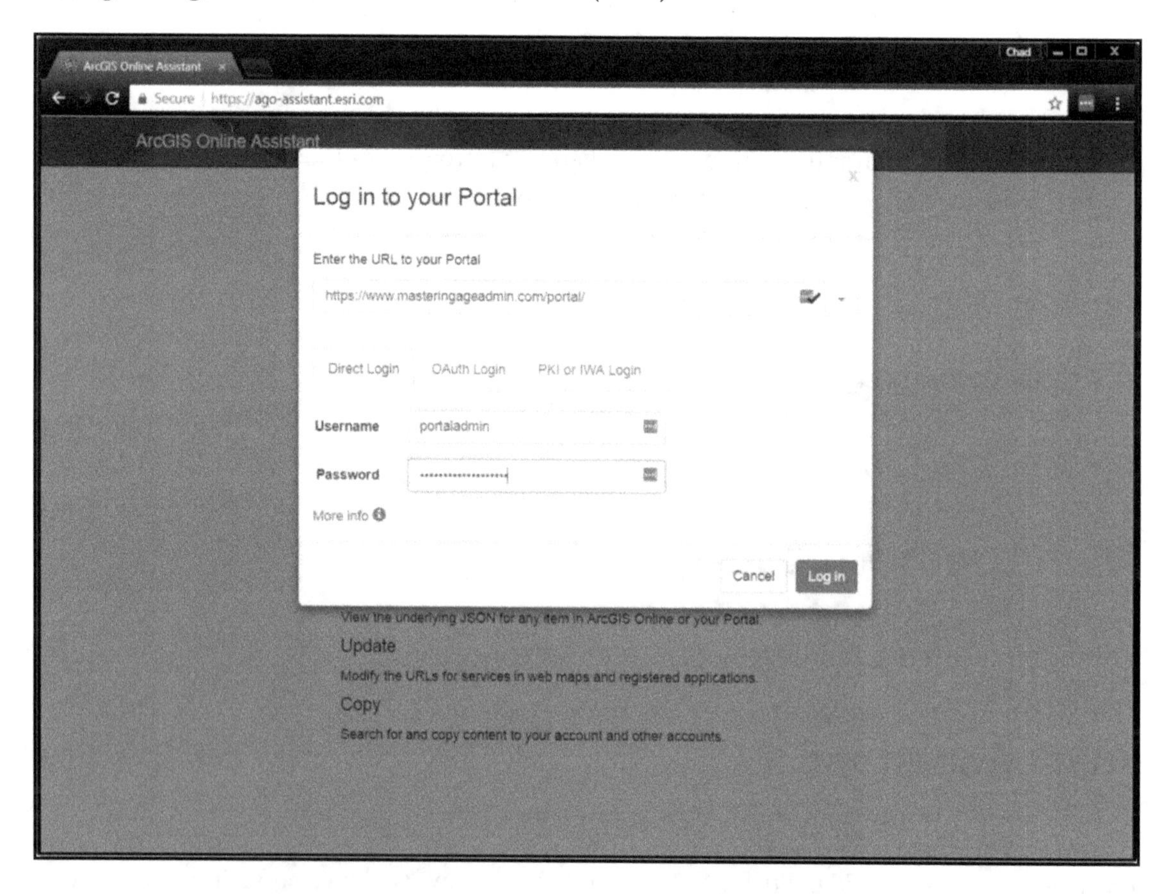

Once you are logged in, you are directed to the root of your Portal for the account you have logged in as. AGO Assistant is a task-driven application, meaning that, to get something done, you have to first choose the task from the **I want to...** drop-down and the AGO Assistant will present highlighted items in the left panel that the chosen task can be performed on.

Viewing an item's JSON

Virtually all items in Portal (and ArcGIS Online, for that matter) have some sort of JSON representation associated with them, even if it is just a description (think metadata). Many items, including Web Maps and hosted Web Mapping Applications, also have data associated with them, and that data, to a certain extent, can be manipulated from right within the AGO Assistant. Yes, you can edit an item's JSON, save the changes, and they are immediately live. To view an item's JSON, log in as the item's owner and under **I want to...**, select **View an Item's JSON**:

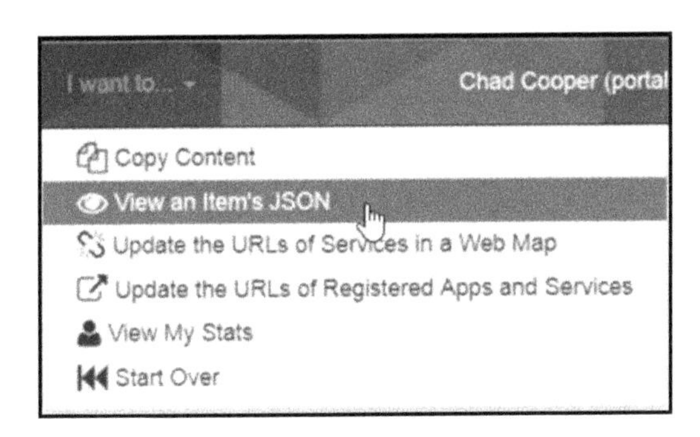

Click on a folder in the left panel to expand it. Any items with JSON to view are highlighted in light blue. Select **Web Map**. Select a hosted **Web Mapping Application** if you have one. Note that it has two sections in the right column: **Description** and **Data**. Take a look at the JSON in these sections. If you have a Basic Web Mapping Application and a Web AppBuilder application, look at the differences between the two; you'll see that the basic app's JSON is, well, more basic than that of the Web AppBuilder app.

Let's look at a simple Basic Web Mapping app in my Portal, aptly named `Basic Viewer`. It was created using only a bare basemap centered on Fayetteville, Arkansas (Go Hogs!) and has a search bar, basemap switcher, print tool, and share tool:

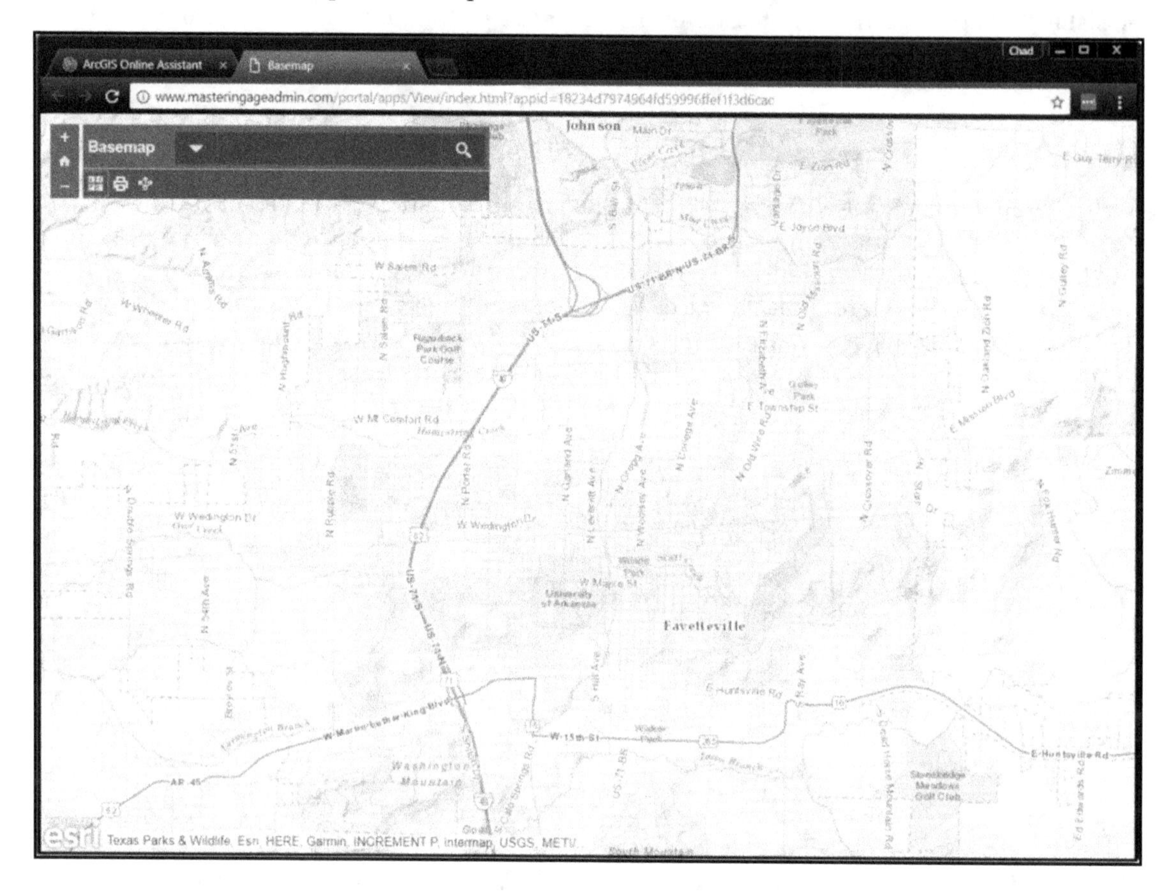

Now let's take a look at this app's JSON, in particular, parts of the **Data** section. In the values object array, there are quite a few objects related to tools, such as the following:

```
"values": {
  "color": "#fff",
  "theme": "#105e78",
  "iconColor": "#fff",
  "activeTool": "",
  "scalebar": false,
  "splashModal": false,
  "toolbarLabels": false,
  "tool_print": true,
  "tool_print_layouts": true,
```

```
        "tool_print_legend": false,
        "tool_share": true,
        "tool_basemap": true,
        "tool_overview": false,
        "tool_measure": false,
        ...
        ...
    }
```

The `tool_print`, `tool_basemap`, and `tool_share` are all `true`, as can be seen in the preceding screenshot; these tools are all present. However, what if we want to add an overview map? Can we do that from here? Let's edit the JSON and find out. At the top of the **Data** window in AGO Assistant, let's click on the pencil icon, as shown here:

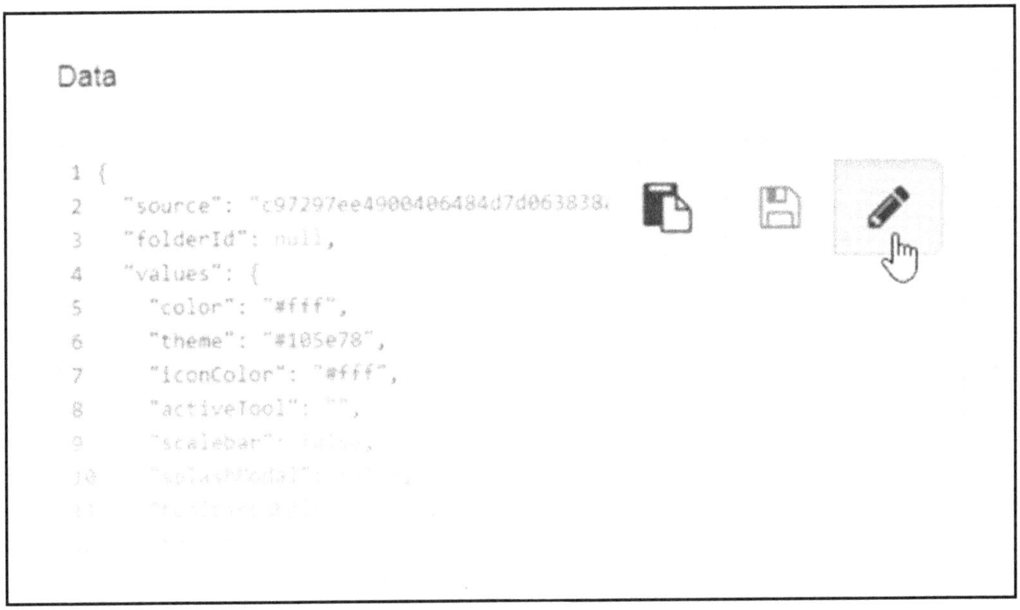

We will accept the **THIS IS UNTESTED AND UNSUPPORTED** warning (my colleagues and I have literally been using this feature for years) by checking the **I understand the risks** checkbox and then the **Proceed** button. We are now in edit mode for the Data JSON of our web app. I'll change the `tool_overview` parameter from `false` to `true`.

However, when I start to delete `false` to replace it with `true`, the line of code is highlighted in pink:

```
7     "iconColor": "#fff",
8     "activeTool": "",
9     "scalebar": false,
10    "splashModal": false,
11    "toolbarLabels": false,
12    "tool_print": true,
13    "tool_print_layouts": true,
14    "tool_print_legend": false,
15    "tool_share": true,
16    "tool_share_embed": false,
17    "tool_overview": fal,
18    "tool_measure": false,
19    "tool_details": true,
20    "tool_legend": false,
21    "tool_layers": true,
22    "tool_sublayers": true,
23    "tool_opacity": true,
24    "tool_layerlegend": true,
25    "locate": false,
26    "locate_track": false,
27    "tool_edit": false,
```

A new feature of the AGO Assistant is that JSON is validated on-the-fly; as soon as I remove the `e` in `false`, the JSON becomes invalid. When I replace `false` with `true`, it becomes valid again, and the pink highlight goes away. I also do not have an active tool, one that is displayed when the application loads. I can change the `activeTool` parameter from an empty string (`""`) to `basemap` to show the `Basemap` widget on application load:

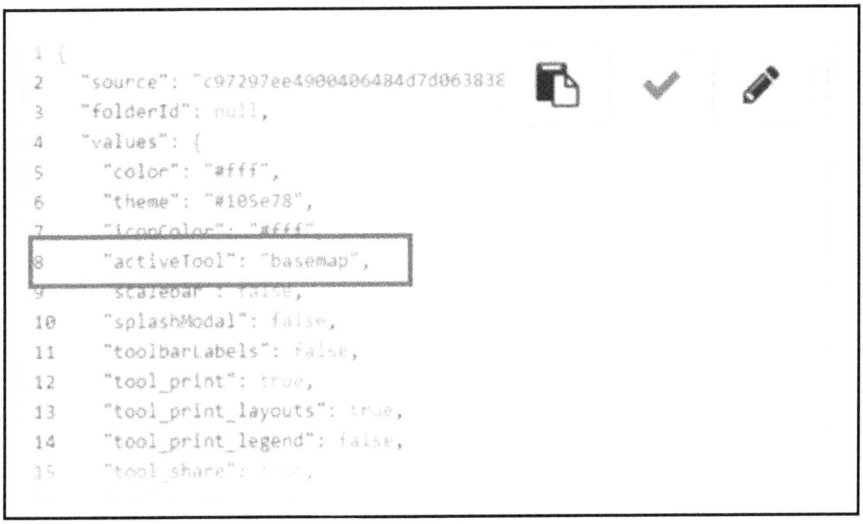

```
"source": "c97297ee4900406484d7d06383E
"folderId": null,
"values": {
  "color": "#fff",
  "theme": "#105e78",
  "activeTool": "basemap",
  "splashModal": false,
  "toolbarLabels": false,
  "tool_print": true,
  "tool_print_layouts": true,
  "tool_print_legend": false,
```

Now, after going back to our web app in the browser and refreshing, we get the **Overview Map** widget icon in the upper-left menu and the **Basemap** widget shown on page load in the upper-right:

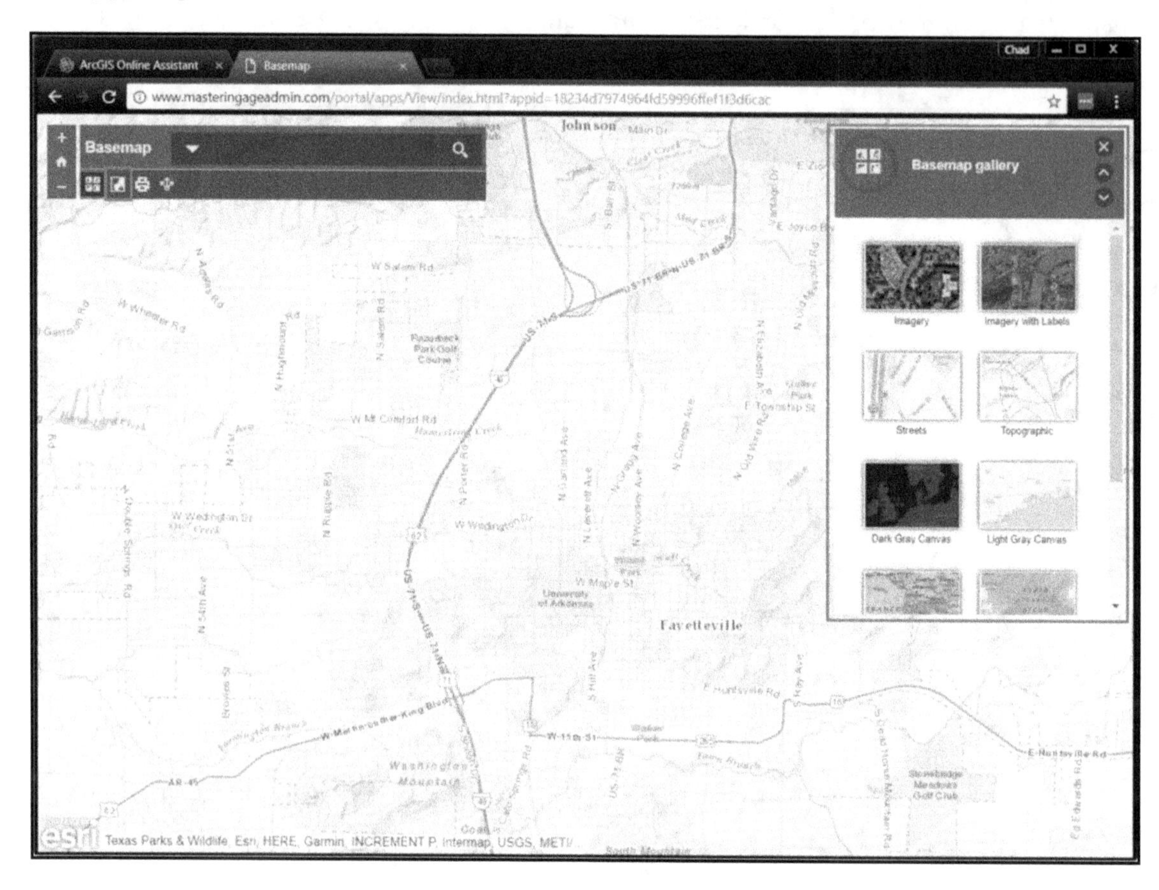

Now, having done all of that, was it a contrived and rather trivial example? Yes. Does it lay the groundwork for Portal item manipulation and demonstrate how powerful and time-saving the AGO Assistant can be? Yes.

Changing URLs

Ever had a Web Map path to a map service that has changed? You probably have. Maybe a third party changed their domain name or enabled HTTPS on their ArcGIS Enterprise instance and now you are left with busted web applications. Never fear, the **Update the URLs of Services in a Web Map** tool in AGO Assistant can fix that. In the **I want to...** menu, select **Update the URLs of Services in a Web Map**; then, in the left column, select **Web Map with broken URLs**. **Operational Layers**, **Tables**, and **Basemap Layers** present in the Web Map will get populated in their sections in the right column. In the **Find/Replace** section, as follows, I have selected to look for all instances of `http://some-domain.com` in my Web Map and replace them with `https://some-domain.com`. In a Web Map with dozens of layers, it is easy to see how much time this tool could save:

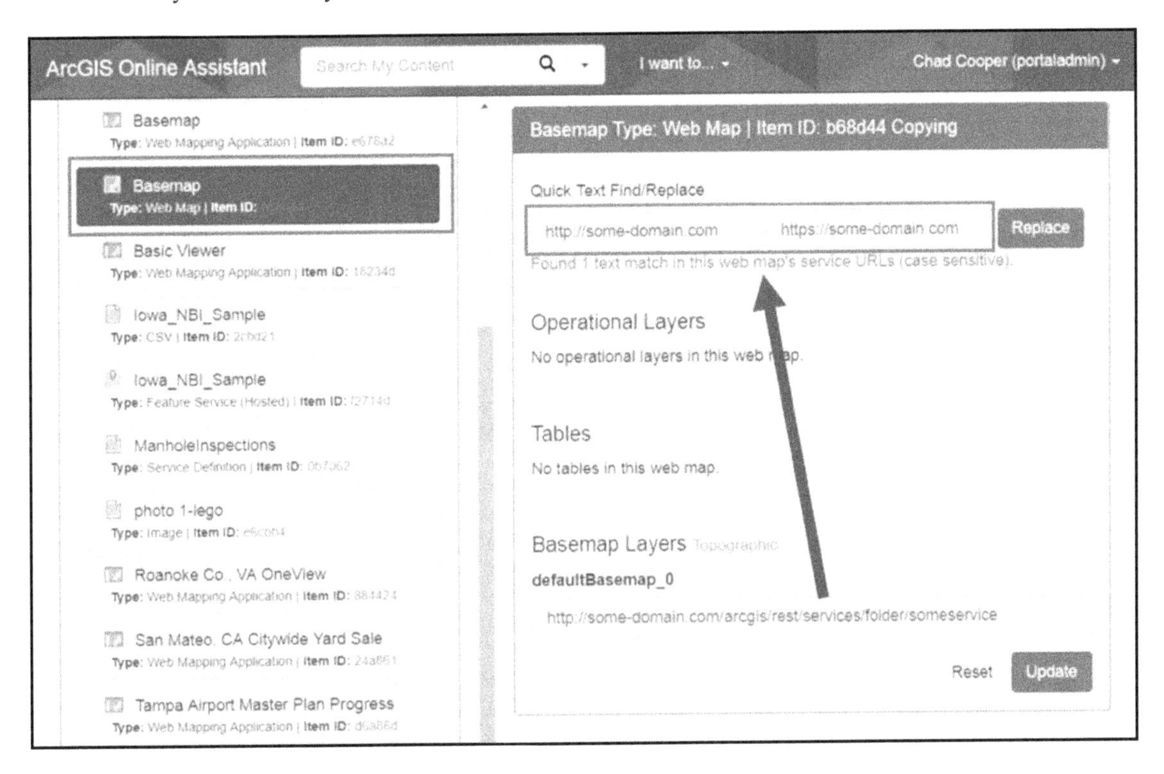

Copying items

Have an item you want to edit the JSON of, but you are just a little bit afraid you might mess up that production item beyond repair? Or maybe you have an application in ArcGIS Online that needs to be migrated to Portal? Sure, if the item you want to edit the JSON of is in your Portal, you can open it in the **Map Viewer** and do a **Save As** and save a copy. Same goes for a hosted Web AppBuilder app in your Portal. However, there is a copy tool in AGO Assistant that is even easier to use. To copy content within your Portal using AGO Assistant, follow these steps:

1. In AGO Assistant, log in to the account containing the content you want to copy. Go to **I want to...** | **Copy Content**. Select **My account** as the account you want to copy into. The items available to copy will be highlighted in light blue in the left column (file-based data such as images, service definitions, CSV files, and more are not available for copy) and a dashed window named **Drop items to copy to the folder** will be created in the **Root** window in the right column.
2. Drag and drop an item from the left column to the dashed drop window in the right column, as shown in the following screenshot:

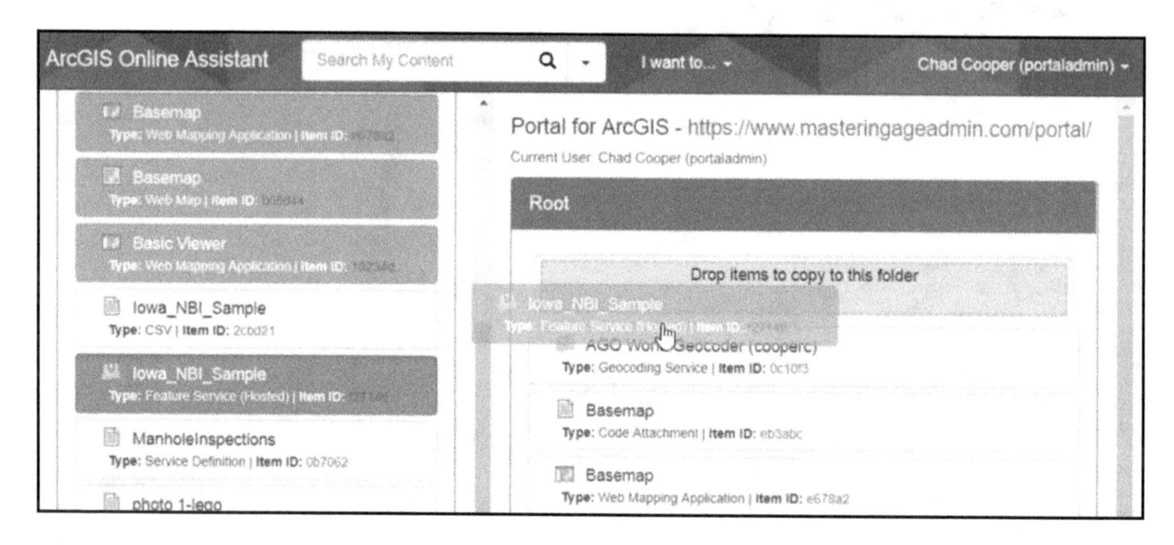

3. Most items do a simple copy and retain the original name, but with a different Item ID. Make note of the Item ID (you can hover over the link of the newly copied item in the right column and the Item ID will be shown), as you will probably want to rename and move the item to another folder. In the case of a feature service, you are given the option to reference the existing data (simple copy--fairly quick) or do a full replication of the original service along with its associated data (full copy--time-consuming, depending on the source data).
4. Once your item is copied, you can make any necessary changes by editing the item's JSON to change a Web Map ID or edit URLs to point to new or updated service URLs.

geo jobe Admin Tools

Admin Tools by geo jobe (`http://www.geo-jobe.com/`) is a suite of tools designed to help streamline ArcGIS Online and Portal for ArcGIS Administration. Much like AGO Assistant, Admin Tools uses the REST API on its backend to perform administrative tasks. Tasks can be performed individually on a single item or multiple tasks can be chained together and performed in bulk on multiple items. Admin Tools is used by over 4,000 users worldwide and comes in three versions:

- **Admin Tools (Free)**: This version can be used free of cost, but offers limited, but still incredibly useful, functionality such as copy, move, and delete items, update owner and sharing properties, and import and export users from CSV and JSON.
- **Admin Tools (Pro)**: This version offers the same functionalities as the free version, but adds additional functionalities for a fee. Additional functionalities here include viewing item dependencies, updating web map URLs, cross-organizational cloning, and importing of groups and items from CSV and JSON.
- **Admin Tools (Portal)**: This version is essentially Admin Tools Pro, but for Portal for ArcGIS instead of ArcGIS Online. For a fee, this version has all the functionality of Admin Tools Pro, but again, for Portal.

Admin Tools is used by many large organizations, such as Esri, Apple, and BP to manage multiple large ArcGIS Online and Portal instances. For more information on these and other products geo jobe offers, visit `http://www.geo-jobe.com`.

Summary

In this chapter, we covered quite a bit of content regarding Portal administration. We showed how to access Portal through both its web interface and Portal Admin. Then, we looked at how to change the public appearance and style of your Portal. Managing content is an especially important topic, which we covered in detail, ranging from customizing basemaps, to configuring the map viewer, to configuring utility services such as printing. The creation of a custom ArcGIS Server print service was also discussed in great detail, as this is an item that almost every organization needs and wants. Next, we moved on to using the Portal REST Admin for administration, covering the Web Adaptor, licensing, and logs. Backing up your Portal is an important administrative task, and we covered it using the `webgisdr` utility, along with a Windows batch script to cover both Portal and non-Portal file-based data that also needs backing up. We then discussed how to change the Portal service account, and finished by talking about the ArcGIS Online Assistant and how it can aid greatly in administration tasks for Portal. Next up is security, in `Chapter 6`, *Security*, where we will discuss user stores, federation, and how to best secure your ArcGIS Enterprise deployment for your organization.

6
Security

Security is quite possibly the most important yet least discussed aspect of any enterprise system, GIS included. The security of your ArcGIS Enterprise system should be a paramount concern warranting significant consideration. As an administrator, the security and integrity of your deployment should always be on your mind. ArcGIS Enterprise has many different security patterns that can be utilized by organizations of all sizes. In this chapter, we will discuss several security patterns ranging from simply utilizing the ArcGIS Server built-in user store to federating ArcGIS Server with Portal, and on to enabling **Integrated Windows Authentication** (**IWA**) and ultimately **Single Sign-On** (**SSO**). Security is a very deep subject, full of an astounding number of details. Covering every aspect surrounding security in ArcGIS Enterprise would constitute a book of its own; however, when we are finished with this chapter, you will understand the following:

- The fundamentals of security and identity stores in ArcGIS Enterprise
- What different security patterns are available with ArcGIS Enterprise and the pros and cons of each
- Why and how to federate ArcGIS Server with Portal
- How to implement IWA
- How to implement SSO
- How to edit, manage, and maintain security settings through different interfaces of ArcGIS Enterprise

Security basics

I'm going to say this again--security is a big deal; it's a big deal regardless of the size or nature of your organization. Whether you are an international organization of 20,000 people or a small business of 10 people, if you discount the security and integrity of your systems, it's not a matter of *if* you will be compromised, but *when*.

Now, I'm not trying to fearmonger, I'm simply stating the facts--there are parties that will compromise your system for no reason other than the fact that *they were able to do it*. As an administrator, it is your job to do everything within your powers and abilities to keep those parties from infiltrating your system.

Password strength

When talking security, little things can make a big difference. Passwords are one of those things. Considered by most to be a necessary evil, passwords are an essential first line of defense to your system and are the most widely used form of authentication throughout the world. We all know our passwords should be *strong*, but what does that really mean?

Password entropy

Password entropy is the measurement of how unpredictable a password is and is based on the character set used (uppercase, lowercase, digits, symbols) and password length. Since password entropy measures unpredictability, it can, therefore, predict how difficult a password is to determine, or *crack*, as it is commonly referred to. Password complexity has often been characterized using the concept of entropy (NIST, 2017). Conventional wisdom has always said that passwords must have some form of complexity, usually in the form of one number, one upper case letter, one lowercase letter, and one symbol (sound familiar?). Enter password length.

Password length

Recent studies by the **National Institute of Standards and Technology** (`https://www.nist.gov/`) have password length be the primary factor in characterizing password strength. Short passwords yield to brute force (guessing) attacks as well as dictionary attacks that use banks of known words and commonly used phrases as chosen passwords (NIST, 2017). So, what *is* a good minimum length to use? NIST recommends a minimum length of 8 characters but also states *Users should be encouraged to make their passwords as lengthy as they want, within reason* (NIST, 2017). The key to users getting the most from a password length without the burden of complexity requirements is for them to create a memorable password.

Generating passwords

Coming up with a memorable, yet somewhat random password on your own might sound easy, but it's not. Lucky for us, the interwebs come to our rescue. Googling **memorable password generator** turns up a plethora of sites dedicated to just that--memorable password generation. Many of these are open source projects with the source code fully and openly available online. A personal favorite of mine and many IT professionals is XKPasswd at `https://xkpasswd.net/s/`. I use XKPasswd almost daily and recommend it to my colleagues and clients; it is highly customizable, and once you find a format you like, you can save a JSON configuration file and easily reload it to get the similar format on every visit to the site without having to go through the configuration process all over again.

Managing passwords

As an administrator, you will have passwords, *lots* of passwords. Sure, we just talked about making your passwords memorable, but when you have dozens or even just several passwords (many of which you don't use on a day-to-day basis), they are still hard to remember. Now that you have all your nice, reasonably-long, non-complex, memorable passwords, how do you, as an administrator, keep track of them? The answer is a password manager or vault, as they are sometimes referred to. A **vault** is a piece of software that stores account credentials for you using some form of known strong encryption. Accessed with a password, shared token, and in some cases, two or three-factor authentication, a vault is a great way to securely manage, safeguard, store, and share your precious account credentials. Tom's Guide (`https://www.tomsguide.com`), a sub-site of Tom's Hardware (`http://www.tomshardware.com`), which has been around since 1996, has a nice review of password managers for 2017 at `https://www.tomsguide.com/us/best-password-managers,review-3785.html`. Tom's Guide recommends LastPass (`https://www.lastpass.com`), which I personally use myself and would recommend. I also use KeePass (`http://keepass.info/`) in my day job at the team level, and I must say that we would not be able to function without a quality password vault.

ArcGIS Server security

Now that we have gotten some general security items out of the way, let's discuss ArcGIS Server security.

Fundamentals of ArcGIS Server security

In an abstract sense, security is a simple concept; by securing an IT system (such as your GIS system), we are protecting it from harm, either accidental or intentional, that could come as the result of unauthorized access. With that said, let's start with looking at how ArcGIS Server is initially configured upon post-installation, and ways we can further strengthen the security of the deployment.

The post-installation scene

After a fresh new installation of ArcGIS Server, you have a simple system:

- There is only one account in ArcGIS Server--the primary site account (PSA) that was specified upon site creation
- All admin and publishing services are secured
- All services are publicly accessible (no security is set up yet, so any service that gets published is open to the public by default)

For an internal-only development or testing environment that is not public-facing, these settings can be sufficient. However, for any production, public-facing or a highly-secured environment, security needs to be configured further. How much further depends on your organization's needs, structure, and security protocols.

Users and roles

Users and roles, along with the permissions granted to roles, form the basis of ArcGIS Server's built-in security framework. Unsecured services are viewable by anyone without any sort of login required. A user is any person or another user (sometimes referred to as a *headless* user, often software, a program, or a script) that will access a GIS server resource. ArcGIS Server keeps track of users in its built-in identity store. Also managed by the identity store are roles, where a role constitutes a set of users. Roles in ArcGIS Server often equate to groups (sometimes informal) within your organization. There could be a group of GIS users that only need to view services, but also a group of analysts that need to edit data in services. You could also have a very small group of administrators that would manage services. These groups could all be roles in ArcGIS Server. Permissions to ArcGIS Server resources are granted at the role level, not the user level. Users inherit the permissions of the roles they are assigned to, allowing efficient and effective access control to GIS resources.

Authentication and authorization

Authentication and authorization are two terms that sound similar, yet have very different meanings. Authentication is the process of verifying who you are; in computing, this is commonly done through logging in with a username and password. With ArcGIS Server, this is done with token-based ArcGIS Server-based authentication or web server authentication. Authorization is about what you are permitted to do; in computing, this is commonly done through permissions, for example, now that you are logged into the system, what do you have permission to access. When talking about security, it is important to keep the concept of authentication versus authorization in mind as it is completely possible to be authenticated into a system, yet have no authorizations to resources within the system once you are there.

Keeping your ArcGIS Server secure

There are many ways to keep your ArcGIS Server deployment secure. In this section, we will discuss several of the most important and more common tasks and concepts. For a full list of security items, search for **Configuring a secure environment for ArcGIS Server** in the ArcGIS Enterprise online documentation.

Using a CA-signed SSL certificate

ArcGIS Server installs with a preconfigured self-signed SSL certificate that can work to initially get your deployment up and running. Using an SSL certificate from a **certifying authority** (**CA**) is important for a couple of reasons:

- **Encryption**: Yes, a self-signed certificate encrypts traffic just the same
- **Trust**: When your end users see the little green lock in their browser address bar, they get a nice, warm, fuzzy feeling knowing that since the CA trusts you, *they* can trust you

With Portal, we are required to use a CA-signed certificate. We'll discuss more about that in this chapter.

Principle of least privilege

In a computing system, a user account should only be granted those privileges that are essential to it performing its intended function. In other words, if a user account or a service account doesn't need access to a certain resource to do its job, then it should not have access to that resource; this is the principle of least privileges.

Only give a user access to what they need and nothing more.

Always keep this principle in mind and if you have a question, ask your IT Systems Administrator (if you have one) or someone that manages Windows accounts; they can probably offer advice, guidance, and probably a scary story or two.

Examples from a GIS system point of view include the following:

- The `arcgisserver` folder and its subfolders: If you have secured services, the MXDs (and possibly data) along with any print service and geoprocessing temporary outputs can be found in the `arcgisserver` directories. Limit access to these directories to only the ArcGIS Server service account and administrators.
- Development/testing/production environments: If you have multiple environments, standards, and protocols in place for publishing (see `Chapter 10`, *Troubleshooting ArcGIS Enterprise Issues and Errors*) to production, lock down the directories where production items such as service MXDs, database connection files, and file-based data are stored, so only those publishers and administrators that need to access those folders may do so. Apply the same principles to your test and development environment data and service directories.

Disabling or modifying the PSA account

Many systems have highly-privileged accounts, often called admin, power user, or sysadmin (`sa` for short). In many cases, these accounts are created during the software installation process and are used to gain initial access to the system after installation. This is the case with ArcGIS Server. In `Chapter 1`, *ArcGIS Server Introduction and Installation*, when we installed and configured ArcGIS Server, we created the **siteadmin** account. Sure, we had the option to name the account anything we wanted, but **siteadmin** has been the go-to standard name for that account for years in the ArcGIS Server world. As such, this creates a vulnerability; in that, anyone who knows ArcGIS Server knows that the first username to try to use to gain access to an ArcGIS Server system is **siteadmin**. This same concept applies to Microsoft SQL Server, as Microsoft recommends disabling the **sa** account.

Esri recommends disabling the ArcGIS Server PSA to help ensure a secure environment. Before doing this, it is important to have assigned an administrator role to one or more users in your identity store (hopefully, you are one of those administrators).

Once the PSA is disabled, changes to the ArcGIS Server identity store are not allowed. The PSA must be enabled again to make these changes.

To disable the ArcGIS Server PSA, follow these steps:

1. Grant administrator privileges to the roles in your identity store that you want to have administrative access.
2. In the ArcGIS Server REST Admin, log in as an administrator and go to **security** | **psa** and click on the **disable** operation.
3. On the **disable** operation page, click the **Disable** button.

Disabling the services directory

The ArcGIS Server services directory, also known as the REST endpoint, provides a browsable HTML interface into all your unsecured ArcGIS Server web services and their respective REST operations. This means that users can browse your public services, discover them in web searches, and perform available operations on them through the available REST endpoint operations (query, find, identify, and more). On the other hand, the REST endpoint is an invaluable tool for your developers, engineers, or analysts that may be building and debugging applications. If you have development/test and production environments, you might consider turning off the services directory in just production. If you only have a production environment, disabling the services directory might be harder to do. Consult with all possible parties involved before disabling this feature. To disable the services directory, follow these steps:

1. Log on to the ArcGIS Server REST Admin as an administrator.
2. Go to **system** | **handlers** | **rest** | **servicesdirectory**.
3. Click on the **edit** operation link.
4. Uncheck the **Services Directory Enabled** checkbox.
5. Click on the **Save** button.

Now, when going to the REST endpoint, a `403 Forbidden` HTTP response is returned and a message stating that the services directory has been disabled:

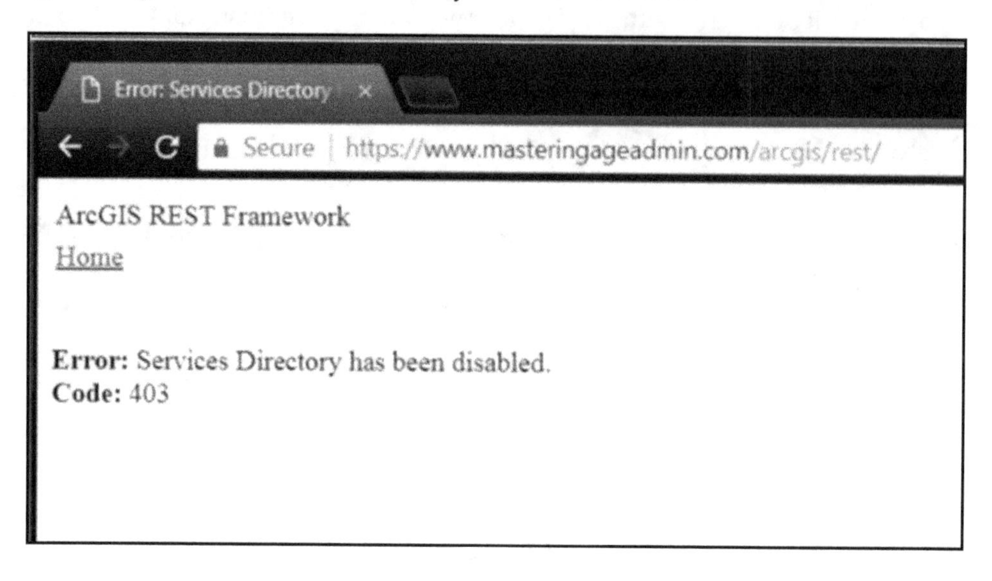

To re-enable the REST endpoint, simply go back to **services** directory and check the **Services Directory Enabled** checkbox.

Scanning your ArcGIS Server instance for security best practices

ArcGIS Server ships with a Python script tool, `serverScan.py`, that can be found in the `\tools\admin` directory of the ArcGIS Server installation. `serverScan` checks for problems based on best practice configurations and generates a report of any issues found.

Configuring security in ArcGIS Server

Server Manager is used to configure ArcGIS Server security settings. To do so, you must log in to **Server Manager** as the PSA.

If the PSA has been disabled, you must first re-enable it in the ArcGIS REST Admin before making changes to your ArcGIS Server security configuration.

Identity stores

Identity stores are how users and roles are managed in ArcGIS Server. Built-in identity stores are maintained on the file system; enterprise stores are maintained in their respective system (Active Directory, for example).

ArcGIS Server built-in store

The ArcGIS Server built-in store is the standard configuration for security with an installation. This store type maintains user and role information in a file-based format in the ArcGIS Server configuration store (remember earlier when we discussed limiting access to the configuration store?) and as such, those users and roles can only be accessed and managed by ArcGIS Server, primarily through **Server Manager**. This store type works well for small to medium-sized organizations that do not need many accounts to access ArcGIS Server. The following three roles exist in the ArcGIS Server built-in store:

- **Administrator**: Full administrative access to the system
- **Publisher**: Ability to read and publish services
- **User**: Read-only access

The existing enterprise system

ArcGIS Server also can leverage an existing external user and role store such as **Microsoft Active Directory** (**AD**) or **Lightweight Directory Access Protocol** (**LDAP**). In this configuration, the AD or LDAP server is a read-only store for ArcGIS Server to access; users and roles from AD or LDAP can be viewed, but not edited or managed, that is still done at that AD or LDAP level. This store type works well in medium to large-sized organizations that have a well-maintained enterprise security store whose users and roles can be used to efficiently enforce security within ArcGIS Server.

Users from the existing enterprise system and roles from ArcGIS Server built-in

Finally, ArcGIS Server can be configured to use an external AD or LDAP server for its user store along with roles from the ArcGIS Server built-in store. Here, AD or LDAP users are also not editable from within ArcGIS Server; they are read-only. Roles, however, can be added, removed, and edited in the built-in store in ArcGIS Server Manager.

Authentication

As discussed earlier, authentication is the process of verifying who you are. With ArcGIS Server, authentication is done one of two ways--at the GIS server tier or at the web tier.

ArcGIS Server authentication

User authentication performed at the GIS server tier utilizes Esri's proprietary token-based authentication. In ArcGIS token-based authentication, when a user provides a valid username and password, ArcGIS Server verifies the supplied credentials, issues a token, and the user, unbeknownst to them, presents the token whenever accessing a secured resource. The important thing to remember about tokens is that they expire. See `Chapter 4`, *ArcGIS Server Administration*, for an in-depth discussion of ArcGIS tokens.

Portal security

Portal is the window into your GIS system and has many settings for keeping it secure.

Fundamentals of Portal security

Security for Portal for ArcGIS is just as important as ArcGIS Server security. Portal is just that, a portal into your data, services, maps, and applications.

Web-tier authentication

Web-tier authentication occurs at the web server tier. If your organization uses Active Directory, you can use IWA to enable an automatic or single sign-on experience through web-tier authentication using the ArcGIS Web Adaptor for **Internet Information Services (IIS)**. Likewise, if your organization uses LDAP, it can be used with ArcGIS Server with your Web Adaptor deployed to a Java application server such as Apache Tomcat or IBM WebSphere.

With web-tier authentication, administration must be allowed through the Web Adaptor. This allows users in the enterprise identity store to publish services from ArcGIS Desktop on their local PCs. To publish, they must connect to ArcGIS Server using the Web Adaptor URL.

The post-installation scene

Like with ArcGIS Server, Portal has a standard security setup after installation, as follows:

- The Portal URL is only known to the person who sets up the Portal deployment
- There is one account in Portal--the initial administrator account that was created during the Portal configuration and setup process
- All administrative operations are initially secured and can only be accessed by the initial administrator account

Again, as it was the case with ArcGIS Server, these initial settings might be fine for an internal-only development or testing environment but for a production environment, you will want to limit access to Portal, control administration and publication privileges, and encrypt communications with Portal.

Keeping Portal secure

There are many different settings available for Portal security. Here, we will touch upon several. You are encouraged to consult the ArcGIS Enterprise online documentation and search for **Portal security best practices** for more information.

Using a CA-signed SSL certificate

Nowhere does Esri state that using a SSL certificate signed by a corporate (internal) or commercial CA is required for Portal for ArcGIS. However, Esri does state that is *imperative* and *very important* that a CA-signed certificate is used to deploy your portal. Along with encryption and trust (see the using a CA-signed certificate with ArcGIS Server section for more information), there are other reasons to use a CA-signed certificate with Portal. Without a CA-signed certificate, users of your Portal will experience the following:

- Warnings in ArcGIS Desktop and web browsers about the site being untrusted. To experienced users, these are annoyances; to untrained users, they are red flags that instill doubt and uncertainty.
- Odd behavior when configuring utility services, printing hosted services, and accessing the portal from client applications (this is a big one).
- Inability to do the following:
    - Open a federated service in the portal map viewer
    - Add a secured service item to the portal

- Log in to ArcGIS **Server Manager** on a federated server
- Connect to the portal from ArcGIS Maps for Office

As can be seen, the list of potential issues far outweigh the investment of a CA-signed SSL certificate. See `Chapter 1`, *ArcGIS Enterprise Introduction and Installation* for more information on acquiring and installing a SSL certificate from a trusted certificate authority.

Enabling HTTPS

When initially configured, and after you configure your CA-signed SSL certificate, all credentials are sent encrypted over HTTPS. However, all other communications with the portal are done over HTTP only and are not secured. Requiring all communications with the portal to be secured is good for your security, but it does have its drawbacks that need to be carefully considered:

- Portal performance may be affected.
- Communications with external web content, such as ArcGIS Server services, OGC services, and so on, will be required to communicate over HTTPS. If HTTPS is not available for these services, they will be blocked, resulting in a message like the following screenshot when trying to add an external service from an HTTP-only ArcGIS Server service:

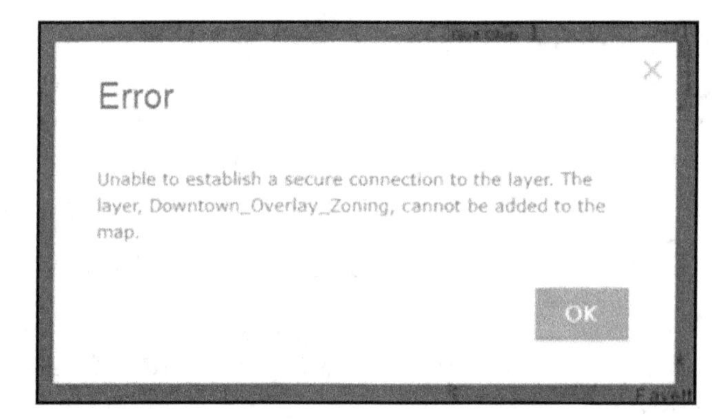

To configure all Portal communications to use HTTPS, follow these steps:

1. Sign into Portal as an administrator.
2. Go to **My Organization** | **Edit Settings** and click on the **Security** link in the left column.

3. Check the **Allow access to Portal through HTTPS only** checkbox.
4. Enable SSL on your web server. Chances are you already did this when you installed your CA-signed certificate on your web server and bound it to port `443` in IIS. See `Chapter 1`, *ArcGIS Enterprise Introduction and Installation* for more information on this topic.

Disable user's ability to create built-in accounts

By default, when a user goes to the sign-in page in Portal, they are presented with a **Create An Account** button that they can use to create a built-in Portal account which is shown as follows:

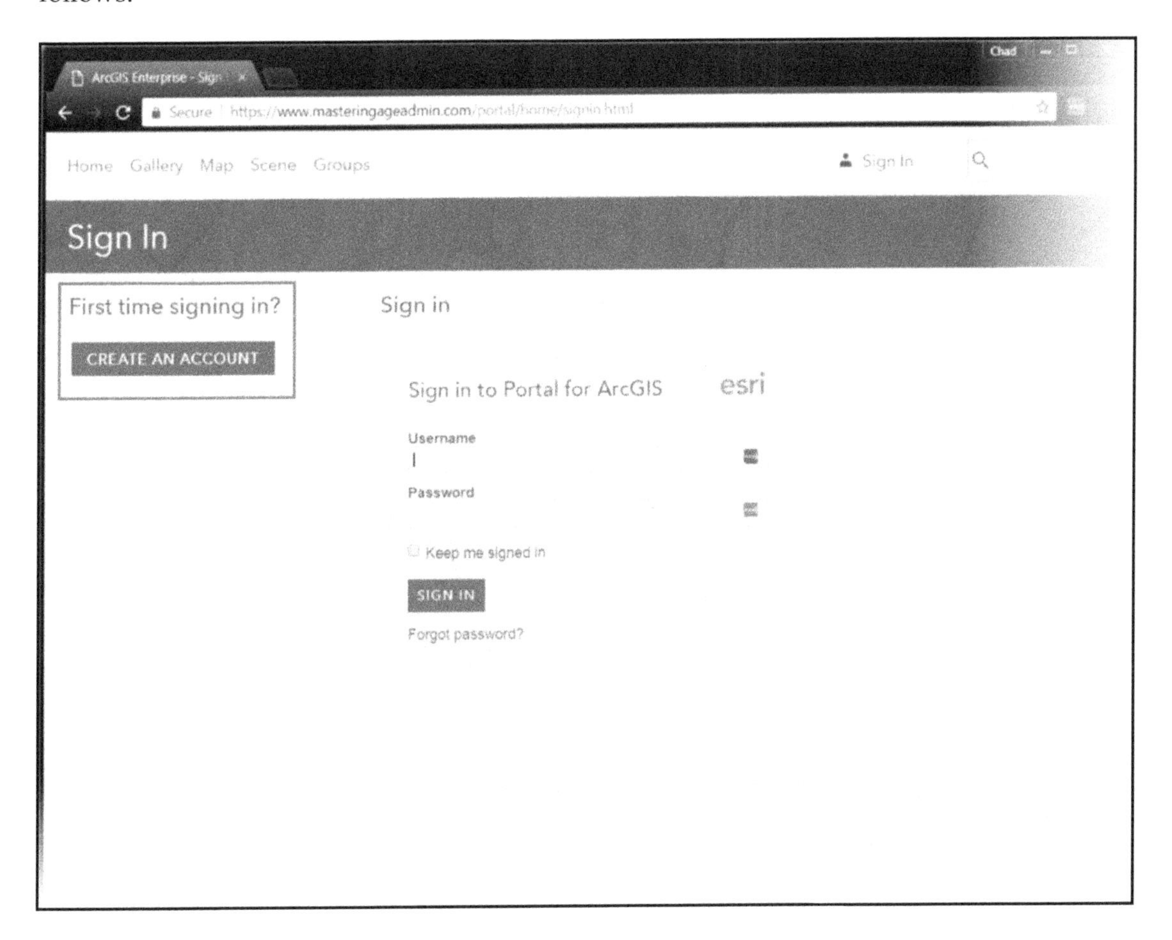

However, if you are using enterprise accounts, you want to create user accounts manually, or if you just want to disable the ability for users to create accounts, this option is easily removed by a simple edit to the system properties in the **Portal REST Admin**. To change this setting, follow these steps:

1. Log in to the **Portal REST Admin** as an administrator.
2. Go to **System** | **Properties** | **Update Properties**.
3. Your **Properties** may be empty. If so, add the following lines of code:

```
{
  "disableSignup": "true"
}
```

4. If your **Properties** has values already, simply add in the `disableSignup` parameter as follows:

```
{
  "WebContextURL":"https://www.masteringageadmin.com/portal",
  "disableSignup": "true"
}
```

5. Click on the **Update Properties** button. Note that this operation restarts your Portal, which may take several minutes to fully complete.

Once your Portal has restarted, go to the **Sign In** page. The **Create An Account** button should no longer be there.

Scanning your Portal instance for security best practices

Like ArcGIS Server, Portal ships with a Python script tool, `portalScan.py`, that can be found in the `\tools\security` directory of the Portal installation. `portalScan` checks for problems based on best practice configurations and generates a report of any issues found.

Configuring security in Portal

Configuring security in Portal is done through the Portal **My Organization** settings and the **Portal REST Admin**, depending on what you are configuring.

Identity stores

As with ArcGIS Server, the source for your Portal users and groups is your identity store. Your security configuration for your Portal is determined by the identity store type you choose. Portal supports two types of identity stores--built-in and enterprise.

Portal built-in identity store

The initial administrator account for Portal uses the built-in identity store. With the built-in store, you can, out-of-the-box, create user accounts in your Portal and groups to manage items.

Like ArcGIS Server services are shared at role level, Portal items are shared at the group level.

Like the ArcGIS Server built-in identity store, this configuration is best suited for small to medium-sized organizations with a limited number of Portal users and no need or capability to leverage enterprise logins.

Enterprise identity store

Through the use of an enterprise identity store, such as AD, LDAP, and identity providers that support SAML, you can use existing enterprise accounts and groups to control access to your Portal. With this approach, no additional user accounts need to be created within Portal; members use their existing Windows domain credentials to log in to Portal, and there is no account credential management done in Portal. In the case of AD, all user accounts are managed within AD; all the Portal administrator must do is add the enterprise users to Portal.

Groups can be created in Portal that leverage existing enterprise groups. When enterprise groups are utilized, access to Portal content for a user is controlled by the rules defined in the enterprise group and group membership management is handled completely outside of Portal, within the enterprise identity store.

Authentication

With Portal, authentication can be handled at the web tier through the ArcGIS Web Adaptor or at the portal tier.

Web-tier

For a portal on Windows with Active Directory configured, Integrated Windows Authentication (IWA) can be used to connect to Portal, enabling a pass-through single sign-on experience for Portal users. To use IWA, the Web Adaptor must be deployed to IIS.

For LDAP, the Web Adaptor must be deployed to a Java application server, such as Apache Tomcat or IBM WebSphere.

To achieve the pass-through sign-on experience, anonymous access must be disabled in both the Portal and ArcGIS Server Web Adaptors, and Windows authentication must be enabled. For this reason, anonymous access to Portal items (even if they are shared with **Everyone**) is not possible with web-tier authentication, as anonymous access is blocked at the web server tier.

Portal-tier

Portal-tier authentication allows access to Portal using both enterprise and built-in identity stores. To achieve this, Windows authentication must be disabled and anonymous access must be enabled on your web server. At the Portal sign in page, a user can sign in using either enterprise credentials or Portal built-in credentials. Pass-through single sign-on will not be available, and users will have to log in every time they visit the portal. Finally, with Portal-tier authentication, anonymous users can access portal resources that are shared with everyone. Portal-tier authentication is oftentimes a win for many organizations as it provides the right mix of enterprise identity store use while still allowing anonymous access to select Portal resources.

Implementing Integrated Windows Authentication and Single Sign-On in Portal

Setting up your portal to use Integrated Windows Authentication is a multi-step process involving Portal, the Portal REST Admin, and for web-tier authentication, IIS.

Configuring Portal to use HTTPS for all communication

To configure Portal to use HTTPS for all communications, do the following:

1. Log in to the Portal website as an administrative user.
2. Go to **My Organization** | **Edit Settings** | **Security**.
3. Check the **Allow access to the portal through HTTPS only** checkbox.
4. Click on the **Save** button.

Update Portal's identity store

This step configures Portal to use an enterprise Windows Active Directory identity store instead of the Portal built-in identity store. This can be one of the trickiest steps in the process. To configure the AD identity store, follow these steps:

1. Log in to the **Portal REST admin** as an administrative user.
2. Go to **Security** | **Config** and click on the **Test Identity Store** operation link. This operation's interface is almost identical to **Update Identity Store**, but **Test** lets you do just that--test your AD identity store connections string. You can then use **Update** to apply it. For **User store configuration**, enter the following, replacing `user` and `userPassword` with your appropriate values. For `user`, you can use any domain account with reading access to Active Directory. I like to use the Portal domain service account for this:

```
{
  "type": "WINDOWS",
  "properties": {
    "userPassword": "somestrongpassword",
    "caseSensitive": "false",
    "userEmailAttribute": "mail",
    "user": "mydomain\\svc_portal",
    "userFullnameAttribute": "displayName",
    "isPasswordEncrypted": "false"
  }
}
```

3. Click on the **Test Configuration** button. If your connection succeeds, you will get a success message below the pink configuration box. If your connection fails, you will get a failed message. If your connection fails, check the password. If it still fails, install (preferably not on one of your production servers) a program such as AdExplorer (`https://docs.microsoft.com/en-us/sysinternals/downloads/adexplorer`) and try to connect to AD as the user account you are using in your connection string. If you cannot connect, there may be an issue with the account. If you still have issues, contact your IT support personnel who are well-versed in AD; there may be an attribute in your connection string that needs to be changed for your organization's AD configuration.
4. Next, if you want to create groups in Portal from existing groups in AD, perform the same configuration test for **Group store configuration**, entering a JSON string like the following code snippet:

```
{
  "type": "WINDOWS",
  "properties": {
     "userPassword": "somestrongpassword",
     "user": "mydomain\\svc_portal",
     "isPasswordEncrypted": "false"
  }
}
```

5. Click on the **Test Configuration** button. Results and troubleshooting for the user connection string in step 3 apply here as well.
6. Once your connection strings test successfully, copy them from the **Test Identity Store** operation over the **Update Identity Store** operation page and click on the **Update Configuration** button.

Add enterprise accounts to Portal

Now that we have connected Portal to the AD enterprise identity store, we can leverage that store to add users to our Portal. There are several ways to add enterprise users to Portal; we will cover the most basic method here, which many of the other methods build upon. Search the ArcGIS Enterprise online documentation for **add members to your portal** for more information on the additional methods of adding enterprise users to your portal.

Add users manually one at a time by following these steps:

1. Sign in to the Portal website as an administrative user.
2. Go to **My Organization** | **Add Members**.
3. Under **Add Members**, select **Add members based on existing enterprise users** option, and click on **Next**.
4. Select the **One at a time** tab. Click on the magnifying glass to search for a user in the **Select User** window. Once you find the user you are searching for, click on the **Select User** button. This is shown in the following screenshot:

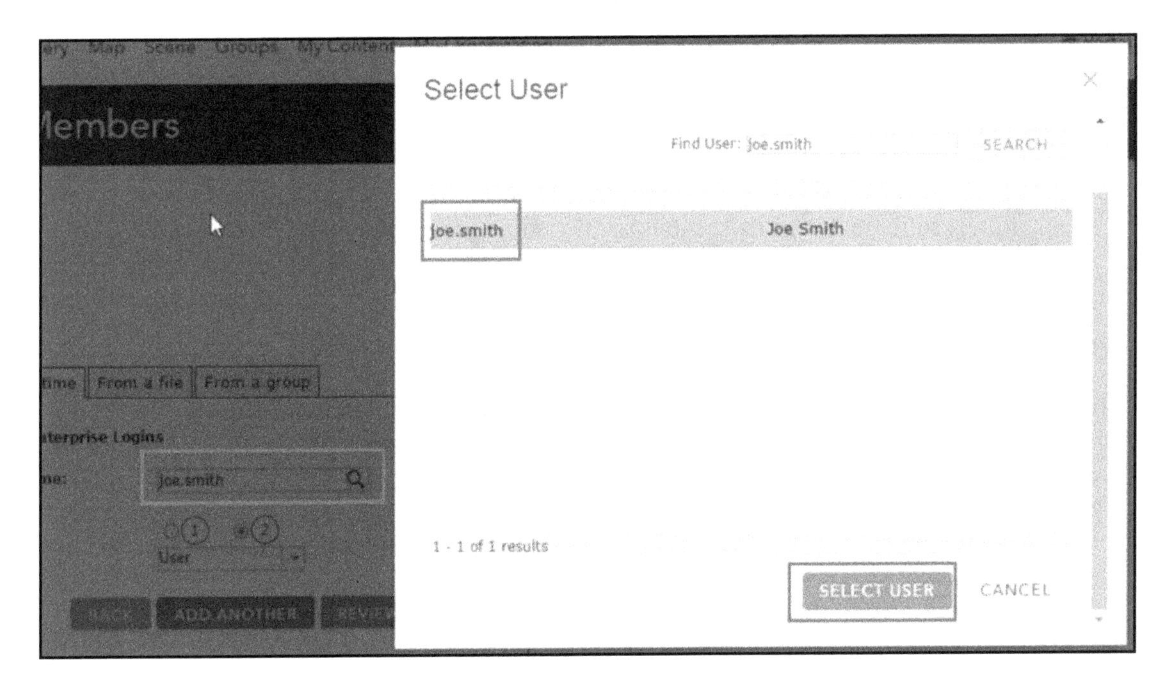

Any domain user account you want to add to Portal must have a well-formed (but not necessarily valid) email address stored in the enterprise identity store. If it does not, you will not be able to add that user to Portal.

5. Apply the proper **Role** and **Level** for the user.
6. Click on **Review**.
7. If the information is correct, click on **Add Members**.

Configure the Web Adaptor to use IWA

If you want to use the web-tier authentication to allow a pass-through single sign-on experience, perform the following steps in IIS to disable anonymous access and enable Windows authentication for both the `arcgis` and `portal` Web Adaptor applications:

1. Open IIS on the **Web Adaptor** server.
2. In the **Connections** panel, find the `arcgis` Web Adaptor.
3. In the **Home** panel, find **Authentication** and double-click on it.
4. Disable **Anonymous Authentication**.
5. Enable **Windows Authentication**.
6. Repeat steps 2 to 5 for the `portal` Web Adaptor.
7. Close IIS.

You may have to refresh the `arcgis` and `portal` applications, or even restart IIS, to get the authentication changes to take effect.

Verify access

While on your domain, go to your Portal website. If you implemented Portal-tier authentication, you should see the **Sign In** link in the upper-right of the Portal website and be able to sign in with your Windows domain credentials. If you implemented web-tier authentication, and your enterprise account has been added to Portal as a user, you should be automatically logged in to Portal and passed through without being prompted or having to log in to Portal.

Using Portal with ArcGIS Server

Portal for ArcGIS became an integral piece of ArcGIS Enterprise starting at the 10.5 release, making it more practical than ever to use Portal with ArcGIS Server.

Benefits

If you've used ArcGIS Server and ArcGIS Online for any amount of time, it's easy to see how combining the powers of Portal with ArcGIS Server can make for easier administration. Using Portal with ArcGIS Server provides the following benefits:

- Portal can help you organize your content and enables discovery within not only your organization, but outside of it as well, using galleries, groups, and searching.
- Portal can help control access to your ArcGIS Server services. This is known as **federation** and is a big deal these days. We will discuss this further later in this chapter.
- Portal can help your organization reach a wider audience by publishing data, maps, and ultimately apps out as web services. Again, discoverability.

Let's look at some of the ways Portal can be used with ArcGIS Server.

Integration

Like many other features of ArcGIS Enterprise, Portal and ArcGIS Server can be integrated at various levels, depending on your organization's needs. There are three common approaches to integration, starting from the least complex and lightly-coupled and going up to the most complex and tightly-coupled.

Registered services

Registering an ArcGIS Server service with Portal (essentially the same as adding the item to Portal) allows your users to easily discover the item and add it to web maps. Since the content you add can be from an external ArcGIS Server instance, this level of integration requires you to only have Portal and not your own ArcGIS Server instance. This is the simplest and most loosely-coupled of the integration methods and requires little to no extra effort from an administrator.

Services from ArcGIS Server 9.3 and higher can be registered with Portal 10.5.1. If your Portal is configured to only communicate over HTTPS, then all external services you add to the Portal must be over the HTTPS protocol.

Federation

Federation is a process whereby more tightly integrating ArcGIS Server with Portal, we can delegate ArcGIS Server security to Portal, effectively eliminating ArcGIS Server-level based security and replacing it with Portal's sharing model. That's right; when you federate ArcGIS Server with Portal, all your security for services is handled as sharing in Portal. As a matter of fact, with a federated ArcGIS Server instance, when you publish a service, it is shared with the portal and shows up in the **My Content** folder of the publishing user's account!

By default, all published services on a federated ArcGIS Server are not shared by default. Note that this is the opposite to a standard ArcGIS Server site where all services published are public by default.

This is an exciting paradigm shift, as ArcGIS Server is now accessed using Portal members and ArcGIS Server users and roles are no longer valid. Portal administrators are now ArcGIS Server administrators, allowing a convenient, trimmed down sign-on experience with one account that can access both resources.

You can federate multiple ArcGIS Server sites with your Portal.

Now that we have bathed in the great virtues of federation, let's consider the drawbacks. The delegation of security to Portal from ArcGIS Server means that any existing ArcGIS Server users and roles are no longer valid and will no longer be used. When you federate, items for all existing ArcGIS Server web services are created in the portal and these items are owned by the Portal administrator who performs federation. This means that existing security in place on those ArcGIS Server services is no longer valid and, after federation, ownership will have to be reassigned to existing Portal members as needed. If you are performing an in-place upgrade of ArcGIS Enterprise and want to federate, carefully consider and plan out all aspects of the security model change to make the transition as smooth as possible.

Federating an ArcGIS Server site with your Portal

To federate an ArcGIS Server site with your Portal, perform the following steps:

1. Log in to the Portal website as an administrative user.
2. Go to **My Organization** | **Edit Settings** | **Servers**.
3. Under **Federated Servers**, click on the **Add Server** button.
4. Enter the following information:
    - **Services URL**: This is the FQDN path to your ArcGIS Server instance Web Adaptor URL, something like `https://www.masteringageadmin.com/arcgis`.
    - **Administration URL**: This is the internal, non-Web Adaptor machine name URL to your ArcGIS Server instance, something like `https://my-machine-name.domain.local:6443/arcgis`.
    - **Username and Password**: These are the ArcGIS Server PSA credentials.
5. Click on the **Add** button. Upon successful federation, your server will show up in the list of validated servers.

Designated hosting server

For even further integration, a federated ArcGIS server can be designated as a hosting server. A hosting server is the most tightly-coupled level of integration and allows the following:

- Items such as feature services, cached maps, and scene services can be published to Portal from other clients or from within Portal
- CSVs and shapefiles can be added from local machines to maps (by drag and drop, no less) in the Portal map viewer
- Addresses can be batch geocoded from a CSV file

When users publish items to Portal (such as CSVs or shapefiles), there must be a place for that data to be stored; therefore, a hosting server must be configured with an ArcGIS Data Store relational data store or a registered enterprise geodatabase acting as the GIS server site's managed database (on a go-forward basis, the ArcGIS Data Store relational data store is the recommended configuration). Finally, if your portal will include federated ArcGIS GeoEvent Servers or ArcGIS GeoAnalytics Servers, your hosting server must also be configured with an ArcGIS Data Store spatiotemporal data store to store the results of those analyses.

Using Portal with the ArcGIS Server REST endpoint

One of the links at the web service MapServer REST endpoint allows the preview of the service in the ArcGIS Online map viewer, as shown in the following screenshot:

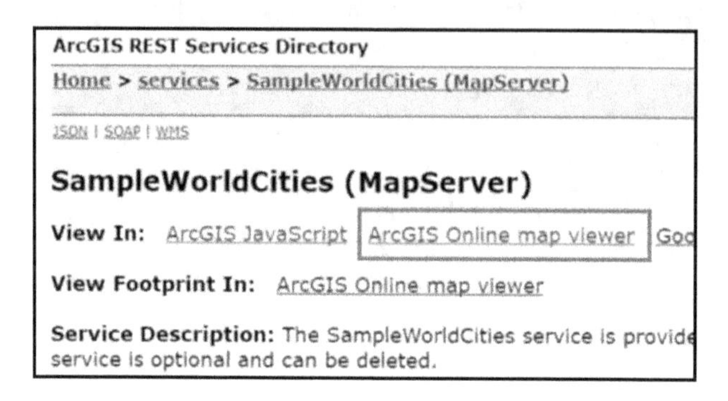

Through a simple configuration change in the ArcGIS Server REST Admin, you can change this to preview the service in your own portal's map view instead. To do this, complete the following steps:

1. Go to the ArcGIS Server REST Admin and log in as an administrative user.
2. Go to **system** | **handlers** | **rest** | **servicesdirectory** and click on the **edit** operation link.
3. The **ArcGIS.com URL** parameter should be something like `http://www.arcgis.com/home/webmap/viewer.html`. Change this to the FQDN path to your portal's map viewer, something like `https://www.masteringageadmin.com/portal/home/webmap/viewer.html`.
4. Change **ArcGIS.com Map Text** from **ArcGIS Online map viewer** to something more appropriate, such as **Portal for ArcGIS Map Viewer**.
5. Click on **Save**.

Go back to your ArcGIS Server REST endpoint and refresh the page. You should now see the map viewer link text changed, as shown here:

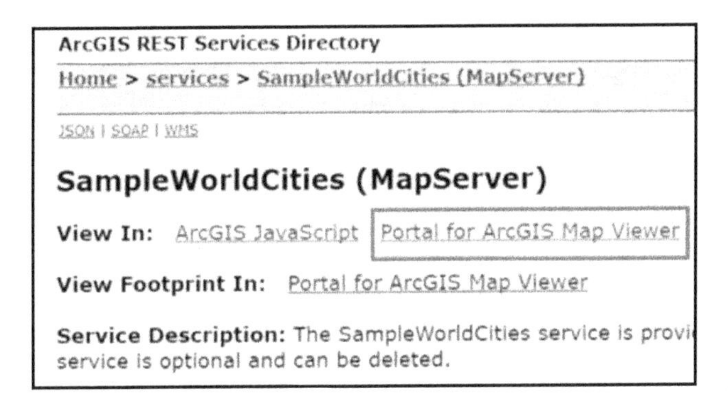

Clicking on that link will launch the service in my Portal map view instead of the standard ArcGIS Online map viewer. This now gives me access to all the custom basemaps and print services that I have configured for my Portal's map viewer.

Updates

Another oftentimes ignored and always annoying security-related task is software and operating system updates. No one likes them, everyone is bothered by them, and most of us put them off like a trip to the dentist. Windows updates especially seem to be the worst about this. I can't tell you how many times I've had Windows updates seemingly start on their own at the worst possible time, like when trying to shut my system down before boarding a plane at the last second. And don't even get me started about Microsoft Office for Mac, but that's a whole other story.

In all seriousness, keeping your software and operating system up-to-date is important for several reasons, as updates do the following:

- Add new features
- Remove old features
- Update drivers
- Deliver bug fixes--important
- Fix security holes--the most important

Running old or even not-so-old but unpatched software can be a dangerous game to play, *trust me*. Remember, earlier in this chapter when I said it's not a matter of *if* you will get compromised, but when, and perhaps even for no real reason? Let me tell you a story.

Years ago, probably a good 13 or so now, I had a personal WordPress site; I think WordPress was on version 1.3 or so at the time (it's on 4.8.1 at the time of writing this). I was just getting into web development and knew just enough PHP (yikes!) to be able to hack together my own highly customized site that looked and worked just like I wanted it to. Problem was since I had customized the code base so heavily, it made upgrading my WordPress installation nearly impossible. So, I did what I do to my dentist, I ignored it. I went several minor versions out of date. No big deal, right? **Wrong**. One day, I went to log on to my site and instead of my site, I found a message from the hacker that had compromised my site. I SSH'd into my site and the root directory was empty. I was dumbfounded, furious, and felt *violated*. I quickly realized though, that it was *all my fault*. I had avoided upgrading my software, vulnerabilities in my old version of WordPress were discovered and exploited, and someone hacked my meager, trivial, unimportant site *just because they could*.

So *please*, set aside a couple of hours once a month to update your servers. Take a few minutes and set up a recurring calendar reminder; otherwise, you will forget (administering an ArcGIS Enterprise environment *is* a lot of work). Set up automatic checks for software updates where you can, and when those annoying notifications pop up, **don't ignore them**.

References

Lefkovitz, N.B. and Danker, J.M., Privacy Authors, NIST Special Publication 800-63B, Digital Identity Guidelines, Appendix A - Strength of Memorized Secrets. National Institute of Standards and Technology, https://pages.nist.gov/800-63-3/sp800-63b.html#SP800-131A, retrieved 9/6/2017.

Summary

Security reigns supreme in any IT system; ignore it and you will pay the consequences. We started this chapter with some security basics on the importance of password strength and management. Next, we dug into how ArcGIS Server security is initially configured and what can be done to further secure it, covering some security best practices and the identity stores and authentication methods that can be employed by ArcGIS Server. We did the same for Portal security next, covering some best practices, identity stores, authentication, and how to implement Integrated Windows Authentication. The different methods to integrate ArcGIS Server with Portal were discussed along with why and how to federate ArcGIS Server with Portal. Finally, we ended the chapter with a short discussion on the importance of applying software updates to your system. Next up in `Chapter 7`, *Scripting Administrative Tasks* we will roll up our sleeves and look at how we can use Python to script ArcGIS Enterprise administration.

7

Scripting Administrative Tasks

As an administrator, you will more than likely be responsible for your entire GIS system--infrastructure, such as application servers, database servers, web servers, and all the software, data, and processes that go along with that infrastructure. Having all of this to deal with, you need to be crafty and come up with as many ways as possible to save time and effort, and this is where scripting comes into play. Python (`https://www.python.org`) has quickly become the de facto standard scripting language of the Esri platform. Considering that Python is literally everywhere in the ArcGIS Enterprise ecosystem (and many other non-Esri systems as well for that matter), knowing how to script with Python is a necessary skill for almost anyone doing any sort of technical work with ArcGIS Enterprise.

This chapter assumes some familiarity and experience with Python. If you are new to Python, there are resources on the internet to help you in the form of tutorials, blogs, and discussion forums. This chapter will cover using Python 2.x with the Esri `arcpy` and `portalpy` modules. We will also use Node.js to work with a REST endpoint. In this chapter, we will cover the following topics:

- Working with geodatabase data with Python and `arcpy`
- Interrogating a REST endpoint with curl and Node,js
- Working with REST endpoints with Python and Node.js
- Inventorying and publishing services with Python
- Using Python to pull error messages from logs and email error reports
- Using Python to administer Portal

Working with data

Data is something that nearly everyone works with daily. As an administrator, you might not work with data as much as others, but there are plenty of administrative tasks that revolve around data.

Loading data into a geodatabase

Loading data into geodatabases may or may not fall under your duties as an ArcGIS Enterprise administrator, but the following script demonstrates a few handy Pythonic methods. This is also a common task that can take on many forms. For our example here, we will simulate the loading of data from an enterprise geodatabase into a publication file geodatabase. Let's say that our web services do not have access to the enterprise geodatabase, so we need a read-only copy of that data that can hydrate our web services. However, we cannot have stale, out-of-date data in our publication geodatabase, so we need the publication data updated every day. This is a perfect example of a need that can be met with Python and `arcpy`.

Before we get into the script, there is a helper input file required for this script. As we see will soon in our script, with Python it is incredibly easy to read a text file into memory and then utilize that input later. Here, we will store the list of publication feature classes in a simple text file, with each new line representing a feature class:

```
WaterDistribution\wFitting
WaterDistribution\wSystemValve
ZoningRegulations
```

Notice in the preceding list that `wFitting` and `wSystemValve` are in the `WaterDistribution` feature dataset. If a feature class is in a feature dataset, just use the feature dataset name with a backslash followed by the feature class name. For any items in the root of the geodatabase, such as `ZoningRegulations`, just simply list its name. This not only makes configuring the script easier, but it ensures that the structure of the data will be the same on the target as it is in the source.

As this script stands, the data must already exist in the target geodatabase.

We start off the script by importing the `arcpy` and `os` modules for use, as follows:

```
import arcpy
import os
```

Next, set our source and target geodatabase connections into the respective variables. Note the `r` in front of the opening quotes on both strings. This tells Python to treat the string as a *raw string literal*. This will allow Python to treat our backslashes in our file paths as simply backslashes and not escape characters, as they normally function in Python:

```
src_gdb = r"C:\connections\owner@sandbox@MSS2014.sde"
tgt_gdb = r"C:\Projects\data\06_02_load_data.gdb"
```

Next, we will create an empty Python list (`layer_list`) that will store contents of our input text file. Each row in the input file will be an item in the list. We will next open the file up for reading, creating an iterable `f` full of lines, each line being a feature class name. We will then loop through `f`, appending each layer name (`line`) to `layer_list`:

```
layer_list = []
with open("06_02_layer_inputs.txt", "rb") as f:
    for line in f:
        layer_list.append(line.strip())
```

Using our input file example, `layer_list` would look like this:

```
["WaterDistribution\wFitting" ,
 "WaterDistribution\wSystemValve",
 "ZoningRegulations"]
```

Now that we have our list of layers, we will loop through that, and, for each one, we will use `arcpy.DeleteFeatures_management` to delete all features in the target object. We use `os.path.join` from the Python Standard Library to join the full path to our target geodatabase with the feature class name:

```
for layer in layer_list:
    arcpy.DeleteFeatures_management(
        os.path.join(tgt_gdb, layer)
    )
```

Next, in a similar fashion, while still in the `for` loop, we will turn around and append all features from the source enterprise geodatabase feature class into the target feature class using the `arcpy.Append_management` :

```
    arcpy.Append_management(
        os.path.join(src_gdb, layer),
        os.path.join(tgt_gdb, layer),
```

```
            "NO_TEST", "", ""
        )
```

Hopefully, this script demonstrates that writing a script doesn't have to be hard nor does the script need to be lengthy. Quite often, what we need to accomplish is a repeatable process, and executing repeatable processes is where any programming language excels.

This script could easily be extended to include logging (consider the `daiquiri` module if you haven't already (`http://daiquiri.readthedocs.io/en/latest/`)), and error handling to make it worthy of nightly runs as a scheduled task.

Modifying field domains

Attribute domains are a great way to enforce data integrity by constraining the values allowed to be used for an attribute in a table or feature class. Domains are also great for your users, making editing easier and smoother by giving them options to easily choose from. Furthermore, attribute domains can be shared and utilized in multiple fields by virtually any feature classes, tables, and subtypes in a geodatabase.

So, what happens when we need to make changes to a domain, say, add a value? If we have a value that is acceptable for all fields utilizing the domain, then we simply need to add a value to the domain. However, if we cannot simply add the value to the existing domain, we will need to create a new domain of the old and new values and set that domain on the appropriate fields. Easy enough, right? Sure, if we have a few fields that need to use the new domain, changing those domains only takes a few minutes in ArcCatalog. However, what if there are dozens or even hundreds (gasp!) of fields that need to be changed? Ain't nobody got time for that. Let's look at how we could use `arcpy` to soften the blow. The following script will loop through all feature classes in a feature dataset and change domains on certain fields if they are present.

1. First, import the `arcpy` and `os` modules:

```
import arcpy
import os
```

2. Next, set the full path to the feature dataset to loop through:

```
input_ds = r"C:\Projects\data.gdb\WaterDistribution"
```

3. Next, create a Python dictionary of key-value pairs where the key is the field we are changing the domain on and the value is the new domain to set on the field. With this configuration, you can change domains for any number of fields, just remember that we are looping through all the feature classes in the feature dataset, so if the field is found, its domain will be changed:

```
field_domains = {"OWNEDBY": "AssetOwner",
                 "MAINTBY": "AssetManager"}
```

4. Let's now set our workspace to our `input_ds` variable and get a list of feature classes, which we will set into the `fcs` variable:

```
arcpy.env.workspace = input_ds
fcs = arcpy.ListFeatureClasses()
```

5. Next, we will start the real work by looping through the feature classes and for each feature class, getting a list of the field names. Finally, we will iterate through the key-value pairs in the `field_domains` dictionary, and if the field name matches a key in `field_names`, we assign that key's value, the new domain, as the field domain using `arcpy.AssignDomainToField_management`. We iterate through `field_domains` using the Python `iteritems()` method, which runs an iterator over the dictionary's key-value pairs, setting the key into the `k` variable, and that key's value in the `v` variable:

```
for fc in fcs:
    fields = arcpy.ListFields(os.path.join(input_ds, fc))
    field_names = [field.name for field in fields]
    for k, v in field_domains.iteritems():
        if k in field_names:
            arcpy.AssignDomainToField_management(fc, k, v)
```

This is a pretty basic script with no messaging, logging, or error handling, but it gets the job done quickly and easily and can be easily extended to do more. That's the beauty of Python, with around a dozen lines of code, we have a repeatable, extensible process. This script could be modified to loop through feature datasets and then all the feature classes in each feature dataset or feature class names could be added into the `field_domains` dictionary, allowing you to filter which feature classes you want a domain changed in, like how we currently limit the field with the `if k in field_names` statement.

Working with ArcGIS Server services

We've talked about services at great length so far, but let's turn our attention to working with those services programmatically. Anyone can go to a REST endpoint and click and pick around; let's look at how we can dig a bit deeper to get more out of our services.

Interrogating a REST endpoint with curl and Node.js

A couple of years back, I needed to interrogate some services at a REST endpoint to get information about the fields and aliases in the service layers. What I really needed was a list of each field name and its alias, preferably separated by a comma; basically, I wanted a CSV file. Something made me think of using curl (`https://curl.haxx.se`) to query the REST endpoint to get at the JSON behind the service, but then I had to parse the JSON. Well, the JSON format is based on a subset of the JavaScript programming language, so I wondered if Node.js could parse the JSON. I eventually ran across the `json` npm package, which is a command-line tool for working with JSON. To follow along, you will need to do the following:

1. Install curl from `https://curl.haxx.se/download.html`. Scroll down to the bottom, to the Windows section, and get the zip for Win64--Generic; curl ships as a zip file. To install, unzip it to wherever you want it to live on your system.
2. Install Node.js from `https://nodejs.org/en/download/`. Download the LTS (long term support) Windows installer MSI.
3. Make sure that the paths to curl and node are both on your PATH environmental variable.
4. Install the JSON package for node:

```
npm install -g json
```

Once everything is installed, we can start parsing a REST endpoint. What we want to do is feed the REST endpoint URL to curl, which will fetch the JSON service. We then take the curl output and pipe it into the Node.js for parsing by the JSON package:

```
C:\Windows\system32>curl -s
http://tryitlive.arcgis.com/arcgis/rest/services/TaxParcelQuery/MapServer/0
?f=pjson | json -a currentVersion
```

Let's break down the preceding commands:

- `curl -s`: This calls the curl program. The `-s` flag puts `curl` in silent mode so it doesn't output anything to the command prompt.
- `http://tryitlive.arcgis.com/arcgis/rest/services/TaxParcelQuery/MapServer/0?f=pjson`: This is the URL to the JSON representation of the first (`0`) layer of a map service.
- `| json -a currentVersion`: This is where it gets interesting. At the Windows command prompt, the bar symbol (`|`) chains commands and redirects the output of the leading command into the input of the trailing command. This takes the output of our curl command and redirects it into the JSON package. The `-a` flag processes the input as an array. We can then pull out the `currentVersion` attribute from the array by calling it.

The output of the preceding command is the following:

```
10.2
```

Indeed, if we go to `http://tryitlive.arcgis.com/arcgis/rest/services/TaxParcelQuery/MapServer/0?f=pjson` in a browser, we will see that the `currentVersion` is 10.2.

Let's take this a step further to getting us the field information we want. In looking at the layer in the preceding link, note that there is a `fields` object:

```
...
"displayField": "LOWPARCELID",
"typeIdField": null,
"fields": [
  {
   "name": "OBJECTID",
   "type": "esriFieldTypeOID",
   "alias": "OBJECTID",
   "domain": null
  },
...
```

We can get to that just as we did with `currentVersion`, as follows:

```
C:\Windows\system32>curl -s
http://tryitlive.arcgis.com/arcgis/rest/services/TaxParcelQuery/MapServer/0
?f=pjson | json -a fields
```

This gives us an array of JSON objects, each one representing a field, as shown in the following code block:

```
[{"name": "OBJECTID", "type": "esriFieldTypeOID", "alias": "OBJECTID",
"domain": null},
 {"name": "Shape", "type": "esriFieldTypeGeometry", "alias": "Shape",
"domain": null},  ...
]
```

Remember how we chained commands together earlier? We can do that again, and pipe the output of our previous command into another attribute parse to get just the name and alias. The JSON package also has a `-d` flag that lets you delimit the output with a character (or whitespace) of your choice:

```
C:\Windows\system32>curl -s
http://tryitlive.arcgis.com/arcgis/rest/services/TaxParcelQuery/MapServer/0
?f=pjson | json -a fields | json -a name alias -d,
```

The preceding command gives the following output:

```
OBJECTID,OBJECTID
Shape,Shape
LOWPARCELID,Low Parcel Identification Number
PARCELID,Parcel Identification Number
BUILDING,Building
UNIT,Unit
CVTTXCD,Tax District Code
CVTTXDSCRP,Tax District Name
SCHLTXCD,School District Code
SCHLDSCRP,School District Name
```

This exercise is perhaps a little unorthodox but shows a different way to get at the REST endpoint and allows us to easily interrogate and inspect it, along with working with the REST endpoint and JSON representation of the service. I also invite you to spend a few minutes looking at curl, as it is a powerful tool.

Publishing services

Publishing services are one of the more common tasks you might do. Having to do a few of them manually usually isn't much of a time killer. However, what about when you have dozens or even hundreds of services in one environment that you need to replicate in another? Maybe you are doing a parallel upgrade of ArcGIS Enterprise on new hardware and you need to stand up your services in the new environment. Or, in the worst case, what if your ArcGIS Server instance crashes and is unrecoverable? How would you ever stand back up all those services?

OnServer

Let me introduce you to OnServer (`https://github.com/CityOfNewOrleans/OnServer`), a Python script with accompanying tools developed by J.B. Raasch at the City of New Orleans, Louisiana. OnServer's tagline is *Track down the data sources and map documents that feed ArcGIS Server Map Services*. OnServer is designed for three primary use cases--information, automation, and restoration. Let's examine how to use OnServer for some administrative tasks.

First, you'll need to download OnServer from the GitHub repository at `https://github.com/CityOfNewOrleans/OnServer`. Look for the **Clone** or **Download** button to get the latest version. While you are there, look at the readme for full usage instructions. OnServer requires no third-party Python libraries to run. Put it in a `scripts` directory on the ArcGIS Server machine you want to run it against.

How OnServer works

OnServer reads the local ArcGIS Server manifest files for every map service in your ArcGIS Server site on the server from which it is run. Your manifests are in the `arcgisinput` directory, for example-- `D:\arcgisserver\directories\arcgissystem\arcgisinput`.

Each service has a folder here or within a subfolder. Within that directory lives the `manifest.json` file that `OnServer` reads:

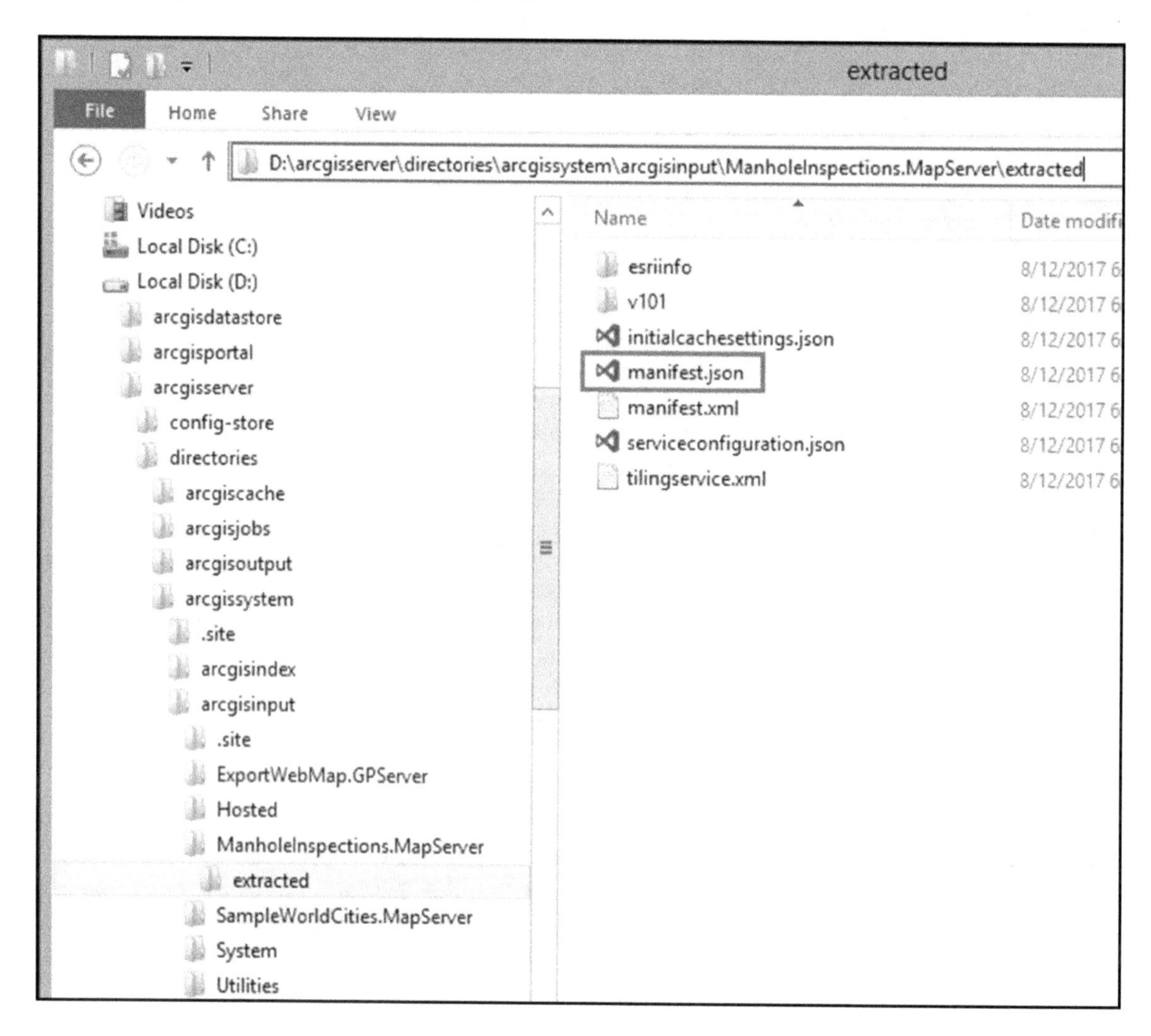

The manifest file is a simple JSON file that lists the database connection strings and data sources for the service, along with the layers and MXD location.

Creating a service inventory

Information is the first primary use of `OnServer`. Ever needed to know just what data sources are behind a map service?

`OnServer` is incredibly simple to run at the command line. The easiest command is the following:

```
D:\Scripts\onserver>D:\Python27\ArcGIS10.5\python.exe onserver.py
```

This will print the following result:

```
ManholeInspections (ManholeInspections)
--------------------------------------
 D:\Services\ManholeInspections.mxd
 - SDEPROD:sde:sqlserver:localhost
   + ManholeInspection
   + ManholeInspection__ATTACH
   + ssManhole
```

We can take this a step further and create an inventory of services in a CSV file that can easily be opened in Microsoft Excel. To do this, we will pass in the optional `-csv` flag and pipe the output to a file:

```
D:\Python27\ArcGIS10.5\python.exe onserver.py -csv >
07_06_onserver_report.csv
```

In a large organization with dozens or hundreds of services, a document such as this can be invaluable. Inventories like this are also great to have for migration or upgrade projects where you need to know what services are there and what data they are consuming:

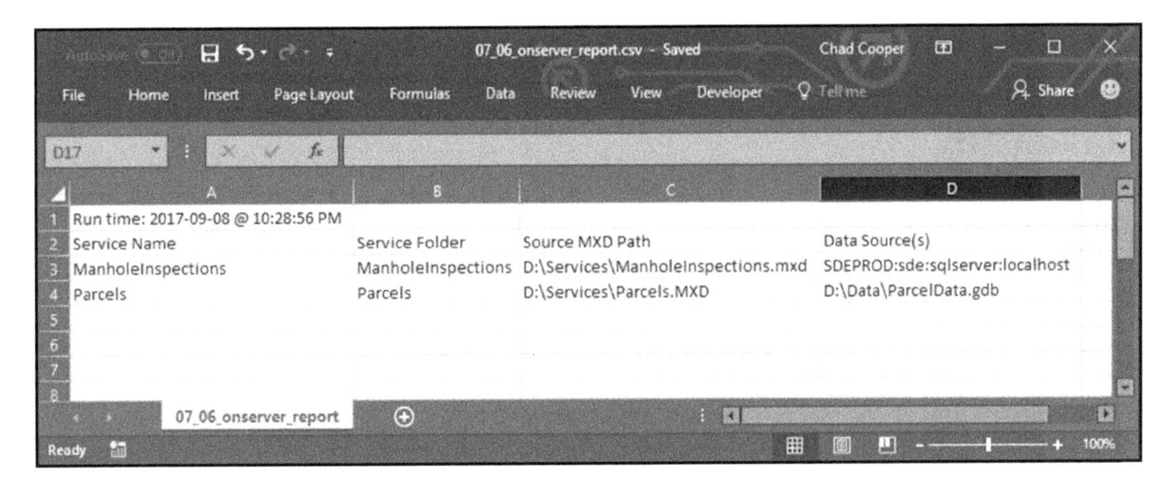

A	B	C	D
Run time: 2017-09-08 @ 10:28:56 PM			
Service Name	Service Folder	Source MXD Path	Data Source(s)
ManholeInspections	ManholeInspections	D:\Services\ManholeInspections.mxd	SDEPROD:sde:sqlserver:localhost
Parcels	Parcels	D:\Services\Parcels.MXD	D:\Data\ParcelData.gdb

OnServer also outputs to Markdown with the optional -md flag. This is handy to add service information into readme files (often written in Markdown), or you could go one step further and use Pandoc (https://pandoc.org/), a fantastic universal document converter, to convert the Markdown file to Microsoft Word to include it in existing pieces of documentation:

```
pandoc -s output.md -o output.docx
```

Since Pandoc is a command-line tool and OnServer can also be run from the command line; it's easy to see how these could be chained together to regularly produce documentation through a scheduled task.

Determining what services a feature class is participating in

Ever needed to make a change to a feature class but you couldn't because a service had a lock on it, but you had no idea what service it is? I know I have. With OnServer, you can search services for a layer name to find out what services it is used in:

```
D:\Scripts\onserver>D:\Python27\ArcGIS10.5\python.exe onserver.py --quiet
ssManhole
```

This gives us the following output:

```
ManholeInspections (ManholeInspections)
```

We now know which service to turn off while we make the change.

MakeMany

Now that we have discussed OnServer, let's discuss its companion process, MakeMany, which consists of two Python scripts--build_remakes.py and make_service.py. The build_remakes.py script takes an input OnServer output file and parses it to generate a text file of service MXD paths and service directories. Run build_remakes.py like this:

```
D:\Python27\ArcGIS10.5\python.exe build_remakes.py
```

The output from `build_remakes.py` looks like the following screenshot:

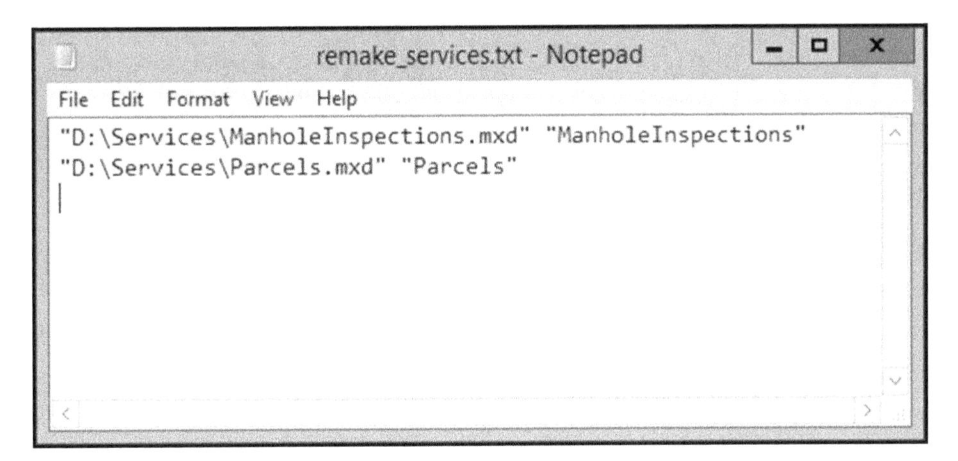

This output file from `build_remakes.py` can then be edited to include a call to `make_service.py` and an ArcGIS Server connection file with at least publisher privileges. Once finished, this file can be saved as a batch file and executed, which will publish all services in the file:

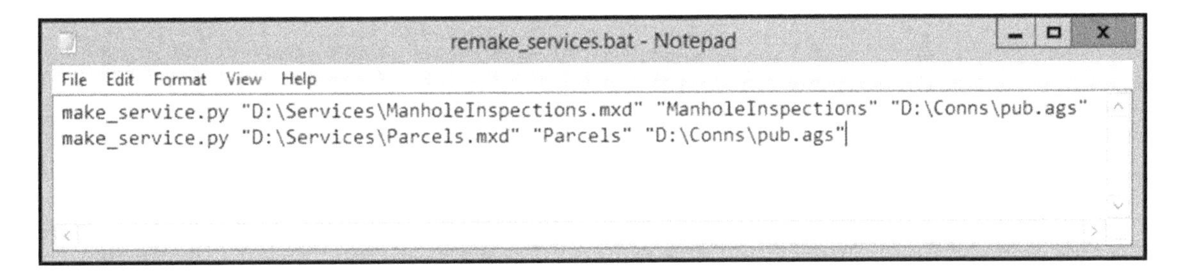

SLAP

The **Simple Library for Automated Publishing** (**SLAP**) of Map Services, SLAP Maps, or simply SLAP, as it can also be referred to as, is a command-line tool to publish map services. It is written entirely in Python, utilizing the ArcGIS Server REST API and `arcpy` and can be found on GitHub at `https://github.com/lobsteropteryx/slap`. Disclaimer--I have contributed to SLAP development; although, SLAP is released under an open source MIT license.

How SLAP works

SLAP uses a configuration file, created either by the user or the SLAP `init` command, along with source MXD files to publish map services to an ArcGIS Server instance. The configuration file lines out the services to be published, it can utilize virtually any service parameter in the REST API, and it can replace workspace paths in the input service MXDs. This makes SLAP ideal for deploying the same set of services to any number of environments (development, testing, and production, for example).

Although it is designed to be utilized as a command-line program, SLAP is written in Python, so it is possible to import SLAP into your own Python programs for further integration.

You are encouraged to look at the SLAP examples (`https://github.com/lobsteropteryx/slap/tree/master/docs`) and explore the source code for further guidance on SLAP usage.

ArcGIS Server error monitoring and reporting

Monitoring and reporting have always been a shortcoming of ArcGIS Server; there just really isn't any sort of out-of-the-box notification system to let you know when things are not going smoothly. We discussed the ArcGIS Server logs in `Chapter 4`, *ArcGIS Server Administration*, and how they can be accessed through ArcGIS Server Manager, but that's the problem--to check the logs, you must log into the Server Manager, query the logs, and view the results. Do you *really* have time to do that every day? I didn't think so. Let's look at a script that queries the ArcGIS Server logs for you and not only reports back the results, but it sends them to you in an email. This script is adapted from an example script by Esri that reports map draw events. The original script can be found at `http://server.arcgis.com/en/server/latest/administer/windows/example-query-the-arcgis-server-logs.htm` or by searching the ArcGIS Enterprise online documentation for **example: query the arcgis server logs**. Our modified script, `query_logs.py`, can be found on GitHub at `https://github.com/chadcooper/mage`. Let's cover the pertinent parts and see how we can use Python to interact with the ArcGIS Server REST API.

The first step required is to gain access to the REST API. To do this, we must provide administrative credentials that will be used to acquire a token for authentication. We set the `username`, `password`, `server_name`, and `server_port` variables for our environment. We also set the `logging_level` that we want to query for.

Remember that, by default, ArcGIS Server logs at the `WARNING` level, so if you want to regularly log anything below `WARNING`, you'll need to change your log configuration accordingly (see `Chapter 4`, *ArcGIS Server Administration*). Here, we want to see only `SEVERE` errors. Next, we will call the `get_token` function and pass in the required inputs:

```
username = "siteadminuser"
password = "somestrongpassword"

server_name = "localhost"
server_port = 6080
logging_level = "SEVERE"

token = get_token(username, password, server_name, server_port)
```

The `get_token` function takes advantage of the `generateToken` REST API operation. First, though, we need to build out several required HTTP objects. We will create the `params` string by passing a Python dictionary of the items we want to pass in to `generateToken` as query string parameters; these are required items, such as the username and password, the response format (`json`), and the client referrer (`requestip`) that binds the generated token to the IP address from where the request originated. The `urllib.urlencode` takes the dictionary of key-value pairs and properly encodes it, replacing spaces with `%20`, for example. We also create the necessary HTTP headers as the `headers` object. HTTP headers are instructions for our upcoming HTTP request. Next, we create a `HTTPConnection` instance to our `server_name` (localhost in my case) over a specified port (`6080` if on localhost) and then call the `request` method, passing in `POST` as the HTTP method, `token_url` as the URL to `generateToken`, and `params` as the string of data to send after `headers` are sent. We then call the `getresponse()` method to get a response back from the server and `read()` that response into the `data` variable. Finally, we call the `json.loads()` method on `data`, effectively deserializing or reconstructing the `data` object into a Python dictionary `token`, which we can then pull our token out of:

```
def get_token(username, password, server_name, server_port):
    params = urllib.urlencode({'username': username,
                               'password': password,
                               'client': 'requestip',
                               'f': 'json'})
    headers = {"Content-type": "application/x-www-form-urlencoded",
               "Accept": "text/plain"}
    http_conn = httplib.HTTPConnection(server_name, server_port)
    token_url = "/arcgis/admin/generateToken"
    http_conn.request("POST", token_url, params, headers)
    response = http_conn.getresponse()
    data = response.read()
    http_conn.close()
```

```
        token = json.loads(data)

        return token['token']</span>
```

Now that we have our token to use for authentication, we will build up another HTTP `POST` request, much like we just did to acquire the token; however, this time, the request will be the query to the `logs` resource `query` operation of the REST API. What we are about to set up is the equivalent of logging on to the ArcGIS Server REST Admin, going to **logs** | **query**, setting up a query, executing it (click on the **Query** button), and getting the response back as JSON. Let's examine how we do that with our Python code. If we want to run this process daily to check for `SEVERE` errors that have occurred, we need a way to filter based on time. Fortunately, the query operation has an `endTime` parameter that will let us do just that and specify the oldest time to include in the result set. Here's the kicker, though--the time must be specified in milliseconds since the Unix epoch (Unix time * 1000) or as an ArcGIS Server timestamp, neither of which just happens to be a standard Python `datetime` format. To get around this, we get a UTC timestamp, subtract 24 hours from it, and set that into the past variable--this gives us a UTC timestamp from 24 hours ago. We then take `past`, convert it to a Python `timetuple` (which looks like `(tm_year=2017, tm_mon=9, tm_mday=8, tm_hour=11, tm_min=44, tm_sec=14, tm_wday=4, tm_yday=251, tm_isdst=1)`), feed that into `calendar.timegm`, which converts it into a Unix timestamp, and then finally multiply it by 1000 to convert it to milliseconds. Next, we encode our query parameters from a dictionary into the `params` string and create our `headers` object. We create an HTTP connection to our server through the specified port and send the `POST` request to `log_query_url`, passing in our `params` and `headers`. We then `read` the response, deserialize it, and close the HTTP connection:

```
    past = datetime.datetime.utcnow() - datetime.timedelta(hours=24)
    unix_stamp = calendar.timegm(past.timetuple())*1000
    log_query_url = "/arcgis/admin/logs/query"

    params = urllib.urlencode({'endTime': unix_stamp,
                               'level': logging_level,
                               'filter': {"codes": []},
                               'token': token,
                               'f': 'json'})
    headers = {"Content-type": "application/x-www-form-urlencoded",
               "Accept": "text/plain"}
    http_conn = httplib.HTTPConnection(server_name, server_port)
    http_conn.request("POST", log_query_url, params, headers)

    response = http_conn.getresponse()
    data = response.read()

    data_obj = json.loads(data)
```

```
http_conn.close()
```

We now have a Python dictionary of the log query results, where the actual JSON result would look like the following:


```
{
  "hasMore": false,
  "startTime": 1504827316822,
  "endTime": 1498327879154,
  "logMessages": [
    {
      "type": "SEVERE",
      "message": "Error getting service.",
      "time": 1504655831704,
      "source": "Rest",
      "machine": "WIN-25FPFGEMUA9",
      "user": "",
      "code": 9016,
      "elapsed": "",
      "process": "4384",
      "thread": "13",
      "methodName": ""
    }
  ]
}
```


And now we get to the fun part: parsing the log messages and sending them in an email. First, we set up our empty `email_body` string to use later. Next, we build up an `email_log_level` message. This inline `if...else` statement does the following: if `logging_level` is `SEVERE`, then the only errors we will get back are `SEVERE` errors, as it is the highest level of logging. But (`else`), if the logging level is not `SEVERE`, as in it is `WARNING` or lower, then we will get back messages at `logging_level` and higher, so we want to include that in our message. Next, we test to see if any messages were returned. If the `logMessages` object of `data_obj` isn't empty, we start looping through the messages. For each message, we build up a string consisting of the error timestamp (`item["time"]` converted back from a Unix timestamp in milliseconds and formatted to something human-friendly), the error level (`item["type"]`), and the item message (`item["message"]`). We keep looping through the results, appending each one to `email_body`, with a line break between each one. When we are done looping through the results, we create our actual `email_msg` consisting of an opening line, stating what log level the errors are followed by the list of errors. We then call the `send_email` function, passing in a subject line and the `email_msg`.

If no results were returned from our query to the logs operation, then we send a nice, short email stating that nothing was returned for the last 24 hours:

```
email_body = ""
email_log_level = logging_level if logging_level == "SEVERE" \
    else "{0} and higher".format(logging_level)

if data_obj["logMessages"]:
    for item in data_obj["logMessages"]:
        msg = "{0} @ {1}: {2}".format(
            item["type"],
            datetime.datetime.fromtimestamp(
                int(item["time"]/1000)
            ).strftime("%Y-%m-%d %H:%M:%S"),
            item["message"]
        )
        email_body = "{0}\n{1}".format(
            email_body,
            msg
        )

    email_msg = "The following {0} errors were logged " \
                "in ArcGIS Server in the last " \
                "24 hours:\n\n{1}".format(
        email_log_level,
        email_body)
    send_email("ArcGIS Server error report", email_msg)
else:
    email_msg = "There were no {0} errors logged in " \
                "ArcGIS Server in the last 24 " \
                "hours.\n".format(email_log_level)
    send_email("ArcGIS Server error report", email_msg)
```

Finally, the `send_email` function connects to the provided SMTP server, logs in with the provided credentials, and sends the email using the `smtplib.SMTP.sendmail` method:

```
def send_email(the_subject, the_message):
    server = smtplib.SMTP("smtp_server", None)
    server.starttls()
    server.login("logon_email_address", "logon_password")
    msg = MIMEText("\n{0}".format(the_message))
    msg["Subject"] = the_subject
    msg["From"] = "logon_email_address"
    msg["To"] = "to_email_address"
    server.sendmail("logon_email_address",
                    "to_email_address",
                     msg.as_string())
    server.quit()
```

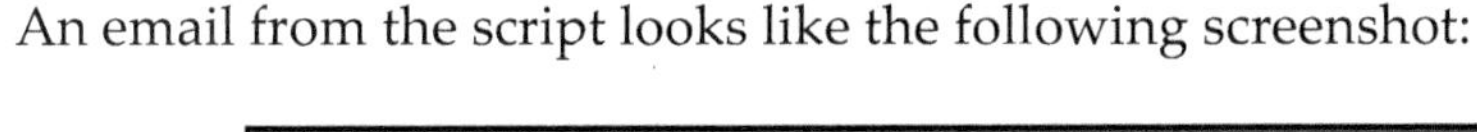
An email from the script looks like the following screenshot:

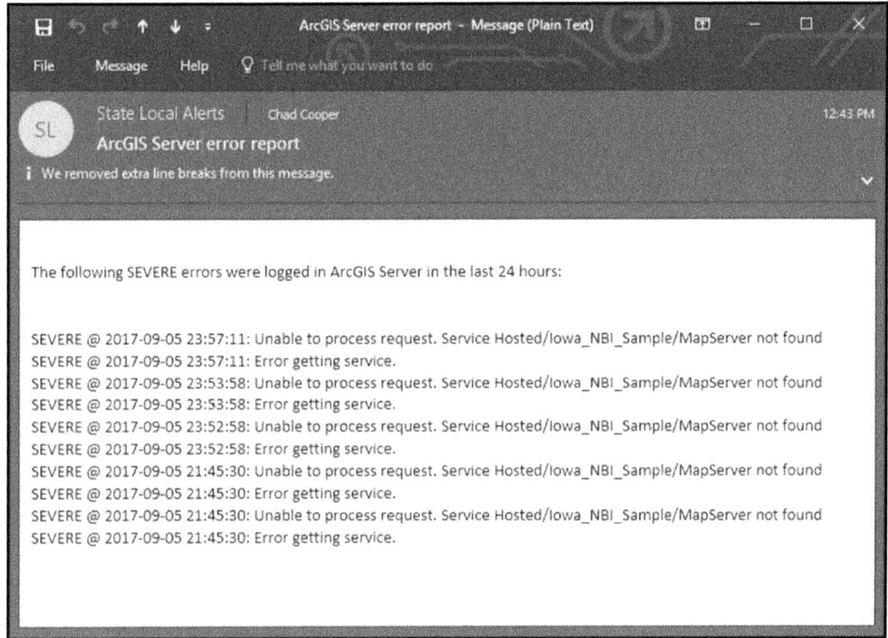

Maybe not the fanciest of emails, but it gets the job done. This script could be scheduled to run daily to poll for `SEVERE` errors and perhaps even weekly to look for other information, such as `WARNINGS`. If you do decide to schedule it to run other than daily, remember to change the `hours` argument in the `datetime.timedelta(hours=24)` call in this line:

```
past = datetime.datetime.utcnow() - datetime.timedelta(hours=24)
```

Also, you will probably want to change the `logging_level`. There is plenty that could be done to improve this script, such as:

- Adding error handling (left out for brevity)
- Adding logging; any script that will be run on a regular schedule that should always have logging
- Making `logging_level`, email `To` address, and the `timedelta hours` value optional arguments that could be passed in at runtime, thus making the script more flexible

Working with Portal through Python

Since the ArcGIS APIs are RESTful, they can be programmatically accessed by any language that can make a web request. Python makes this very simple to do, and for that reason, virtually, all code samples that you will find from Esri are in Python, as are their wrapper libraries.

PortalPy

PortaPy is a Python 2.7 module written by Esri that installs with Portal into the Portal installation directory on your Portal server but is also available on Esri's GitHub account (`https://github.com/Esri/portalpy`). PortalPy can run from any machine with Python 2.7 installed, meaning you can configure this on your local PC that more than likely has ArcGIS Desktop installed, which comes with Python 2.7.x.

Installation and configuration

To work with `PortalPy` from your local PC, you'll have to install the module. To do so, complete the following steps:

1. On your PC, create a directory called `portalpy` (mine is `C:\Projects\portalpy`).
2. Go to the `Esri GitHub portalpy` repository and download `PortalPy`. Unzip the archive so `portalpy.py` sits at the root of the `portalpy` directory you just created on your local PC.
3. On your PC, create an environmental variable with a variable name of `PYTHONPATH` whose variable value of the full path to your `portalpy` folder; in my case, `C:\Projects\portalpy`.
4. Create a file called `test.py` in the same directory as `portalpy.py`, and then enter the following code snippet into `test.py`:

```
import portalpy
url = "https://portal_url.com/portal_web_adaptor_name"
portal = portalpy.Portal(url)
print portal.get_version()
```

The `url` for my Portal is this:

```
https://www.masteringageadmin.com/portal
```

5. Save and close `test.py`. Run `test.py`. You should get back a version number such as `5.1`.

PortalPy usage

Let's start out small here and list all the users in our organization:

```
import portalpy
url = "https://www.masteringageadmin.com/portal"
portal = portalpy.Portal(url,
                         "someadminuser",
                         "strongpassword")

all_users = portal.get_org_users()
for au in all_users:
    print au["username"]
```

Note here that in order to get a list of the users, we need to log in, which is easy enough; just pass in an administrative username and password. Executing this code gives us the following output:

```
esri_boundaries
esri_demographics
esri_livingatlas
esri_nav
joe.schmoe
portaladmin
system_publisher
```

Hmm, I recognize `joe.schmoe` and `portaladmin`, but what are those other accounts? They don't show up in my list of users in my organization. These are internal Portal members used to own built-in groups, even if you don't have those groups. In other words, don't worry about them, they are system users that are supposed to be there.

Let's do the same for groups using the `search_groups` method. There is no method to get all groups like with `get_org_users()`, but we can use `search_groups` with a wildcard query string to return all groups just the same:

```
import portalpy
url = "https://www.masteringageadmin.com/portal"
portal = portalpy.Portal(url,
                         "someadminuser",
                         "strongpassword")
all_groups = portal.search_groups("*")
for ag in all_groups:
```

```
        print ag["title"]
```

This code returns the following output:

```
Esri Boundary Layers
Esri Demographic Layers
Featured Maps and Apps
Killer GISinc applications
Living Atlas
Living Atlas Analysis Layers
Navigator Maps
Our Org Basemaps
```

As was the case with users, we get more groups than might be expected returned to us. Here, I recognize `Featured Maps and Apps` and `Killer GISinc applications`, as those both show up in my groups, and `Killer GISinc applications` is *obviously* a group I created. The other groups are system groups, owned by the system users we just discovered when we listed all our organization's users earlier.

Portal for ArcGIS command-line utilities

Just as with ArcGIS Server, there is a suite of Java-based tools provided by Esri in the `Portal installation directory` on your Portal server at `<Portal installation directory>\tools`. In my case, that directory is `D:\Program Files\ArcGIS\Portal\tools`.

Incidentally, `PortalPy` is installed in the `\tools` directory as well.

Here, there are several tools to scan for security best practices, transferring item ownership, recovering a portal when no administrator accounts are available (oops), removing members, and adding members. Let's look at doing some user administration with one of these tools. A full list of these tools and their usage can be found at the ArcGIS Enterprise online documentation by searching for "**portal for arcgis command line utilities**".

Adding built-in users in bulk

The `CreateUsers` tool, which can be found at `<Portal installation directory>\tools\accountmanagement`, allows you to add built-in or enterprise accounts to your portal in bulk from an input text file.

This tool can only be executed by a built-in administrator-level account on the Portal for ArcGIS machine. An enterprise identity store administrator account cannot be used.

The first step in this process is to create the text file of user accounts, which must follow this format:

```
<account>|<password>|<email address>|<name>|<role>|<description>|
<first name>|<last name>|<level>
```

A separate line is used for each account and values are delimited with pipes (`|`). Several of these parameters need not be defined, but let's define some of the others:

- `account`: This is the username for the account. Account names must contain alphanumeric ASCII characters or underscores can be up to 128 characters long.
- `name`: This is the alias for the account and what is listed near the top right of the Portal site when a user is logged in.
- `role`: This is the role the account will have in the organization, where valid roles are `viewer`, `user`, `publisher`, `admin`, or a custom role name.
- `description`: This is an optional, up to 250-character long description of the account.
- `level`: This is the membership level of the account, where `1` is a Level 1 user that can use existing content but not create or share, and `2` is a Level 2 user with the ability to create, share, edit items, and so on.

Let's look at the content of a file I created to add a publisher user:

```
gis_pub|how-bomb-IS-coffee.6|pub@myorg.com|GIS
Publisher|publisher|GIS publisher account|GIS|Publisher|2
```

Executing the script and passing in the `--idp builtin` flag along with my administrator credentials when prompted, creates the one user:

```
CreateUsers.bat --file 07_05_create_users.txt --idp builtin
CREATE USERS using...
File: 07_05_create_users.txt
Portal admin url: https://WIN-25FPFGEMUA9:7443/arcgis/portaladmin
Admin username: portaladmin
password:
Creating users ...
SUMMARY: 1 of 1 user(s) successfully created.
D:\Program Files\ArcGIS\Portal\tools\accountmanagement>
```

If you do not pass in the `--idp` flag, enterprise accounts are registered by default.

Summary

We covered a lot of ground in this chapter. Knowing how to use scripting to your advantage is quickly shifting from being a non-essential, nice to have talent to a required skill for technical GIS positions. Hopefully, after reading this chapter and seeing just a few of the many ways Python can be used to script within the ArcGIS ecosystem, it's easy to see why Python skills are so important. From geodatabase administration to ArcGIS Server to Portal, all have administrative tasks that can be scripted with Python. With the REST API, if it can be reached through a URL, it can be scripted. The Python Standard Library has everything needed to query our ArcGIS Server logs and send emails with error reports (my favorite part of this chapter). We also looked at the Portal for the ArcGIS command-line utilities and how to utilize them to create Portal users programmatically. Next, in `Chapter 8`, *The ArcGIS Python API*, we will use the ArcGIS API for Python, a very powerful library to access and work with your GIS. I feel the ArcGIS API for Python is one of the most exciting things to come out of Redlands in quite a few years, and it's only going to get better. That said, I think you better get familiar with it, so let's get to it.

8
The ArcGIS Python API

The beta 2 version of the ArcGIS API for Python was released in September 2016. Since then, there have been several more releases leading up to the 1.2.4 version that is available at the time of this writing. Esri describes the API as "*a powerful, modern, and easy to use Pythonic library to perform GIS visualization and analysis, spatial data management and GIS system administration tasks that can run both in an interactive fashion, as well as using scripts*". The API was designed with analysts, developers, power users, content publishers, administrators, and data scientists in mind and allows full access to your web GIS. In this chapter, we will cover how to install, configure, and use the ArcGIS API for Python. We will look at how to use the API to do the following:

- Changing web map service URLs
- Creating a web map inventory
- Replicating content
- Managing users and groups
- Working with features

What is the ArcGIS API for Python?

The ArcGIS API for Python was conceived at the 2015 Esri International User Conference. The project, codenamed **Geosaurus**, was initiated to design and implement a Pythonic web GIS API that would be powerful, modern, and easy to use. So, what exactly does that mean?

Let's look at some of the defining terms and how they relate to the usability of the API:

- **Powerful**: The API is powerful, in that it allows you to work with all aspects of your web GIS, where a web GIS can be either ArcGIS Online or Portal for ArcGIS. With the API, you can create, manage, and use GIS resources such as web layers, web maps, users, and groups.
- **Modern**: The API is modern, in that it is built for Python 3 and integrates easily with libraries such as pandas, NumPy, and the SciPy ecosystem of libraries.
- **Easy to use**: Ease of use is one of Python's greatest virtues, and Esri went to great strides to build the API with usability in mind.
    - **Interactivity**: The API comes ready to work well within the Jupyter Notebook (`http://jupyter.org/`), which is an open source web application where you write and execute live code, visualizations (web maps!), and embedded documentation.
    - **Pythonic**: The API follows modern Python conventions, making it readable, easy, and natural to use and learn. The API is a Pythonic representation of a GIS and is implemented on top of the REST API, thus abstracting away many of the complex and tedious REST operations we have covered in the previous chapters, such as acquiring tokens to log in to your GIS.
- **Modular**: The API is broken down into modules, each with types and functions focused toward one aspect of your GIS (more on this in the next section).
- **Samples and documentation**: The API has a full suite of documentation (`http://esri.github.io/arcgis-python-api/apidoc/html/#`) and rich, interactive samples, and demo code that can be downloaded as Jupyter Notebooks and easily customized to your needs.

Now that we have discussed what the API is, what it can do, and what makes it different, let's look at what the API looks like.

How the API is structured

The API is contained in and distributed as the `arcgis` package. The `arcgis` package is composed of 13 (at the time of this writing) modules, each one focused on a different aspect of the GIS. The following diagram shows the modules and how they can be grouped together by functionality:

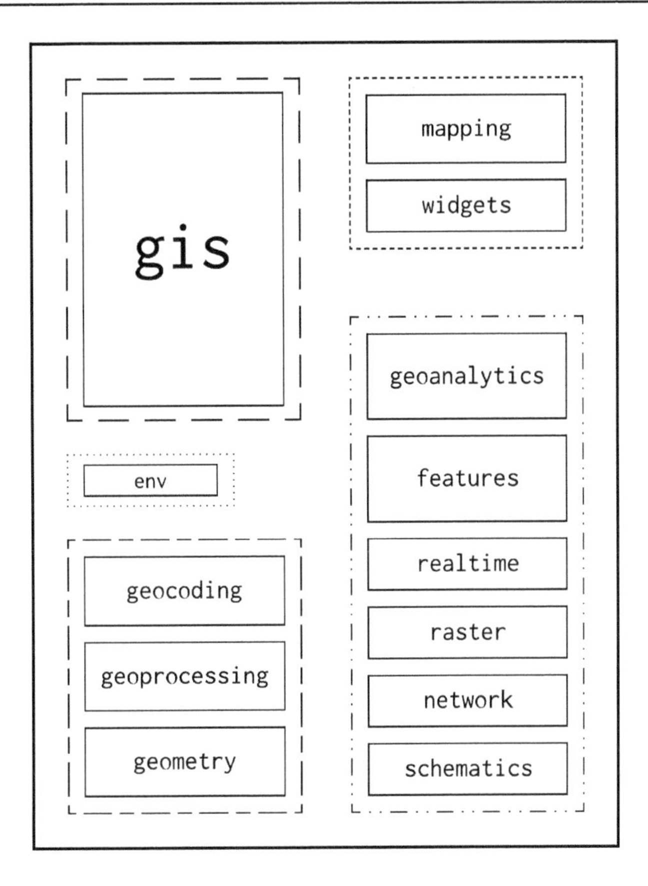

Let's briefly discuss what each of these modules does for us:

- **gis**: This provides an entry point in the GIS hosted within ArcGIS Online or Portal for ArcGIS. It manages users, groups, and content. It's a very important module for administrators.
- **env**: This stores environmental and global settings, such as the currently active GIS, the default geocoder, and output spatial reference.
- **geocoding**: This provides functionality for geocoding.
- **geoprocessing**: This allows us to import geoprocessing toolboxes as native Python modules, making the functions in the tools callable.
- **geometry**: This provides functions for working with geometries and converting them between different representations.
- **mapping**: This provides components for visualization and analysis with web maps, web scenes, and map image layers.

- **widgets**: This is also for visualization and analysis, but with the MapView Jupyter Notebook widget. Once you create a MapView you can add items such as layers or graphics to it.
- **geoanalytics**: This provides types and functions for working with large datasets such as big data in GIS datastores or feature layers.
- **features**: This provides types and functions for working with feature data, feature layers, and feature layer collections in the GIS.
- **realtime**: This includes support for receiving, processing, and analyzing real-time data feeds and streaming sensor data.
- **raster**: This provides classes and functions for working with raster data and imagery.
- **network**: This provides classes and functions for network analysis such as, routing, service area generation, or closest facility finding.
- **schematics**: This provides types and functions for working with simplified representations of networks or schematics.

Getting set up to use the API

There are several ways to use the ArcGIS API for Python, either through an interactive live Sandbox or by installing it in your environment in one of several ways. Let's discuss a few of the more common installation methods.

The ArcGIS API for Python has no Esri software dependencies (such as is the case with `arcpy`, where you must have Desktop or ArcGIS Server installed), so it can be installed on Windows, macOS, or Linux.

Try it live

Many of you may be familiar with the ArcGIS Solutions Gallery (`http://solutions.arcgis.com/gallery`), formerly known as *Try It Live*, a collection of live sample sites and applications provided by Esri. The ArcGIS Solutions Gallery is a great place to find and sample hundreds of ready-to-use applications, products, and solutions from Esri. The ArcGIS API for Python has a similar offering, a live Sandbox in the form of a temporary Jupyter Notebook browser session environment that can found at `https://notebooks.esri.com/`. Here, you can browse through sample code, view guides, and view API presentations and their content. For anyone getting started with the API, this Sandbox is a fantastic resource.

Installing using Conda

Conda (`https://conda.io`) is a package dependency and environment management system for many popular programming languages, such as Python, R, Ruby, and JavaScript, to name a few. As such, the ArcGIS API for Python is distributed via Conda. To use Conda to install the `arcgis` package, you will first need to download and install the latest version of Anaconda for Python 3.x (`https://www.anaconda.com/download/`). After you install Anaconda, in your terminal, execute the following command to download and install the API:

```
conda install -c esri arcgis
```

The API package can also be updated through the Conda command line with the following command:

```
conda upgrade -c esri arcgis
```

Installing using ArcGIS Pro

The API can also be installed via the **Python Package Manager** (**PyPM**) in ArcGIS Pro 1.4 or higher.

If you plan on using the ArcGIS API for Python and `arcpy` in the same Python environment (that is, the same scripts), you must install the API through ArcGIS Pro 1.4 or higher.

To install the API from within Pro 1.4 or higher, launch Pro and go to the backstage (if you have a project open, go to the **Project** tab to get backstage), and then click on the **Python** left tab. Next, as the following screenshot shows, select **Add Packages** and search for **arcgis**, first clicking on the green refresh button to ensure that you are getting the latest version. Finally, when the API is found, click on **Install** to the far right to install the module. You will be notified of dependencies that will also be installed and asked to accept the licensing terms. Select **OK**. The API will install quickly, more than likely in less than 1 minute.

This is shown in the following screenshot:

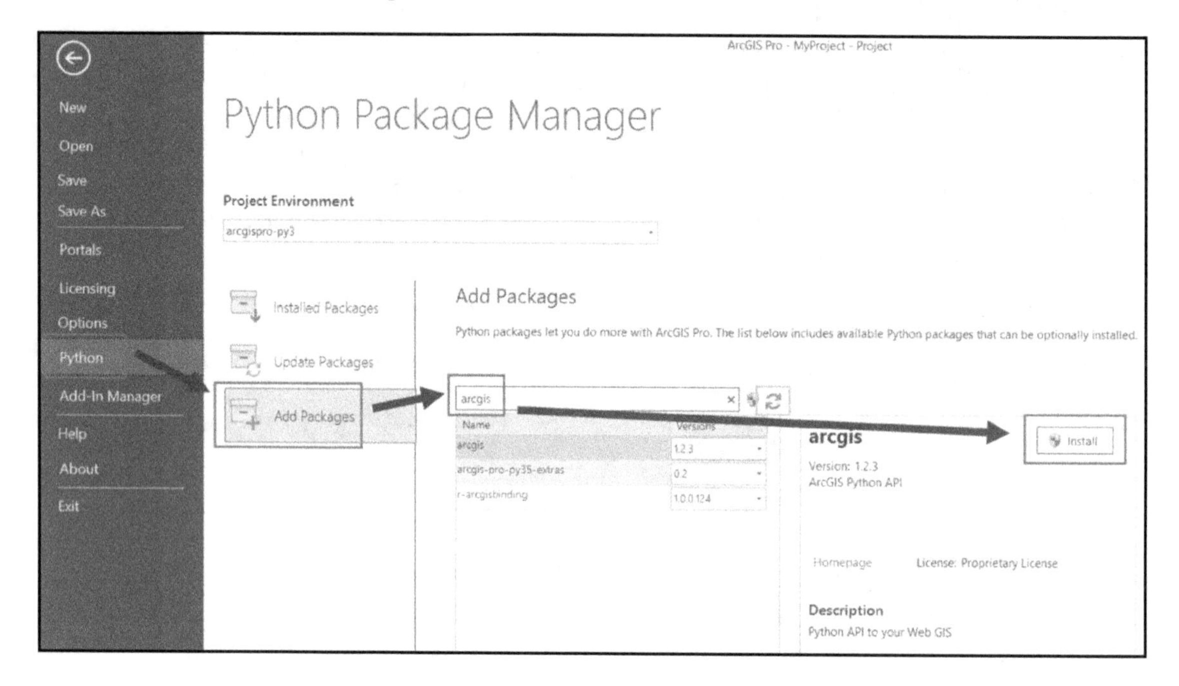

Packages can also be updated through **Update Packages** in the **Python Package Manager** of Pro. For the remainder of this chapter, examples will use the API as installed through **Pro Package Manager**.

Testing the API installation

Once you have the API installed either through Conda or ArcGIS Pro, start a Jupyter Notebook with the following command at your terminal:

```
jupyter notebook
```

Now, if you're on Windows, and most of you will be, chances are that you will get a `jupyter is not recognized as an internal or external command, operable program or batch file` error when you enter the preceding command. You get this error because Windows has no idea where `jupyter.exe` is, unless you tell it explicitly or you have the path to `jupyter.exe` set in your `PATH` environmental variable. One way around this is to create a batch file with the following command in it:

```
"C:\Program Files\ArcGIS\Pro\bin\Python\envs\arcgispro-
py3\Scripts\jupyter.exe" notebook
```

Here, I have entered the full path to `jupyter.exe` on my system (yours may be slightly different), followed by the `notebook` command. Double-clicking on this batch file from Windows Explorer will launch a **Jupyter Notebook** dashboard in a new browser tab in your default browser. On the upper-right of the notebook window, choose **Python 3** from the **New** drop-down to create a new empty notebook. On line 1 of the new notebook, enter the following lines of code:

```
from arcgis.gis import GIS
gis = GIS()
gis.map()
```

Next, on the **Cell** menu, select **Run Cells and Select Below** (*Shift* + *Enter* is the shortcut), which will execute the code. It may take a few seconds to run, and you should end up with a map, as follows:

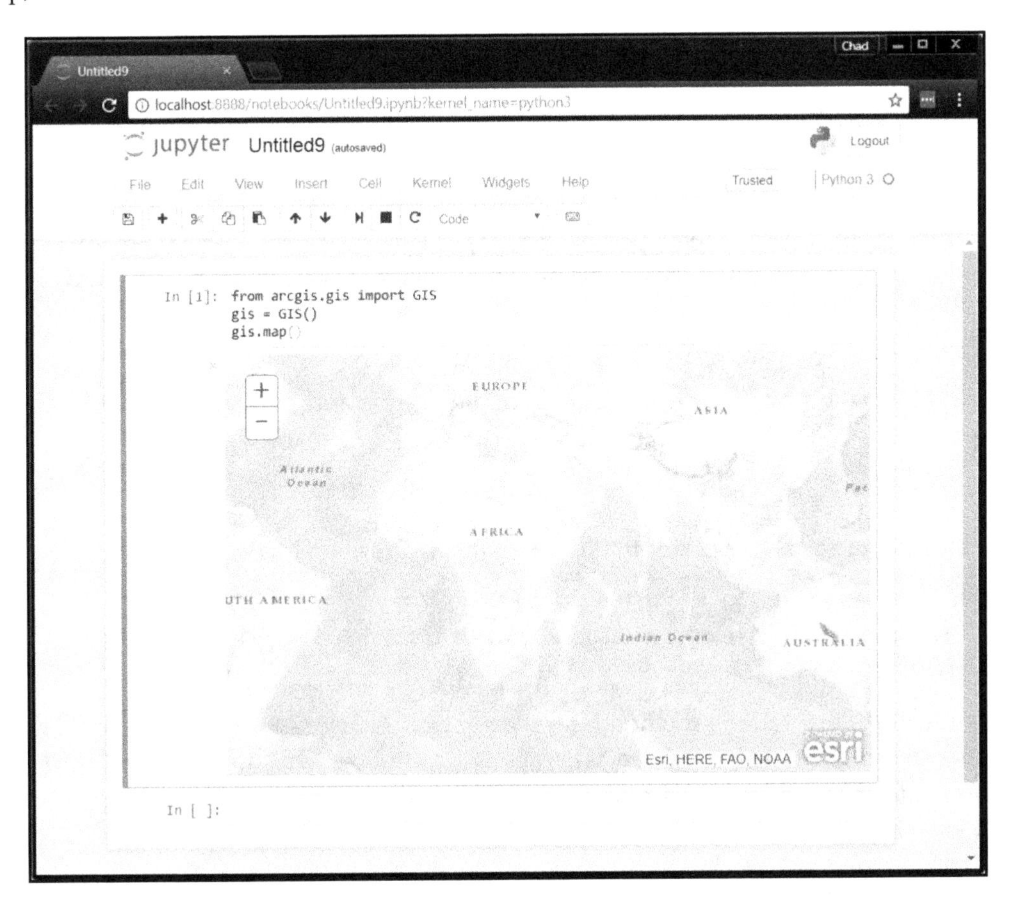

Congratulations, you've just successfully installed and used the ArcGIS API for Python!

Working with services

As an administrator, you will need to manage services of all sorts. We've already covered methods of administering services using ArcGIS Server Manager and with the REST API. In this section, we'll look at ways to work with services and items through the ArcGIS API for Python.

Changing web map service URLs

The web is a dynamic environment, there's no denying that. Consequently, URLs to resources you reference in your applications are going to change. When they are your URLs, you usually know about it in advance and can have time to plan for the change. The big surprises come when a URL to an item that isn't yours changes. You usually find out about these through a phone call or email from a user telling you that your application is busted. So, what's an admin to do in a case like this? We've looked at ways to handle URL changes with ArcGIS Online Assistant. However, accomplishing this same task with the ArcGIS API for Python is easy to do as well.

Let's say that we have some older web maps that reference externally-facing ArcGIS Server services that, at one time, were served out over HTTP, but the provider has since acquired an SSL certificate and is now serving out over HTTPS. This is a common issue, and, as a result, you need to update those service references in some web maps. Let's walk through the code that changes these URLs for us.

First, we will start off by creating a new Jupyter Notebook and importing `GIS` from the `arcgis.gis` module:

```
In [1]: from arcgis.gis import GIS
```

We will then create a `GIS` object in the `gis` variable by passing in a Portal URL and credentials:

```
In [2]: gis = GIS("https://www.masteringageadmin.com/portal",
                  "portaladmin", "pass")
```

Next, we will use the `search` method on the `arcgis.gis.ContentManager` class. When we do this, we pass in the `query` and `item_type` arguments. Our query is for tags of `city-maps` on items of type `Web Map`. Consult the API documentation for all available types and query syntax, as it is very rich and powerful:

```
In [3]: search_results = gis.content.search(query="tags:city- maps",
                                            item_type="Web Map")
```

We will then print out the search results as follows:

```
In [4]: print(search_results)
```

This outputs a list of the following `Web Map` objects:

```
[<Item title:"LandUse" type:Web Map owner:portaladmin>,
 <Item title:"Parcels" type:Web Map owner:portaladmin>,
 <Item title:"EOC" type:Web Map owner:portaladmin>]
```

Let's dig down into those `Web Map` objects a little further. First, we will import the `arcgis` module from the API. Next, we will look through the list of search results and using the `WebMap` class create an instance of the `Web Map` as a JSON object. This allows us to access `operationalLayers` in `Web Map` and print out the `url` of each:

```
In [5]: import arcgis
        for search_result in search_results:
            web_map = arcgis.mapping.WebMap(search_result)
            layers = web_map['operationalLayers']
            for layer in layers:
                print(layer['url'])
```

Printing each `url` outputs as follows:

```
http://www.masteringageadmin.com/arcgis/rest/services/
LandUse/LandUse/MapServer/0
http://www.masteringageadmin.com/arcgis/rest/services/
Parcels/Parcels/MapServer/0
http://www.masteringageadmin.com/arcgis/rest/services/
EOC/USNationalGrid/MapServer/0
```

As you can see, the services URLs reference the HTTP protocol, and we need to change those to HTTPS. To do so, we will loop through the layers in the `Web Map` of each search result like we just did in the preceding code, but then we will build `new_url` by taking existing `url` and replacing `http:` with `https:`. Finally, we will set `operationalLayers` to the updated layers list of `https` URLs and call the `update` method on our `Web Map` object:

```
In [6]: for search_result in search_results:
            web_map = arcgis.mapping.WebMap(search_result)
            layers = web_map[''operationalLayers'']''
            for layer in layers:
                new_url = (layer['url'].replace('http:','https:'))
                layer['url'] = new_url
            web_map['operationalLayers'] = layers
            web_map.update()
```

To see the result of our URL replacement, we will loop through the layers of each operational layer in our `Web Maps` again, printing out those service URLs as follows:

```
In [7]: for search_result in search_results:
            web_map = arcgis.mapping.WebMap(search_result)
            layers = web_map['operationalLayers']
            for layer in layers:
                print(layer['url'])
```

Just as we wanted, we now have service URLs over HTTPS:

```
https://www.masteringageadmin.com/arcgis/rest/services/
LandUse/LandUse/MapServer/0
https://www.masteringageadmin.com/arcgis/rest/services/
Parcels/Parcels/MapServer/0
https://www.masteringageadmin.com/arcgis/rest/services/
EOC/USNationalGrid/MapServer/0
```

This was a nice introduction to the API. You are encouraged to experiment with examining item content using the API, especially using the `search` method, as you will use that extensively to access content.

Creating a Web Map inventory

How many times have you had to open a Web Map in the map viewer or view its properties just to try and get an idea of the services and layers that are hydrating it? I know I've personally done that more times than I'd care to, and it can be very frustrating and time-consuming. Wouldn't it be nice if you could have an inventory in say, an Excel spreadsheet, which could be automatically created and updated for you? Fortunately, The ArcGIS API for Python makes doing something like this possible with around 30 lines of code. Let's look at how we might go about creating a small script that queries a subset of Web Maps and updates an Excel workbook with URLs of all the operational and basemap layers in the web maps.

First, we import the modules we'll need for our script. `arcgis` imports the `arcgis` module from the ArcGIS API for Python. The `collections` module implements specialized container types that are alternatives to Python's general purpose built-in containers such as `dict` and `list`. Here, we will use the `dict` subclass, `OrderedDict` (more on that in a minute). Thirdly, we will import the `GIS` class from the `arcgis.gis` module. Finally, we will import the `pandas` library. Pandas is a Python library that provides easy to work with data structures and data analysis tools:

```
import arcgis
```

```
import collections
from arcgis.gis import GIS
import pandas as pd
```

For the export to Excel functionality of this script to work, you will need to install the `pandas` and `openpyxl` modules. The easiest way to install these is through the **Python Package Manager** in ArcGIS Pro.

Next, we will create an instance of our Portal GIS as `gis`, and perform a search that looks for `Web Maps` that are tagged as `city-maps`:

```
gis = GIS("https://www.masteringageadmin.com/portal","
          "portaladmin"," strongpass")
search_results = gis.content.search(query="tags:city-maps",
                                    item_type="Web Map")
```

Now that we have connected to our Portal and returned search results of our web maps of interest, let's start going through those results and pulling out the parts we are interested in reporting on. The first thing we will do is create an empty list `l` that will be used as a container for what we will ultimately be our rows in our report, where each row, a dictionary object, represents a layer that is referenced in our web maps. The `l` list will get passed into pandas as a list of dictionaries that pandas can work with:

```
l = []
```

Next, we will loop through our `search_results` list, where each `s` is a web map. We will pass each `web map` object into `arcgis.mapping.WebMap()`, which converts it into web map JSON. Once we have the web map JSON, we can start pulling bits of information out, such as the `operationLayers` and `baseMapLayers`:

```
for s in search_results:
    wmo = arcgis.mapping.WebMap(s)
    ops_layers = wmo['operationalLayers']
    basemap_layers = wmo['baseMap']['baseMapLayers']
```

Once we have the operational and basemap layer objects, we can loop through each of them, adding bits of information from each, such as Item ID, URL, and title, to an `OrderedDict` object. What's an `OrderedDict` object, you say? Well, by definition, standard Python dictionaries are unordered.

An `OrderedDict` object, however, remembers the order in which entries were added, and this is important because we want all our fields to always be in the same order so we can export them to Excel. Once we get all the bits of information we want about the operational layer, we add `d` to the list `l`:

```
for op_layer in ops_layers:
    d = collections.OrderedDict()
    d["Web Map Name"]" = s.title
    d["Web Map Item ID"]" = s.itemid
    d["Layer Type""] = "Operational Layer
                        ({0})"."format(op_layer["layerType"])
    d["Layer URL"]" = op_layer["url"]
    l.append(d)
```

An example of an `OrderedDict` object from the preceding code will look like this:

```
OrderedDict([('Web Map Name', 'LandUse'),
             ('Web Map Item ID', '0931bb9d370b4cbb39de45179f'),
             ('Layer Type', 'Operational Layer (FeatureLayer)'),
             ('Layer URL', '<some_url>')])
```

In the preceding code, the `OrderedDict` object is a list of Python tuples, with each tuple representing a key and a value pair, where the `key` is the field name and `value` is the value of that field for that web map. Remember that tuples are ordered, so the order of `key` and `value` will never be changed. Think of each instance of an `OrderedDict d` as a row in our upcoming Excel workbook.

Next, we will do the same thing for basemaps, keeping in mind that a web map can have more than one basemap:

```
for base_layer in basemap_layers:
    j = collections.OrderedDict()
    j["Web Map Name"] = s.title
    j["Web Map Item ID"] = s.itemid
    j["Layer Type""] = "Basemap Layer
                        ({0})".format(base_layer["layerType"])
    j["Layer URL"] = base_layer["url"]
    l.append(j)
```

Now we get to the good stuff using pandas, and only four lines of code to write out our results to an Excel file. The first thing we will do here is to create an instance of a pandas `DataFrame`, a two-dimensional labeled data structure.

When we call the DataFrame method, we will pass in our l list we populated earlier with dictionaries representing rows. The l list may look something like the following. Remember, each instance of an OrderedDict here will be a row in our Excel workbook:

```
[OrderedDict([('Web Map Name', 'LandUse'),
            ('Web Map Item ID', '0931bb9d370fb9c4bb39de45179f'),
            ('Layer Type', 'Operational Layer (FeatureLayer)'),
            ('Layer URL',
             '<some_url>')]),
 OrderedDict([('Web Map Name', 'LandUse'),
            ('Web Map Item ID', '0931bb9d370b49c4bb39de45179f'),
            ('Layer Type', 'Operational Layer (FeatureLayer)'),
            ('Layer URL',
             '<some_url>')]),
 OrderedDict([('Web Map Name', 'LandUse'),
            ('Web Map Item ID', '0931bb9d370fb9c4bb39de45179f'),
            ('Layer Type', 'Basemap Layer (TiledMapLayer)'),
            ('Layer URL',
             '<some_url>')]),
 OrderedDict([('Web Map Name', 'Parcels'),
            ('Web Map Item ID', '70031f46a9788235b2dd31b4f68b'),
            ('Layer Type', 'Operational Layer (FeatureLayer)'),
            ('Layer URL',
             '<some_url>')])]
```

Think of a DataFrame like a spreadsheet or a database table, both with columns and rows:

```
df = pd.DataFrame(l)
```

Next, we will create a writer object for a workbook called output.xlsx. This is the file we will soon write our results to:

```
writer = pd.ExcelWriter('output.xlsx')
```

Finally, we will call the pandas.DataFrame.to_excel method on our DataFrame df, passing in our ExcelWriter object and a sheet name. Calling writer.save() saves the Excel workbook:

```
df.to_excel(writer, sheet_name='Sheet1')
writer.save()
```

The result of this script is an Excel workbook, where each record represents an operational or basemap layer in a web map:

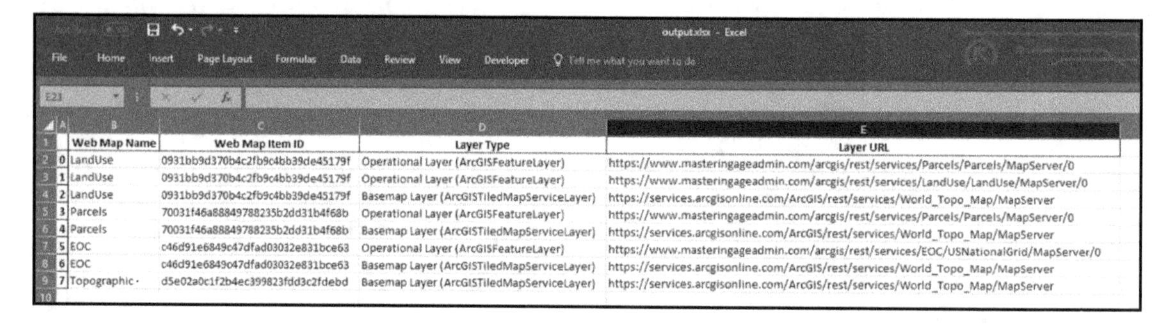

	Web Map Name	Web Map Item ID	Layer Type	Layer URL
0	LandUse	0931bb9d370b4c2fb9c4bb39de45179f	Operational Layer (ArcGISFeatureLayer)	https://www.masteringageadmin.com/arcgis/rest/services/Parcels/Parcels/MapServer/0
1	LandUse	0931bb9d370b4c2fb9c4bb39de45179f	Operational Layer (ArcGISFeatureLayer)	https://www.masteringageadmin.com/arcgis/rest/services/LandUse/LandUse/MapServer/0
2	LandUse	0931bb9d370b4c2fb9c4bb39de45179f	Basemap Layer (ArcGISTiledMapServiceLayer)	https://services.arcgisonline.com/ArcGIS/rest/services/World_Topo_Map/MapServer
3	Parcels	70031f46a88849788235b2dd31b4f68b	Operational Layer (ArcGISFeatureLayer)	https://www.masteringageadmin.com/arcgis/rest/services/Parcels/Parcels/MapServer/0
4	Parcels	70031f46a88849788235b2dd31b4f68b	Basemap Layer (ArcGISTiledMapServiceLayer)	https://services.arcgisonline.com/ArcGIS/rest/services/World_Topo_Map/MapServer
5	EOC	c46d91e6849c47dfad03032e831bce63	Operational Layer (ArcGISFeatureLayer)	https://www.masteringageadmin.com/arcgis/rest/services/EOC/USNationalGrid/MapServer/0
6	EOC	c46d91e6849c47dfad03032e831bce63	Basemap Layer (ArcGISTiledMapServiceLayer)	https://services.arcgisonline.com/ArcGIS/rest/services/World_Topo_Map/MapServer
7	Topographic ·	d5e02a0c1f2b4ec399823fdd3c2fdebd	Basemap Layer (ArcGISTiledMapServiceLayer)	https://services.arcgisonline.com/ArcGIS/rest/services/World_Topo_Map/MapServer

This script could be run on demand for inventories or set to run as a nightly scheduled task, giving insight into what services are used in web maps within your organization.

Displaying pandas DataFrames

It is important to note that many things can be done with the pandas `DataFrames`, the simplest of which is to display `DataFrame` in a cell in the Juypter Notebook. To do this, once you have an instance of the populated `DataFrame`, just call it like this:

```
df = pd.DataFrame(l)
df
```

The `DataFrame` panda will be displayed as shown in the following screenshot:

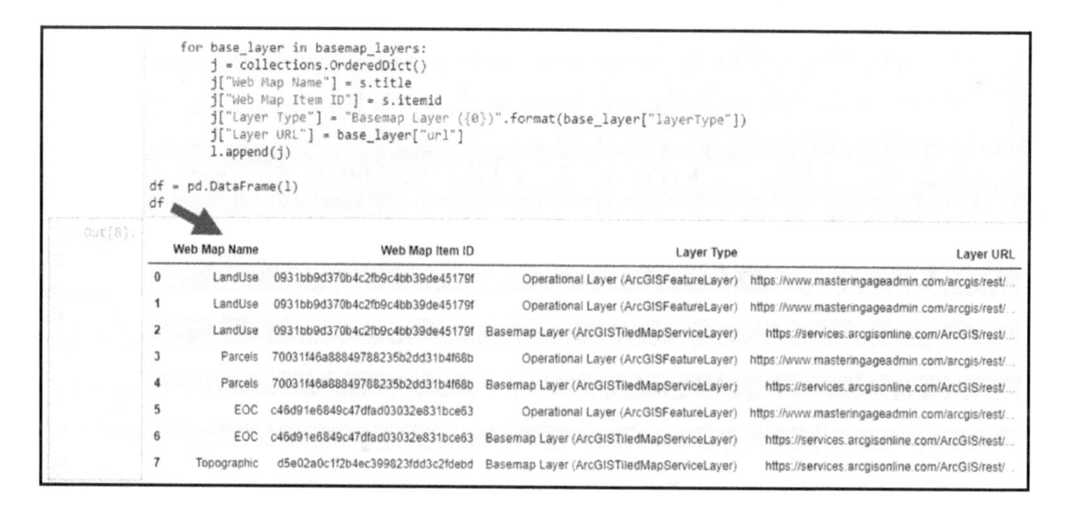

```
    for base_layer in basemap_layers:
        j = collections.OrderedDict()
        j["Web Map Name"] = s.title
        j["Web Map Item ID"] = s.itemid
        j["Layer Type"] = "Basemap Layer ({0})".format(base_layer["layerType"])
        j["Layer URL"] = base_layer["url"]
        l.append(j)

df = pd.DataFrame(l)
df
```

Out[8]:

	Web Map Name	Web Map Item ID	Layer Type	Layer URL
0	LandUse	0931bb9d370b4c2fb9c4bb39de45179f	Operational Layer (ArcGISFeatureLayer)	https://www.masteringageadmin.com/arcgis/rest/...
1	LandUse	0931bb9d370b4c2fb9c4bb39de45179f	Operational Layer (ArcGISFeatureLayer)	https://www.masteringageadmin.com/arcgis/rest/...
2	LandUse	0931bb9d370b4c2fb9c4bb39de45179f	Basemap Layer (ArcGISTiledMapServiceLayer)	https://services.arcgisonline.com/ArcGIS/rest/...
3	Parcels	70031f46a88849788235b2dd31b4f68b	Operational Layer (ArcGISFeatureLayer)	https://www.masteringageadmin.com/arcgis/rest/...
4	Parcels	70031f46a88849788235b2dd31b4f68b	Basemap Layer (ArcGISTiledMapServiceLayer)	https://services.arcgisonline.com/ArcGIS/rest/...
5	EOC	c46d91e6849c47dfad03032e831bce63	Operational Layer (ArcGISFeatureLayer)	https://www.masteringageadmin.com/arcgis/rest/...
6	EOC	c46d91e6849c47dfad03032e831bce63	Basemap Layer (ArcGISTiledMapServiceLayer)	https://services.arcgisonline.com/ArcGIS/rest/...
7	Topographic	d5e02a0c1f2b4ec399823fdd3c2fdebd	Basemap Layer (ArcGISTiledMapServiceLayer)	https://services.arcgisonline.com/ArcGIS/rest/...

Replicating content

The ability to create and replicate content within the GIS is a powerful feature of the ArcGIS API for Python. Let's say we have a zoning web map that only shows one zoning code. The zoning department likes this web map and would like to see one just like it for all other zoning codes. That's easy enough to do by hand when there are maybe only a few zoning codes, but what about when there are dozens? We've already looked at the JSON that makes up a web map, so surely there is a way to manipulate this JSON and use it to create a new web map, right? Indeed, there is a way to do just that. Let's look at how we might take the one existing web map and use it as a template to create other web maps just like it, but with different filters for zoning codes.

We begin, as usual, by importing the modules we will need. We have already used and discussed the `arcgis` and `arcgis.gis.GIS` module. In this script, we will also be using the `json` module to help us work with the web map JSON and the `re` module, or regular expression, to help us parse and manipulate our layer filter:

```
import arcgis
import json
import re
from arcgis.gis import GIS
```

Next, we will create a Python list with the zoning codes we need to create new web maps for, keeping in mind that we are only doing a handful here, but you could do this with dozens or hundreds of codes:

```
codes = [11, 14, 30, 40]
```

Next, we will create a `GIS` object and execute a search on our GIS for a web map titled **Landuse-12**. We will then create a `web map` object from that sole search result, as follows:

```
gis = GIS("https://www.masteringageadmin.com/portal",
          "portaladmin", "somepassword")
search_result = gis.content.search('title':LandUse-12',
                                   item_type = 'Web Map',
                                   outside_org = False)
web_map_object = arcgis.mapping.WebMap(search_result[0])
```

Since we are creating one web map for each zoning code, we will loop through the code lists and do everything in the loop for each code:

```
for code in codes:
```

Once we are in the loop, we will dig down into the operational layers, and using the Python dictionary `get()` method, we will look for `layerDefinition` (our zoning code filter). If we find `layerDefinition`, we will look to see if there is `definitionExpression`. Think of `definitionExpression` as a `where` clause. The `definitionExpression` in this web map is `LANDUSECODE = '12'`. Next, we will set up a regular expression search to look for any two-digit occurrences surrounded in single quotes at the end of the string query. In other words, we are looking for the two-digit zoning code. If we find one, we will replace it with code--the zoning code for the map we are currently creating. We will then set that `new_dq` as `definitionExpression` of `layerDefinition` in the current `web_map_object` JSON that we are working with:

```
for layer in web_map_object['operationalLayers']:
    def_query = layer.get('layerDefinition')'
    if def_query:
        query = def_query.get('definitionExpression')
        if query:
            search_obj = re.search(r"'[0-9]+''$"",'" query)
            if search_obj:
                if search_obj.group().replace("''", "").isdigit():
                    new_dq = query.replace(search_obj.group(),
                                           "'{0}'".format(code))
                    layer['layerDefinition'] \
                        ['definitionExpression'] = new_dq
                    dq = layer['layerDefinition'].get(
                                     'definitionExpression')
```

Next, we will set up the JSON to be used to create the actual web map. We will create a title using the current code, add any tags we see fit, and set the text of the web map by serializing `web_map_object` to a JSON-formatted string. We will then pass in `web_map_props` to the `gis.content.add()` method, setting the folder to store the web map in. Finally, we will share the new web map with everyone like this:

```
web_map_props = {"title": "LandUse-{0}".format(code),
                 "type": "Web Map",
                 "tags": "arcgis-api",
                 "text": json.dumps(web_map_object)}
web_map_item = gis.content.add(web_map_props,
                               folder="Replicating Content")
web_map_item.share(everyone=True)
```

Before running this script, the **Replicating Content** folder in Portal had only one web map in it--**LandUse-12**:

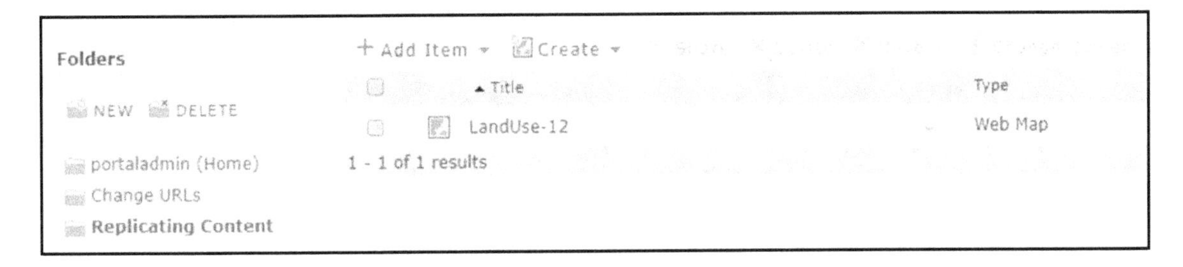

After running our script, we now have five web maps, as shown in the following screenshot:

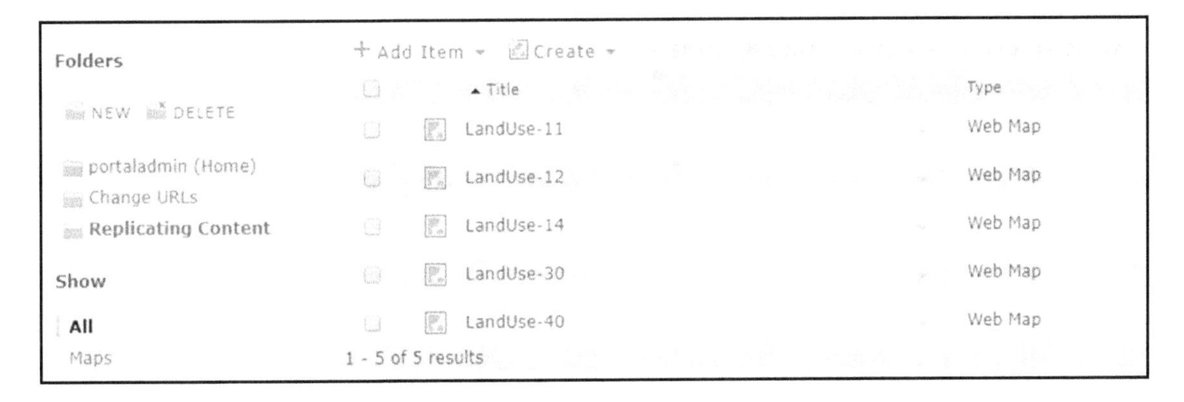

If we look at **LandUse-11**, for example, we can see that only land use code of `11` is being shown on the map and our filter that we created using the API is in effect:

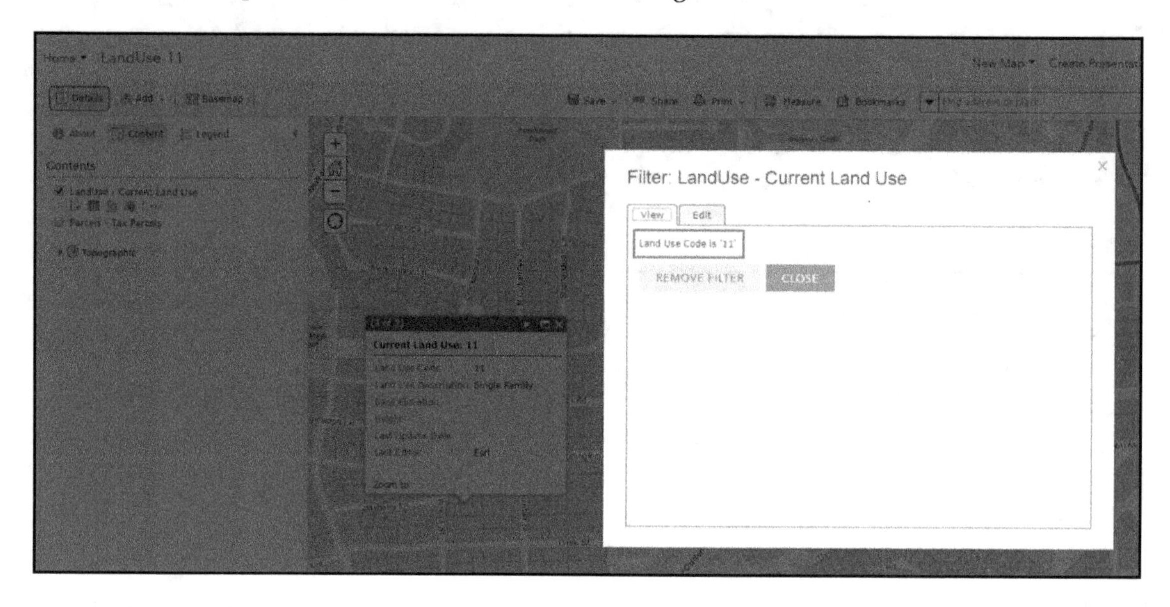

What we just wrote was another example of how easy it is to connect to your GIS and work with content within it. Now that we've worked with services and web maps, let's shift our focus to user management.

Working with users and groups

We've covered working with services and items such as web maps, but you'll also need to manage users and groups as well. Examples we have utilized so far in this chapter have shown how the ArcGIS API for Python abstracts away much of the minutiae of connecting to Portal or ArcGIS Online and allows you to just get to work. User and group management is no different; we connect just as we have in previous examples to gain access.

Managing users

In Chapter 7, *Scripting Administrative Tasks,* we used PortalPy to pull user information out of our Portal. We can do the same thing with the ArcGIS API for Python, but in a modern, Pythonic fashion. Let's look at how we can easily pull user information from Portal:

```
import collections
import time
import pandas as pd
from arcgis.gis import GIS
```

After we've imported the libraries we will need, we create a connection to our GIS and then perform a search for users whose username does not begin with `esri_`. Note how we use the bang (`!`) symbol as a `NOT` operator:

```
gis = GIS("https://www.masteringageadmin.com/portal",
          "portaladmin", "somepass")
all_my_accounts = gis.users.search('!esri_')
```

What we do next should look familiar, as we did something very similar in our last example. We will create an empty list to store records in, and then we will loop through our search results, putting attributes for each user into `OrderedDict`. We will then append each `OrderedDict` to the list of records, as follows:

```
l = []
for user in all_my_accounts:
    d = collections.OrderedDict()
    d['User Name'] = user.username
    d['First Name'] = user.firstName
    d['Last Name'] = user.lastName
    d['Email'] = user.email
    d['Role'] = user.role
    d['Provider'] = user.provider
    d['Level'] = user.level
    date_created = time.localtime(user.created/1000)
    d['Created'] = "{0}/{1}/{2}".format(date_created[1],
                                        date_created[2],
                                        date_created[0])
    l.append(d)
```

Finally, we will create a pandas `DataFrame` with the results list and display it, as follows:

```
df = pd.DataFrame(l)
df
```

	User Name	First Name	Last Name	Email	Role	Provider	Level	Created
0	gis_pub	GIS	Publisher	pub@myorg.com	org_publisher	arcgis	2	9/8/2017
1	portaladmin	Portal	Administrator	supercooper@gmail.com	org_admin	arcgis	2	7/10/2017

Managing groups

Working with groups is a very similar experience to content and users; perform a query to get the objects you are looking for, then read or manipulate them. Let's look at some groups in my Portal and see who owns and administers them:

```
from arcgis.gis import GIS
gis = GIS("https://www.masteringageadmin.com/portal",
          "portaladmin", "somepass")
```

After we create a connection to our GIS, we will search for all groups owned by the `portaladmin`:

```
groups = gis.groups.search('owner:portaladmin')
```

Next, we will iterate through `groups`, printing out information about the group members:

```
for group in groups:
    print(group.title)
    mems = group.get_members()
    print("\"tOwner: {0}".format(mems['owner']))
    print("\"tAdmins: {0}".format(", ".join(mems['admins'])))
    print("\"tUsers: {0}".format(", ".join(mems['users'])))
```

This gives us the following output:

```
Featured Maps and Apps
    Owner: portaladmin
    Admins: portaladmin
    Users:
Killer GISinc applications
    Owner: portaladmin
    Admins: portaladmin
    Users:
```

```
Our Org Basemaps
   Owner: portaladmin
   Admins: portaladmin
   Users:
```

Working with features

Earlier in this chapter, we briefly covered all the modules in the ArcGIS API for Python. Many of these modules are geared toward analysts and data scientists, but, as an administrator, you will still occasionally get your hands in some data processing. The ArcGIS API for Python has capabilities to both update and overwrite feature layers.

Publishing and overwriting a feature layer

In this example, we will use an Excel workbook to keep track of project locations and statuses. We will then push a worksheet from that workbook out to CSV and update a hosted feature service with that CSV. A scenario like this allows end users to update the feature service using an existing (and very common) workflow, keeping track of data, assets, and so on, in an Excel workbook. Users will be updating the feature service without even knowing they are doing so, simply by keeping the Excel worksheet up-to-date. Let's look at how this code will be laid out.

Publishing the initial feature layer

First, we will import the libraries we need and connect them to our GIS instance, as follows:

```
import pandas as pd
from arcgis.gis import GIS
gis = GIS("https://www.masteringageadmin.com/portal",
          "portaladmin", "password")
```

Next, we will use the `ExcelFile` method of pandas to load our Excel workbook into a pandas object. We will then call `parse` on the sheet with our data; `Sheet1` in this instance:

```
xls_file = pd.ExcelFile(r"C:\data\station-data.xlsx")
df = xls_file.parse("Sheet1")
```

We can print out the `DataFrame` as follows by calling the `df` variable:

```
df
```

	StationID	Long	Lat	Status
0	1	-94.200326	36.028075	Active
1	2	-94.213327	36.034187	Inactive
2	3	-94.218176	36.021193	Active

We will now call `to_csv` on our `DataFrame` object to export it to a CSV that we then turn around and add it to Portal as an item and publish:

```
df.to_csv(r"C:\data\station-data.csv")
csv_item = gis.content.add(item_properties={"title":"mt kessler"},
                           data=r"C:\data\station-data.csv")
stations_item = csv_item.publish()
```

Now, draw a *blank* map of the Fayetteville, Arkansas area, as follows:

```
map1 = gis.map("Fayetteville, AR")
map1
```

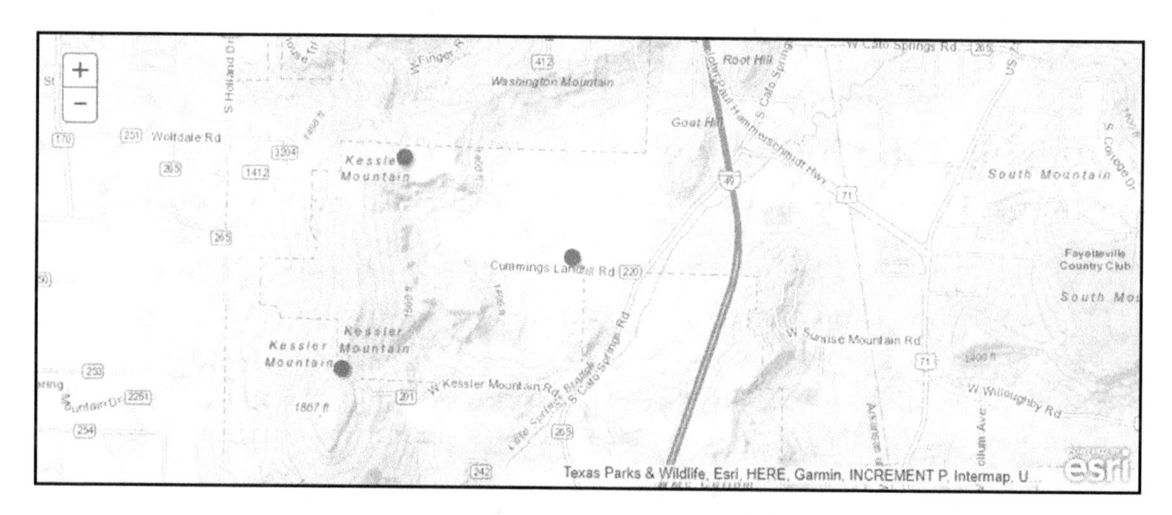

After we have our map, we can use `add_layer` to add our `stations_item` to the preceding `map1`. This will update the existing map, which, to be honest, is very cool to watch happen:

```
map1.add_layer(stations_item)
```

We can also interrogate our stations item to get its URL:

```
stations_item.url
'https://www.masteringageadmin.com/arcgis/rest/services/
   mt_kessler/FeatureServer'
```

That's it; in roughly 10 lines of code, we just published an Excel worksheet to a feature service. **Ten lines**.

Overwriting the feature layer

The Excel spreadsheet of our stations is updated daily, so the feature service needs to be as well.

First, import our usual libraries, but this time we will also import the `FeatureLayerCollection` class from the `arcgis.features` module. This class will allow us to update our existing feature layer:

```
import pandas as pd
from arcgis.gis import GIS
from arcgis.features import FeatureLayerCollection
```

Connect to our Portal as follows:

```
gis = GIS("https://www.masteringageadmin.com/portal",
          "portaladmin", "password")
```

We know the item ID of our feature service and it doesn't change, so let's use that to search for the item. We then take that item and pass it into `FeatureLayerCollection.fromitem()` to create a feature layer collection:

```
flayer_item = gis.content.search('f017aa0e71e74987be6cdb48f192aff3')
flayer_collection = FeatureLayerCollection.fromitem(flayer_item[0])
```

Next, let's bring in the Excel workbook again; parse the sheet of interest into a `DataFrame`, then export that to a `CSV` file:

```
xls_file = pd.ExcelFile(r"C:\data\station-data.xlsx")
df = xls_file.parse('Sheet1')
df.to_csv(r"C:\data\station-data.csv")
```

Finally, we call the `overwrite` method of `FeatureLayerCollectionManager`, passing in our CSV as the input data file. This overwrites all the features in the hosted feature layer collection with the contents of the `CSV` file. Once that finishes, we are informed of our success:

```
flayer_collection.manager.overwrite(r"C:\data\station-data.csv")
{'success': True}
```

To see our changes in the feature service, create another blank map of Fayetteville, as follows:

```
map2 = gis.map("Fayetteville, AR")
map2
```

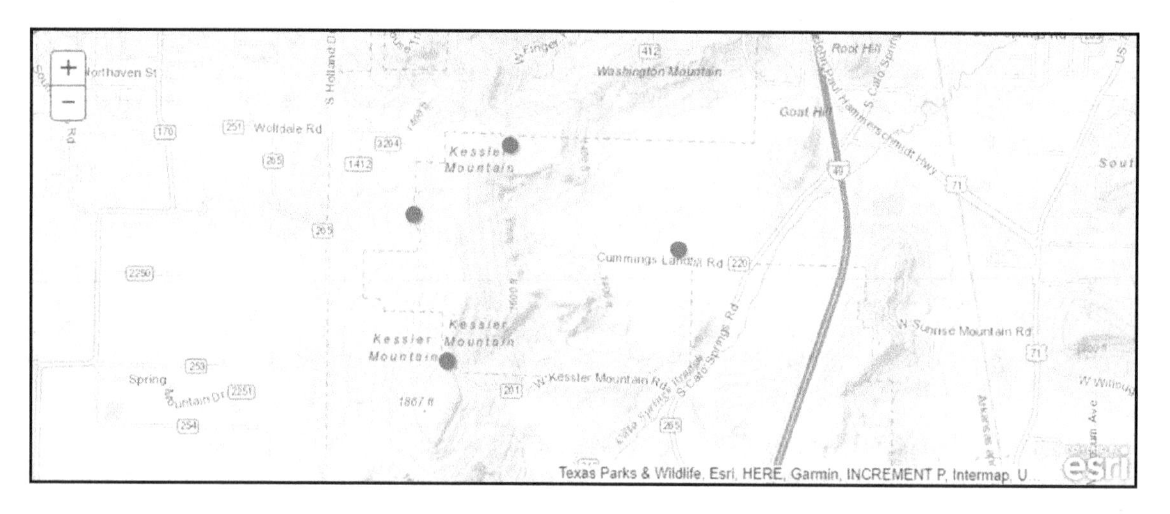

Add the updated feature layer item to map, and it will redraw, showing our new westernmost point location:

```
map2.add_layer(flayer_item)
```

Here, we updated an existing feature service with an Excel workbook in roughly 10 lines of code. So, in around 20 lines of code, we published a feature service from an Excel workbook and then updated the feature service with changes that we made to the workbook. This example truly shows the power and ease of use of the ArcGIS API for Python and how it abstracts away so much of the chatter involved in communicating over the REST API.

Summary

For anyone working with Python on the Esri platform, the ArcGIS API for Python is one of the most exciting things to come along in quite a few years. The API is well-structured, easy to set up, and even easier to use. By abstracting away much of the overhead typically involved in working with the REST API, the ArcGIS API for Python allows you to get more done with less code in less time. In this chapter, we looked at how to get set up to use the API. We also wrote code to change map service URLs and create a web map inventory. We scratched the surface of the pandas library, showing how to use DataFrames, one of the most prevalent data structures in pandas, to not only display data in a Jupyter Notebook but also how to use DataFrames as methods to move data into the GIS. Next, we looked at ways to interrogate users and groups in the GIS and methods to work with and manage them. Finally, we looked at how the ArcGIS API for Python lets us work with features and how easy it is to translate data from an Excel workbook to a feature service in the GIS. The ArcGIS API for Python is the future of Python on the Esri platform--become familiar with it. Next, in `Chapter 9`, *ArcGIS Enterprise Standards and Best Practices*, we will look at standards and best practices and how they can help your GIS enterprise system run smoothly and efficiently.

9

ArcGIS Enterprise Standards and Best Practices

Standards and best practices could easily be the second and third most important topics around ArcGIS Server, after security, of course. In fact, many of the standards and best practices we will discuss in the chapter impact security. Regardless of industry, standards and best practices are topics that no one typically wants to discuss or implement, but that everyone could benefit *greatly* from.

In this chapter, we will cover many topics, some of them briefly, some a bit more in depth. Regarding standards, we will talk about the following topics:

- Storage locations for your data and how to keep things tidy and neat
- Naming conventions for items such as the following:
    - Database connections
    - Folders
    - Services
    - Map document internals

Best practices will be a longer section, covering topics such as the following:

- Accounts and credentials
- Map documents and how to optimize settings for performance
- Map service settings and how to optimize settings for availability
- Print services' do's and don'ts
- How to make scripting easier
- Storage best practices

Why are standards and best practices needed?

Standards and best practices are often misinterpreted as rules and let's face it, no one liked rules as a child, and no one likes them as adults. Instead of rules, think of them as guidance; guidance that can help you and your team (if you have one) work more efficiently with less frustration and confusion. A system set up and maintained with standards and best practices over the years is a much simpler system to upgrade once the newest version of ArcGIS Enterprise comes out. Also, with standards and best practices in place, it is easier to bring new employees on board and get them familiar with the environment.

Standards

Before going any further, let's define just what exactly a standard is. A **standard** is a level of quality or attainment. When a standard is in place, it provides a target to shoot for or an expected way that something should be done. By enacting, having, and, most importantly, enforcing standards, you can make administration and management of your ArcGIS Enterprise environment all that much easier by providing consistency. Let's explore some ideas for standards and how they can affect your environment.

Storage locations

Where and how you store your data is important. No one likes to have to hunt for data, dig through directories, and no one especially likes the dreaded broken data source. With a little bit of planning and diligence, your data can be accessible and easy to get to (for those with access). The following are some things to keep in mind when storing data on disk:

- Never, ever store data in **My Documents** or anywhere under a user profile on Windows (for example, **C:\Users\ccooper\...**). Profiles get corrupted, people leave and their profiles get deleted, and, most importantly, the profile is only available to that user when they log in.
- Have a set location on an accessible file share or server drive (preferably a fixed non-operating system fixed drive) to store file-based data for ArcGIS Enterprise. Remember that real users aren't the only ones that will need access to this data, but service accounts may as well.

- Do not use spaces in folder or filenames. Now, in most cases, ArcGIS Enterprise seems to do just fine consuming data from folders or filenames with spaces. However, one area that this can be problematic still is in scripting.
- Keep folder and filenames as short, yet meaningful, as possible.

Naming conventions

Utilizing standardized naming conventions for all items in your GIS system is an easy and effective organizational tool. By having set naming standards and patterns throughout your system, it will be easier to locate items and establish relationships between them. Let's look at some ways to name things--the key takeaway from many of these standard suggestions is to keep things consistent.

Enterprise database connections

Name the enterprise database connections according to the username, database name, and server name. For example, if the SQL user `webeditor` is connecting the `GISPROD` database on the `GISDB` server, the SDE connection filename would look like the following:

```
webeditor@GISPROD@GISDB.sde
```

This naming scheme provides for uniform identification of which user is connecting to which database on which server.

Operating system-level directories and files

When it comes to naming folders for storing data on disk, developing a standard and using it consistently is paramount. Some simple rules to start with are as follows:

- No spaces in directory or file names. No exceptions. I know, it's 2017 and we aren't running Windows NT, but spaces in directory and file names can still cause problems.
- Keep names short, yet descriptive. For example, if you need folders for both ArcGIS Server connection files and Enterprise geodatabase files, you can name them **ags_connections** and **db_connections**, but you can also just have a **connections** folder with the **ags** and **db** folders underneath connections. Same for files; keep them as descriptive as possible, but only as long as it is necessary. If you find yourself having to have long file names, you may need to rethink your directory structure and add subdirectories to group like files together.

- Either capitalize the first letter of words or use all lowercase. Pick one and stick with it. This sounds easy in theory, but it's very easy to get careless here. For example, you could do something like the following:
    - `County_Maps`
    - `county_maps`
    - `CountyMaps`
    - `countymaps` (not really recommended, for readability)
- What you don't want to do is something such as this:
    - `County_maps`
    - `Countymaps`
    - `countyMaps`
    - `county_Maps`
- If you must use a separator between words, use an underscore (_).

Services and their sources

Developing and applying a clean, consistent, and standardized nomenclature as early as possible for your ArcGIS Server services will lead to a tidy environment that is easy to navigate and maintain. I've done quite a few ArcGIS Server (now Enterprise) upgrades over the years, and one thing that can make the migration of services from the old environment to the new go smoothly is standardized ArcGIS Server folder and service names that match the operating system folder and source MXD files names exactly. Let's look at this a little closer.

For illustration, let's say an organization has administrative access to the ArcGIS Server enabled through the Web Adaptor, and two publishers in the organization can publish services from their PCs. They publish a service each:

- User A has an MXD stored in `C:\Users\usera\Documents\Projects\MXDs\CountyBasemap.mxd` on their PC with data sources in an enterprise geodatabase that is registered with the ArcGIS Server. They publish the service to the ArcGIS Server, putting the service in a folder called **Basemaps** and naming the service **Terrain**.
- User B has an MXD stored on their personal network drive (which only they have access to) at `Z:\My Projects\maps\temp\Untitled.mxd`. Their data source is also registered with the ArcGIS Server. They publish `Untitled.mxd` to ArcGIS Server, putting it in the root folder and naming the service Roads.

The preceding two publishing workflows are extremely flawed and will eventually cause problems.

If you are thinking there is nothing wrong with the preceding publishing workflows, then read on very carefully. In both cases, the publisher references a source MXD that is in a location that only they have access to. This means that no one else in the organization, in particular, their publisher counterparts, will be able to access that MXD to make changes to the service. Secondly, they both named the service different from the source MXD. This means the only way to find out what the source MXD is for the service and where it is located is to, either in ArcCatalog or ArcGIS Server Manager, look at the service properties.

Wouldn't it be nice to be able to look at a services' name and folder in ArcGIS Server and instantly know from those where you could find the source MXD? Well, this is easy to do. To make your services easily discoverable, follow these steps:

- Have the main storage location in a shared directory location that is only accessible by those needing access to it (publishers and admins). Let's say this location is on a file server named `fs1` on a share named `gis`, so the UNC path to the share would be `\\fs1\gis`. In the `gis` folder, create a folder named `services`. Think of the `services` folder as being the root folder of your ArcGIS Server instance.
- Design a subfolder plan for your service MXDs under the `services` folder. These folders will be replicated as folders under the root folder of ArcGIS Server. This is how you want your subfolders to also be structured at your ArcGIS Server REST endpoint. For example, I work in state and local government, so it's common for services to be stored in folders such as `Planning`, `Cadastral`, `Parcels`, `Water`, `Sewer`, and so on. This folder structure can only be one directory deep under the root services folder, as ArcGIS Server only allows folders to one deep under the root folder.
- Name your MXDs exactly what you want your services to be named. If you want your parcels service to be named `Parcels`, then name the source MXD `Parcels.mxd`. If you want your roads service to be named `Transportation`, then name the source MXD `Transportation.mxd`, not `Roads.mxd` or `Streets.mxd`.

Now that we've laid the foundation for this plan, let's execute it by creating a folder structure with MXDs like the following:

```
\\fs1\gis\services
   Administration
      AdministrativeAreas.mxd
   Basemaps
      City.mxd
      County.mxd
   Boundaries
      CityLimits.mxd
   Cadastral
      Cadastral.mxd
   Parcels
      Parcels.mxd
   Sewer
      Sewer.mxd
   Transportation
      Roads.mxd
      Signs.mxd
      TrafficSignals.mxd
   Water
      Water.mxd
```

Nice and tidy, isn't it? The MXDs are organized, well-named, and accessible in the well-known location by only those who need access. Any organizational publisher with access to the **gis** share can make changes to the service MXD and republish a service.

Now, for the ArcGIS Server services, when publishing any of the MXDs in the preceding directory structure, the service needs to reside in an ArcGIS Server service folder that is named the same as its MXD directory. For the preceding MXDs published to the ArcGIS Server, the structure will look like this:

```
ArcGIS Server root directory
  Administration
     AdministrativeAreas
  Basemaps
     City
     County
  Boundaries
     CityLimits
  Cadastral
     Cadastral
  Parcels
     Parcels
  Sewer
     Sewer
```

```
Transportation
   Roads
   Signs
   TrafficSignals
Water
   Water
```

See how the services and their folders are identical in structure to the MXDs and their folders? Since we know the root storage location of all service MXDs (**\\fs1\gis\services**), determining where the MXD for any service resides is easy. The MXD for the **TrafficSignals** service is at **\\fs1\gis\services\Transportaiton\TrafficSignals**. Having a structure such as this makes it incredibly easy for both new and existing staffers to find their way around the system.

The preceding organization system works for services of all types, not just map services. Use the same system for image services, geoprocessing services, and so on.

Map service MXD standards

There are several naming standards that can be utilized in map service MXDs as well. First, give your layers names in the **Table of Contents** that you would want your end users to see; don't accept the default (typically the layer name) that gets applied when you add the data to the map. With enterprise geodatabase layers, this is quite often the fully qualified layer name, such as the following:

This layer would typically get named `Manholes` in the **Table Of Contents**. To set the alias for a feature class at the geodatabase level, go to the feature class properties by right-clicking the feature class in the **Catalog** window and selecting **Properties**. On the **General** tab, enter the **Alias** you would like for the feature class and click the **OK** button:

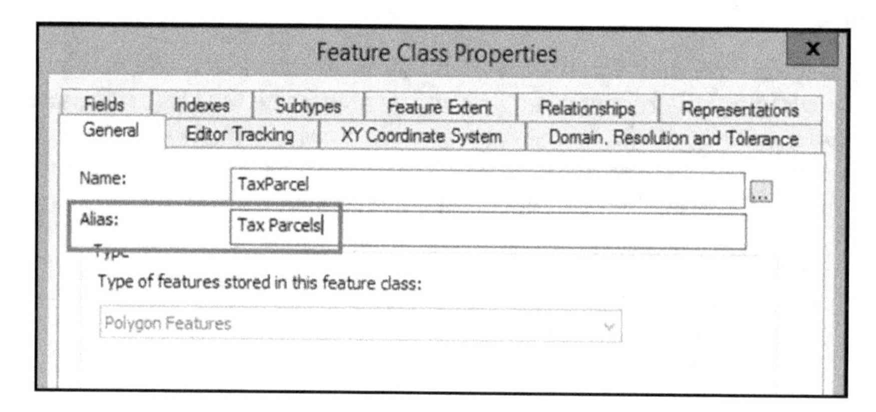

Setting aliases for feature class (and other geodatabase objects) names in the geodatabase will ensure that every time that object is added to an MXD, the alias will be used in the **Table Of Contents**, and not the name.

Another naming item of concern in service MXDs is field aliases. Setting proper field aliases will present your end users with field names that are human-friendly and will make sense to them. Many field names make perfect sense to someone who designed the database schema or someone who works with the data every day; however, to the average end user, it might not make any sense at all. As with feature class name aliases, field aliases can be set at the database level for feature classes and tables.

Hiding fields in the data behind a map service is a common practice, but be careful which fields are hidden. To hide fields in a service, you turn them off in the source MXD; therefore, not making them available at all in the map service.

In the following screenshot, we are hiding the **Invert Elevation**, **Rim Elevation**, and **Cover Type** fields by turning them off in the **Fields** tab of the **Layer Properties**:

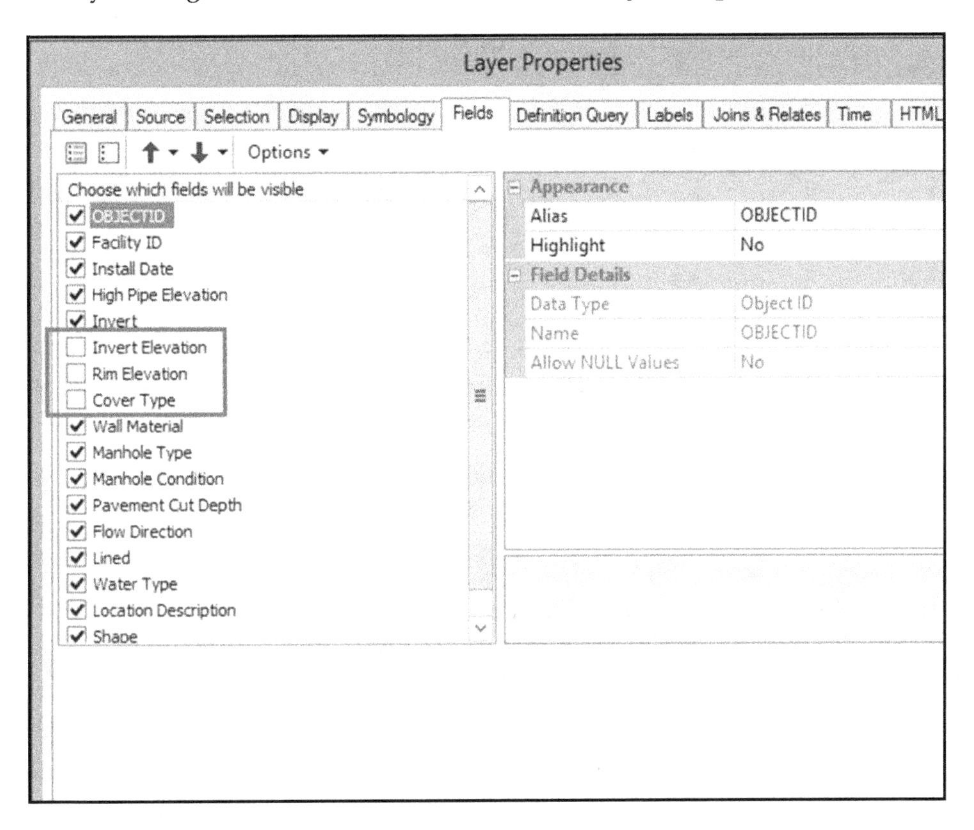

Turning off fields is fine, and quite often necessary to protect sensitive data. However, turning of fields such as **OBJECTID** and geometry fields can have detrimental impacts on some web mapping applications. In some cases, **OBJECTID** fields could be needed for querying data, and geometry fields (**Shape**) are needed to return geometries from searches or queries. Keep these points in mind when using layers searches and queries.

Best practices

Best practices can be defined as professional procedures that are accepted or prescribed as being correct or most effective. We've talked about best practices throughout this book without even knowing it. Let's cover some specific best practices that can help your ArcGIS Enterprise system run smoothly and efficiently.

Credentials

We covered password strength at length in `Chapter 6`, *Security*. Let's talk more about service accounts and some best practices around them.

Service accounts

ArcGIS Server, Portal for ArcGIS, and Data Store for ArcGIS all run as Windows services, but what exactly is a service? A service is a program that runs in the background and executes with no user interaction. Most services are typically configured to start automatically with Windows. A service account is a Windows user account that exists solely to provide a security context for a Windows service and determines the service's ability to access local and network resources. In other words, the service account that a service runs under controls what the service can and cannot access.

Esri recommends using domain-level accounts as the ArcGIS Server, Portal, and Data Store service accounts for production systems. With a domain user account, the service's actions are limited by the access rights and permissions associated with the account. In a site with multiple GIS servers, the ArcGIS Server service on each machine can run under the same domain service account. If your ArcGIS Server configuration store is located on a network share, you are required to use a domain account, as a local service account will not be able to access resources outside of the local machine.

Since service accounts are only to be used to run services, naming them accordingly is beneficial. A typical practice is to prepend or append `svc` to the account name. An ArcGIS Server service account may be named `svc_arcgis`, and a Portal account, `svc_portal`.

Service accounts should always be granted access to necessary resources and no more. For more information on resources that the ArcGIS Server account requires access to, search the ArcGIS Enterprise online documentation for **the arcgis server account**. Keep in mind that in most instances, ArcGIS Server, during the installation process, will grant the ArcGIS Server account access to any required resources. The **Configure ArcGIS Server Account** tool performs the same task.

Map documents

I've seen quite a few map services over the years, and there are a few golden rules you do not want to break.

Scale dependencies are one of the most important and easily implemented performance and usability settings that can be implemented on a map service. Scale dependencies are rules set on layers in the map document that determine at what scales layers will and will not draw. Do users need to see parcels at a 100,000 scale? Absolutely not, but they do need to see county boundaries, for instance. At a 24,000 scale, do they need to see address points? Probably not, but they do need to be able to see parcels. Drawing unnecessary layers, such as address points, at the wrong scale can be a huge performance hit that will do nothing but clutter up your map and aggravate your end users. Carefully plan and set your scale dependencies.

For ideas on proper scale dependencies, go to `http://solutions.arcgis.com/` and look for templates specific to your industry. These templates are a great starting point for ideas on how to configure many aspects of your map services, not just scale dependencies.

Just as with layers, labels can be set to draw at only certain scales. Keep this in mind when designing your services, only draw labels when they need to be there. Also keep in mind that with label classes, each class can draw at its own scale ranges, allowing you to perhaps label more as the user zooms in on a layer, for example.

For dynamic map services, it is best to keep symbology and symbols as simple as possible. Overly complex symbols can impact draw times and clutter up your maps.

Database connections

Regardless of whether you are on a production server or your personal PC, when you create a database connection in ArcCatalog or ArcMap, the `.sde` connection file gets stored in the user profile of the currently logged on Windows user. In my case, for example, connections in the **Database Connections** section of ArcCatalog are stored at **C:\Users\Administrator\AppData\Roaming\ESRI\Desktop10.5\ArcCatalog**. Now, while this is a major annoyance of mine, I fully understand why this is like this by design; the software needs a well-known location to be able to store user content, and it has control over its own directories in the user profile, a well-known location. That said, do not store enterprise database connection files for any MXD connections that will get published to ArcGIS Server in the default user profile location. We talked earlier in this chapter about the naming of enterprise database connection files, but where should they be stored? When we discussed service MXD storage locations earlier, it was recommended to store MXDs in a commonly accessible location, such as a file share, something such as `\\fs1\gis\services`.

For connections, create a folder on the same share, named `connections`, then in that folder, create another folder named `db`. Connect to this folder in ArcCatalog and copy (properly named) `.sde` connection files there. Now, users with access to this directory can utilize these connection files for their publication services.

Storing the credentials in `.sde` connection files is a common practice and is completely acceptable under almost all circumstances. However, be very cautious to not store the credentials on connections with elevated privileges, such as `sa` (sysadmin), `sde`, or any member of `dbo`. As these accounts have elevated privileges, if an unauthorized user was to gain access to these connections (more often by accident than intent), they could potentially do major damage to your system, as they could have the rights to alter and/or delete objects.

ArcGIS Server

As you know by now, ArcGIS Server is a crucial piece of the ArcGIS Enterprise system. That said, there are several best practices that can be utilized to help ease the burdens of administration and help your system perform smoothly and efficiently.

Registered data sources

We discussed registering data sources in `Chapter 3`, *Publishing Content*. As an administrator, you have control over what data sources are registered with the ArcGIS Server and if publishers can copy content over to the server during the service publication process (refer to `Chapter 4`, *ArcGIS Server Administration* and the `blockDataCopy` parameter). Armed with these tools, you can effectively lock down the data sources that your publishers can utilize in their map service MXDs, forcing them to only use approved data stores. Keep in mind that enterprise geodatabases are not the only data stores that can be configured; you can also register data folders and databases. By communicating these approved, known locations with your publishers, you not only keep your environment clean and tidy for yourself, but you are also eliminating permissions problems for your publishers.

Print services

Actual map products have always been a core part of any GIS system. Users may not physically print as many maps as they did 10 years ago, but the need for digital map products in the form of PDFs or other image formats has increased. Web cartography is a topic that could constitute an entire book all on its own, so we won't go in depth here.

Instead, let's discuss some general dos and don'ts when it comes to printing templates:

- Keep the number of layers turned on by default in your applications as low as possible. Use scale dependencies to help with this. For example, if a user prints a map at 100,000 scale, you might want Public Land Survey System township lines visible. However, at 10,000 scales, township lines probably aren't necessary in general, but parcel lines are. This is just a hypothetical example and your use cases will vary. Know your users and find out what layers they need to see; this will help keep their maps from being a cluttered mess.
- Include a north arrow. Yes, most of us in the geospatial industry know that usually, unless otherwise stated, north is up on a map. However, be aware that not everyone knows that. Also, there are instances where north arrows are required for regulatory purposes. Unless you have a specific reason for not including a north arrow on your print templates, put one on them; they take up very little space and it is simple to do.
- Include a dynamic scale bar. As with the north arrow, adding these in your templates is a painless task for you that can greatly benefit your users. Design it for readability and ensure that if it lies within the map window of the print template, it can be read no matter what basemap the map ends up having.
- Legends in web map print exports have been a topic of contention for many years; some people love legends on their print products and demand they are there, others are just fine without them. A common issue with legends in web application printouts is that, in most cases, maps are small, often letter-sized, and therefore have very limited real estate for a legend to fit in. Also, with all ArcGIS Server service types other than feature services, a legend will include all legend entries for all features regardless of the map extent. In other words, with feature services, the legend in a printed document will only include legend entries for features found in the printed map extent; whereas, if using a standard map service, all layers will be shown in the legend even if they are not present in the printed map extent. When it comes to legends in web application printouts, know your users and what they are looking for and experiment heavily with exports to ensure that legends act, function, and look appropriately.

Tuning services

Like many topics we have covered, tuning services is one that could constitute its own book. Tuning is typically concerned with performance and availability, where settings for availability are configurable in ArcGIS Server service settings and many performance-related issues can be resolved at the map-document level for map services.

Availability

Availability settings, such as pooling and timeouts, are set at the service level in the **Service Editor** during publishing, but can also be changed any time after a service has been published. With pooling, multiple connections can feed on a single pool of instances. A connection uses the instance for a period of time, gets a result, and then the instance is free to execute another request from any user. Depending on the request, execution can take milliseconds (a map pan or zoom) to minutes (a network trace or geoprocessing task). With pooling, you set a minimum number of instances that should start when the service itself starts. Likewise, you can set the maximum number of instances that can be available with pooling settings; not that these are per machine, so, if you have two GIS servers and specify a minimum of one instance, then 1 will be spun up on each GIS server for a total of 2. For a service that only gets viewed, panned, zoomed, and possibly lightly queried on, you could go so far as to set the minimum instance number to 0 and maximum to 1. Panning, zooming, and light querying are fast operations, so when requested, an instance could be spun up, used quickly, and then released for future requests. On the contrary, if you have a service that is used by field crews for emergency network tracing operations, you want that service to be instantly available to multiple users at the same time. In this situation, you may set your minimum number of instances to 2 and your maximum to 4. The following screenshot shows the pooling settings for a map service:

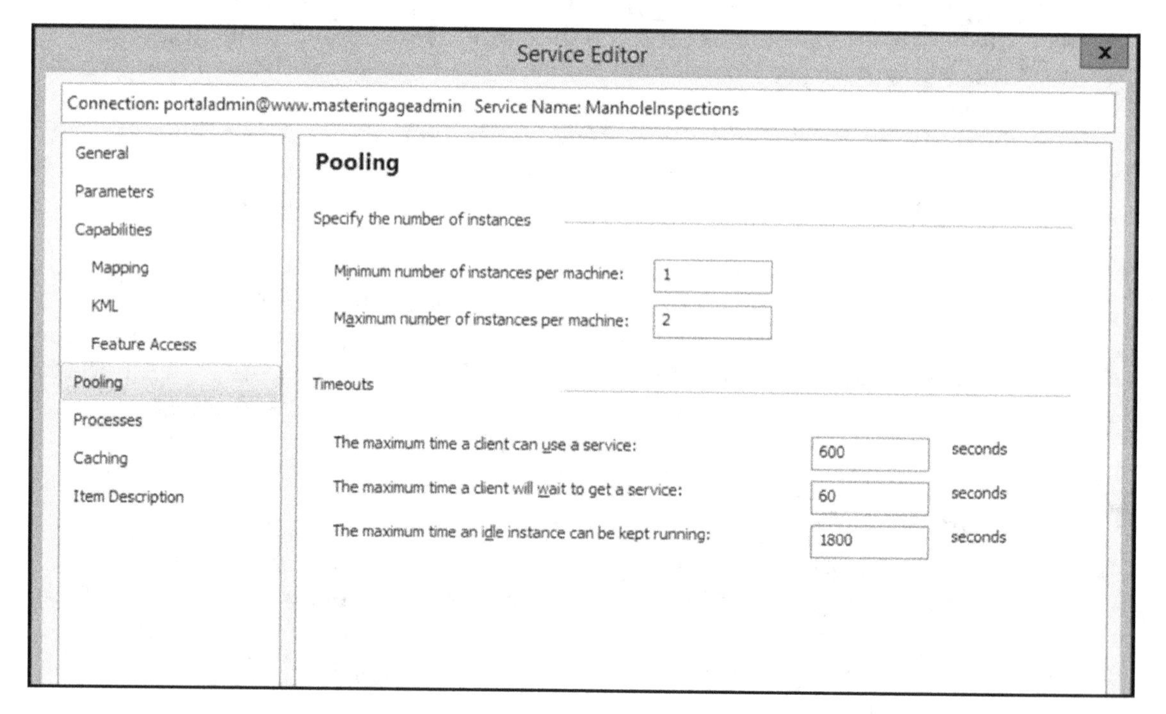

Timeouts are a second availability setting on the **Pooling** tab of the **Service Editor**. With timeouts, there are three different settings, as follows:

- **The maximum time a client can use a service**: This is the maximum amount of time a client can get a reference to a service instance and use it before it is automatically released and the client loses its reference to it. Referring to our earlier examples, panning and zooming a parcels service is quick work; we could lower this value to 10 seconds for parcels. For our network tracing service, those operations are longer, so we may want to set that at 60 seconds or longer if we know those processes are time intensive.
- **The maximum time a client will wait to get a service**: Wait time is the amount of time it takes between a client requesting a service and the client getting the service. When all instances (set by the maximum number available in pooling) of a service are in use, a client requesting the service gets put into a queue to wait. If the wait time exceeds the maximum wait time for the service, a timeout occurs. For services with quick operations, set this value lower than those that are time intensive, such as our network tracing example.
- **The maximum time an idle instance can be kept running**: When a client finishes with a service, it is kept running on the server until it is requested by another client. While it is still running, it is consuming memory. The default value here is 30 minutes (1,800 seconds), and for simple services such as parcels, this number can be reduced substantially to a matter of minutes at most. Our crucial networking service, however, is different. We want that service immediately available, so we will leave it at 30 minutes or maybe even bump it up to an hour.

Performance

We discussed a few best practices earlier for map service MXDs; now, let's examine some more settings and configurations that can affect map service performance:

- **Joins and relates**. This is a big one; many times, enterprise data is kept in systems outside of GIS. It's tempting to add that data and join or relate it based on a primary field. While this may be fine for desktop applications and daily use, it doesn't scale well and tends to perform poorly when used on larger datasets. In these cases, it is often best to develop a script to be run as a scheduled task that will join the enterprise data onto the GIS data which can then be consumed by your web services. For more information, search the ArcGIS Enterprise online documentation for the `AddJoin_management` geoprocessing method.

- **Detailed versus generalized datasets**: You have a map view that encompasses the entire world and a beautiful highly-detailed country's polygon layer. The problem is, when zoomed out to the entire world, those highly-detailed polygons take forever to draw. In an instance like this, use a simplified version of your countries with much fewer vertices along with scale dependencies to draw the generalized layer when zoomed out and detailed layer when zoomed in.
- **Queries**: For map layers that will be queried from web applications, keep attribute and spatial indexes up-to-date. Attribute indexes will speed up attribute queries, whereas spatial indexes will speed up spatial queries. The faster your application can perform queries and present results back to the user, the better. For datasets that get updated regularly, updating indexes can be challenging. Consider using a scheduled Python script and `arcpy.RemoveIndex_management()` along with `arcpy.AddIndex_management()` to update existing attribute indexes.
- **File geodatabases for static data**: In instances where you have static data that either never gets updated or maybe only gets updated once or twice a year, you could see performance gains by storing that data in a file geodatabase. Remember, file geodatabases are files on the file system, so they can be read much faster than an RDBMS can. Want to get even better performance? Store the file geodatabase on a **solid-state drive** (**SSD**), which will provide even better I/O than a traditional magnetic spinning hard drive.

Portal for ArcGIS

When it comes to Portal, organization and housekeeping are crucial to keeping a well-maintained environment. It is important to always keep in mind that with Portal, just as with ArcGIS Online, every item has an Item ID, and IDs are how items are referenced. This is how and why you can change ownership of items and move items from one folder to another without impacting any other items that might be hydrated by the moved item. Use these capabilities to your advantage by organizing items into folders and changing names when necessary to keep your content organized.

Ownership of items is a common hang up for those new to Portal administration. To publish items to Portal, use a publisher-level account, not an admin-level (principle of least privileges, remember?). Furthermore, restrict the ability to share to the public to just a few administrators. When a content creator is ready to have an item published to the public, an administrator can take ownership of the item and perform the sharing, thus preventing accidental release of information and controlling sharing to the outside.

Keep your production Portal items owned by one single account so they are viewable in one place. As a security practice, only log into Portal with an administrative-level account when you truly need to do administrative tasks.

Python scripting

Pep 8 (`https://www.python.org/dev/peps/pep-0008/`) provides exhaustive and detailed standards, best practices, and coding conventions to be used in your Python code. There are even online tools that can be used to analyze your code for common errors (or *lint* your code, as it is referred to) and report back PEP 8 violations; `http://pep8online.com/`, for example. Instead of covering coding conventions that are already covered in PEP 8, let's focus on ways to improve usability and durability of your scripts.

Script storage

Just as with ArcGIS Server map service documents, it is important to have a standard location in your environment to store your scripts, such as `\\fs1\gis\scripts`. This provides a location where scripts can be accessed from any one of your GIS servers on the network.

Connection files

Earlier in this chapter, we talked about database connection files and how to best name and store them. Those same principles apply here:

- Use standard naming conventions for your SDE connection files, such as `webeditor@GISPROD@GISDB.sde`
- Store connection files in a standard, yet secured, well-known location, such as directly beneath your scripts directory; for example `\\fs1\gis\scripts\connections`

Logging

Logging is a crucial aspect of any good script, especially one that gets run on a regular basis. Having a log to fall back on in the event of a failure of a script can mean the difference in figuring out in two minutes what went wrong or having to manually run the script again, hope that the error occurs again, and wait for the error to happen--not an ideal method of troubleshooting.

For Python scripts, the Python standard library ships with the logging module, which is typically sufficient for most logging tasks. Other logging libraries exist, and the `daiquiri` module (`http://daiquiri.readthedocs.io/en/latest/`) is a great library that enables simple logging configuration setup, handlers, and formatters. The `daiquiri` module makes it incredibly painless to configure multiple loggers in your scripts. It can be installed via `pip` by the following command:

```
pip install daiquiri
```

Let's look at how we may implement logging with `daquiri`. First, we will import the necessary modules:

```
import daiquiri
import datetime
import logging
```

The next step is to set up our logging through the `daiquiri.setup` function. Here, we can specify multiple different outputs, each with their own formatter. Our first output is a `TimedRotatingFile` output, which is a rotating log file output triggered by a fixed interval. In our setup here, we use an instance of `datetime.timedelta` to log to the current log file for 7 days. After that, the log file gets archived (in the same directory next to the current log file, with a name such as `log.txt1`) and a new `log.txt` gets created and logged to. A total of 10 backup log files are kept on disk, and when it comes time for the 11th archived log file to be created, the oldest archive file is deleted. For the `TimedRotatingFile` output, we also pass in a path to the log file and set up a `daiquiri.formattter.TEXT_FORMATTER` formatter that will print out the time, file name, error/message line number, log level, and log message:

```
daiquiri.setup(
    outputs=(
      daiquiri.output.TimedRotatingFile(
          r""C:\scripts\log.txt"",
          formatter=daiquiri.formatter.TEXT_FORMATTER(
              fmt=""%(asctime)s - %(filename)s - %(lineno)d - "" \
                  ""%(levelname)s - %(message)s""
          ),
          interval=datetime.timedelta(days=7), backup_count=10
```

Our second output goes to a generic `Stream` that will get printed out in a console if we are running the script at a command prompt or in the console output of an IDE while running there or debugging. Here, we specify the output to go to `sys.stdout`, the interpreter's standard output. We also set up a `daiquiri.formatter.ColorFormatter`, which, if your console supports it, colorizes the log output.

Finally, for all our outputs, we set the logging default level to `INFO`:

```
        ),
        daiquiri.output.Stream(
            sys.stdout,
            formatter=daiquiri.formatter.ColorFormatter(
                fmt=""%(asctime)s - %(filename)s - %(lineno)d - "" \
                    ""%(levelname)s - %(message)s""
            )
        ),
    ),
    level=logging.INFO)
```

To use our `logger` object, we will get an instance of it by calling `daiquiri.getLogger()`:

```
    logger = daiquiri.getLogger(__name__)
```

Now that we have our logger instance, we can freely use it to simply log info, warning, and error messages to both standard output at the console and to our log file simultaneously:

```
    logger.info("Info message")
    logger.warning("Warning message")
    logger.error("Oops, something went really wrong")
```

Setting up a `logger` in Python is incredibly easy to do and provides great benefits for keeping track of processes, seeing how efficiently they run, and troubleshooting when errors occur.

Scheduled tasks

Chances are, you will have the need to run something as a scheduled task on your servers, more than likely, a script. On Windows machines, Task Scheduler is the standard out-of-the-box way to run tasks at predefined times and intervals. Task Scheduler has been around in some form since Windows 95; unfortunately, it really hasn't changed much since then either. Task Scheduler is rudimentary, but, for most cases, when configured properly, it gets the job done. Let's look at some best practices that can make using Task Scheduler easy.

All scheduled tasks in Task Scheduler must run under a Windows account, which is set under the **Security options** section on the **General** tab when creating a task:

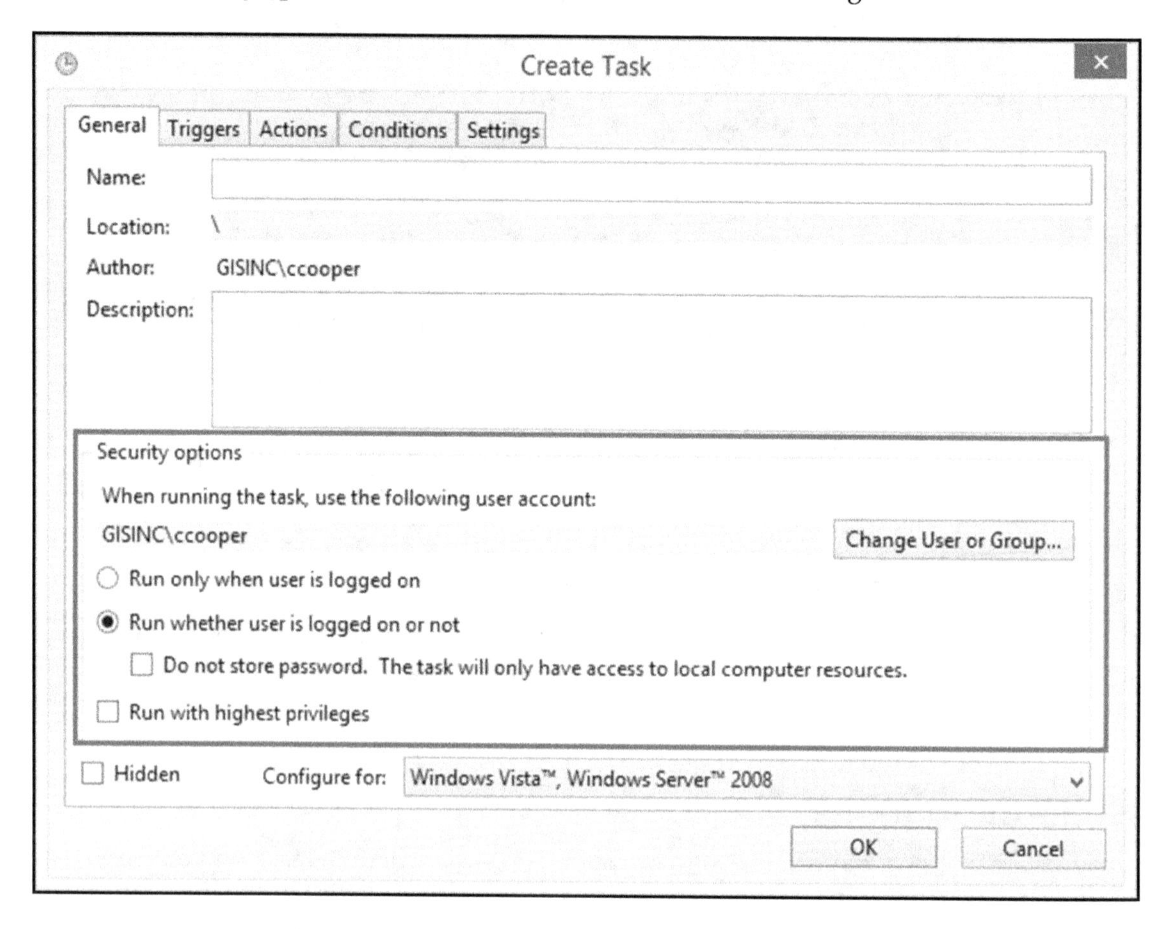

In the preceding task, my personal domain account is set as the default as I am the user that initiated the task creation. There are considerations to be aware of when selecting the account that a task runs under:

- **Access and permissions**: Does the account have access to all resources the job relies on to complete successfully? If the task is run under a local Windows account, it will not have access to network resources. If using a domain account, does that account have access to any required network resources? If the account used doesn't have access to required resources, it will fail.
- **Password expiration**: I've seen this too many times, a job fails and no one can figure out why. In looking at the settings for the task, a domain user account (typically the person who created the task) was used to run the task under, and, in most organizations, user account passwords eventually expire. I'll ask the person *Did you change your domain password recently?* The answer is usually *Yes*. We change the password on the scheduled task, and it runs as intended. Also, if someone leaves the organization, their domain account is more than likely going to be disabled. When any of these events occur, the task will not be able to execute as the credentials being used are no longer valid.

To avoid issues with access and password expiration, use a Windows service account to run your scheduled tasks. Service accounts can be set to have passwords that never expire and they can be explicitly granted privileges to any required network resources. Talk to your IT systems administrator about issuing you a service account; tell them what you are trying to accomplish and they will probably be accommodating to your needs.

Finally, when creating your scheduled task, name it properly and give it a detailed description on the **General** tab. Doing so will not only help others know what a task is and what it does, but it will help you two years down the road as well.

Storage

When it comes to file system storage, your two primary concerns should be security and storage quotas/limitations. In other words, what accounts can access what resources and how much disk space are your resources consuming (that is, are your drives filling up?).

Lock resource access down

Limiting access to resources in your GIS system goes back to the principle of least privileges that we discussed in `Chapter 6`, *Security*. For example, if your ArcGIS Server configuration store is on a network share, ensure that only the ArcGIS Server account and any GIS administrator accounts have access to that location. One wrong (even accidental) move by a user and a directory in the config store could be deleted, potentially bringing down your entire system. Likewise, we discussed earlier in this chapter having a well-known location for your service map documents. A location such as this would only need to be accessible by GIS publishers and administrators.

Measures such as these are not only intended to keep malicious users from accessing secured resources but are possibly even more important to have in place to keep accidental access and subsequent unintentional harm from occurring within your system.

Moving the IIS web root

On your web server, one measure that can be taken to ensure that the operating system drive doesn't fill up accidentally with IIS log files is to move the storage location of the IIS **inetpub** directory, commonly referred to as the web root, onto a secondary data drive, if one is available. To accomplish this, Microsoft provides a configurable script available at `https://support.microsoft.com/en-us/help/2752331/guidance-for-relocation-of-iis-7-0-and-iis-7-5-content-directories`.

Even though Microsoft provides the preceding-referenced script, they do not support it. Also note that after moving the **inetpub** directory, Windows servicing events require that the original location (typically in **C:\inetpub**) be kept intact and not removed.

Storing ArcGIS Enterprise logs off the operating system drive

In `Chapter 4`, *ArcGIS Server Administration*, we talked about moving the ArcGIS Server logs off the operating system drive. This same best practice can be applied to Portal logs as well. Keep in mind that you don't want to store logs from any piece of ArcGIS Enterprise on a network share, so moving the logs is only viable if you have a secondary fixed attached drive on the server.

As shown in the following screenshot, to change the location of your Portal logs directory, log in to the Portal Admin as an administrator and go to **Logs** | **Settings** | **Edit**. Change **Log Location** accordingly:

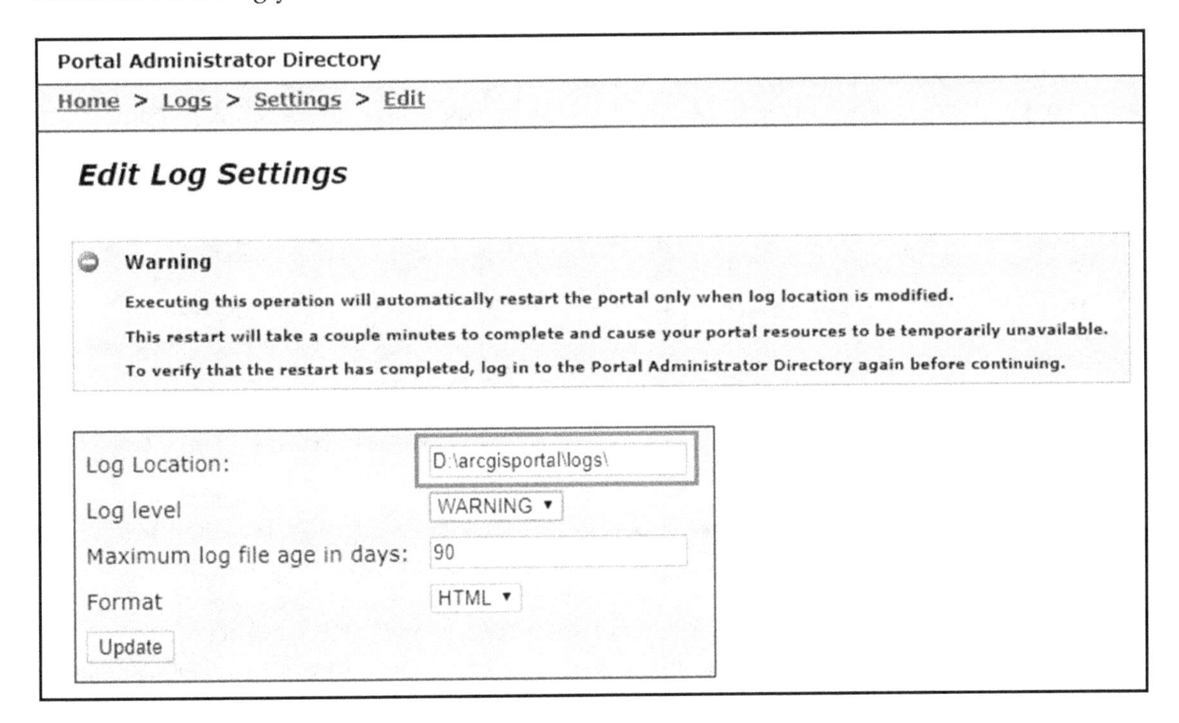

Documentation

This is as good a place as any to discuss the topic of documentation, because, just like with standards and best practices, most people (except me) don't like to write documentation. The biggest obstacle to getting documentation written is almost always time--no one has the time to dedicate to writing documentation because there is real work to get done. Trust me, I get it. However, by not documenting your processes and workflows, you are only hurting yourself in the long run.

The bus factor

Ever had someone tell you something *in case they get hit by a bus?* They want you to have some knowledge that they possess in case they are no longer around. The bus factor revolves around this principle and is a measure of the risk resulting from work that is undocumented, not shared, encrypted, obfuscated, or just plain incomprehensible to others. Let's say you have an enterprise GIS team of five people. Of those five people, two know the credentials and have access to the production servers. If those two people win the Powerball Lottery and disappear, no one will be able to access production to fix it when it breaks, so your team's bus factor is 2. Ultimately, the higher the bus factor, the better. With your five-person team, the highest and the best your bus factor can get is 5.

One of the key ways to increase your bus factor is to document all processes and communicate and cross-train within your team. Documentation doesn't have to be anything fancy and doesn't require any special tools (think Word document). Generating workflow diagrams doesn't have to be hard or require fancy tooling either; a basic graphics package or even PowerPoint can work just fine. However, if you do want to consider some tools for creating documentation, here are a couple:

- **Markdown + Pandoc**: Markdown (`https://daringfireball.net/projects/markdown/syntax`) is a lightweight markup language that is everywhere in the software and development communities. If you've ever read a README file on GitHub, chances are you were reading Markdown that was converted to HTML. Markdown is simple and easy to learn. There are even Markdown editors and online converters available. When used in conjunction with Pandoc (`https://pandoc.org/`), a universal document converter, Markdown can easily be converted to PDF or Word documents.
- **Sphinx**: Sphinx (`http://www.sphinx-doc.org`) is a Python module originally developed for the Python online documentation (`https://docs.python.org/3/`). Sphinx takes input written in reStructuredText markup and can output into a wide variety of formats such as HTML, plain text, and LaTeX (for PDF output).

Summary

Throughout this book, we have discussed ArcGIS Enterprise at length, and how to install, configure, and secure it. In this chapter, we discussed standards and best practices, which are often seen as rules, and no one likes to be told what to do. The key to working with standards and best practices is to think of them as guidance brought to you by those before you who have learned things the hard way. This is the knowledge that is being passed down to you to make your system run smoother and more efficiently. Many standards are common sense, especially those related to naming; it's enacting and sticking to the standard that requires work. Best practices are the same way; enacting and sticking with them is the only way to ensure that they work. Finally, the importance of documenting your system and workflows cannot be overstated enough; documentation is not only for others, but it is also for yourself, intended to help you remember why something was set up the way it was, for example. In the next chapter, the final chapter of this book, we will look at troubleshooting ArcGIS Enterprise issues when they arrive. Many of the standards and best practices suggested here in this chapter, when utilized, can make troubleshooting issues much easier.

10
Troubleshooting ArcGIS Enterprise Issues and Errors

Sooner or later, issues are going to come up in your ArcGIS Enterprise system. Knowing how to effectively and efficiently troubleshoot errors and issues and put out fires when they arise is an important and necessary skill of an ArcGIS Enterprise administrator. Effective troubleshooting is a fine art that can truly only be mastered with years of experience. Always keep in mind that Enterprise GIS systems are complex and intertwined, often reaching out beyond the borders of the GIS system itself. Knowing what to look for where in the system will go a long way in helping resolve issues. In this chapter, you will learn how to roll up your sleeves, dig in, and methodically, calmly, and patiently determine what is causing the issue at hand and how to best fix it.

In this, our final chapter, we will focus on the following topics:

- Using logs to help troubleshoot issues
- Different issues you may come across during installation and configuration
- Permissions issues
- Troubleshooting scripts
- Tools available to help with troubleshooting and testing

Keeping your cool

One of the first things you must do while working on an issue is to remain calm. I fully realize that in many cases, this is much easier said than done. However, remaining calm and collected will help you focus on the task at hand. Panic, or anger for that matter, is only going to clutter your brain and cloud your judgment, decreasing your effectiveness to properly work on the issue. Likewise, limit distractions; if the issue is important enough, ignore all other emails, set your phone to silent, set your instant messenger status to do not disturb--do anything you can to improve your focus. Don't be afraid to tell (or ask politely, if the case may be) people to leave you alone as you are currently working an important issue. I work with a project manager who, when I am working on an issue on a project, will ask me "*Do you need anything from me or should I just leave you alone?*", knowing that I can work better if left to focus and I will reach out to him if I do indeed need anything from him. Remember that the key here is that you need to be able to focus on the issue to be able to resolve it as quickly and effectively as possible, and keeping your cool will help you do just that.

Gathering information

It's not working.

You've probably heard this a few times from users. Very rarely will a user send you a message regarding an issue that tells you what they are truly experiencing with an application, dataset, and so on. Keep in mind that they aren't as intimately familiar with all the aspects and functionalities of your system; they might only use one tool in one application, so when that one facet isn't working right for them, they see it as broken, and that's fine. Part of your job is to calmly and politely walk through their workflow with them and find out exactly what they were doing when things didn't go as they expected. Sometimes, this can be done over email; sometimes, it might be best to do it over a phone call; and, sometimes you might need to look over their shoulder to watch them reproduce the issue. However you do it, you need to translate *it's not working* into some actionable information that you can then use to start working on the issue. Some things to find out from the user reporting the issue are mentioned here:

- What exactly were they doing when the error/issue occurred? Get them to describe or show their workflow to you.
- If they are using a web application, what browser are they using and what version? Have they tried another browser?

- If they are working in a desktop application, what version?
- Were any error messages presented to them? If so, can they show you or tell you what the messages were?
- These are general questions, but they get the dialog started. Depending on what the user is experiencing, from here, the questions and conversation can go in a multitude of different directions, and it is up to you to guide that. Gathering this information of what is leading up to the issue will help you develop a plan to address and, hopefully, resolve it.

Using available resources

I've been in this line of work for quite a few years now and work on issues on regular basis. Very rarely do I come across an issue that, after a few minutes of Googling, returns no results. In other words, chances are, you aren't the first person to have this problem, so if you don't immediately know what the issue is or have a resolution, Google any error codes or messages to see what first comes up. With Esri error codes, you can usually search for something like `esri 001369`, where `001369` is the error code you are dealing with. Be careful here though, as you can riffle through dozens and dozens of help forum posts for certain errors (such as `001369` mentioned earlier) and get absolutely nowhere, as some error codes and messages can be ambiguous or even a red herring for something else that is actually going on. If you start to see lots of help forum posts on the error code or message, but the cases reported seem to be all over the place and there are no real solutions, step back and reassess. It's very easy to get sucked into reading these and it ends up being a complete waste of time as you get no resolution.

We've seen throughout this book how ArcGIS Enterprise relies on many aspects of IT that are quite often outside the purview of most GIS professionals (network shares, SSL certificates, permissions, domain accounts, and so on). A solid working relationship with your IT staff can be indispensable in a time of need, such as an outage. You need your IT staff, so get to know them; build a working relationship with your systems administrator, network engineer, database administrator, and others. Unfortunately, in too many organizations, the IT department is quite often forgotten about, so buy them a cup of coffee and take them out to lunch occasionally.

Using the logs

We have discussed this already, but I'll reiterate--the most important thing to remember about using ArcGIS Server and Portal for ArcGIS logs is to *remember to check them*. Now, just because an issue or error is being experienced doesn't necessarily mean that a `SEVERE` event will get logged, or even a warning for that matter. When this happens, change the log level to `VERBOSE` or perhaps even `DEBUG`, repeat the workflow that is causing the issue, and see if anything interesting gets logged that could be a precursor to the actual error you are trying to track down.

Always remember to change your log level back to your default setting after you change it for troubleshooting purposes.

In `Chapter 4`, *ArcGIS Server Administration*, and `Chapter 5`, *Portal for ArcGIS Administration*, we covered logging administration. Let's discuss using the logs a little more.

ArcGIS Server logs

The best way to view the ArcGIS Server logs is to log in to **ArcGIS Server Manager** as an administrator and go to **Logs** in the header menu. This takes you to the **View Log Messages** page where you can query the logs by level, time, source, and machine.

Let's look at the different options and features available in **View Log Messages**. In the following list, we will explain each one:

1. **Columns**: This brings up the **Manage Columns** window where you can select log parameters to view in the log listing. The defaults are **Level**, **Time**, **Message**, and **Source**. In a multi-machine ArcGIS Server site, **Machine** can be helpful in determining which machine in the site is throwing an error:

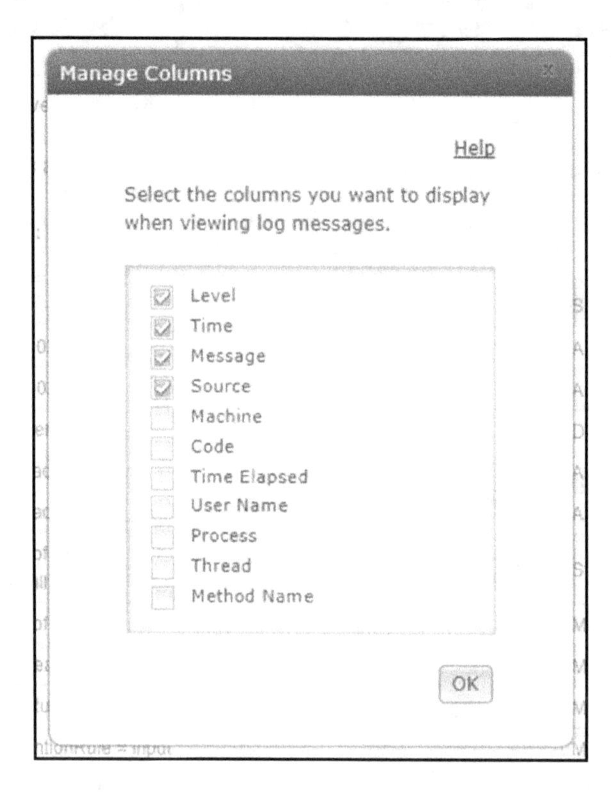

2. **Delete Logs**: This function allows you to delete all log messages that have been created. Be careful with this one, as once you delete the logs, you cannot recover them.
3. **Settings**: We discussed this in `Chapter 4`, *ArcGIS Server Administration*, but it is worth mentioning again:

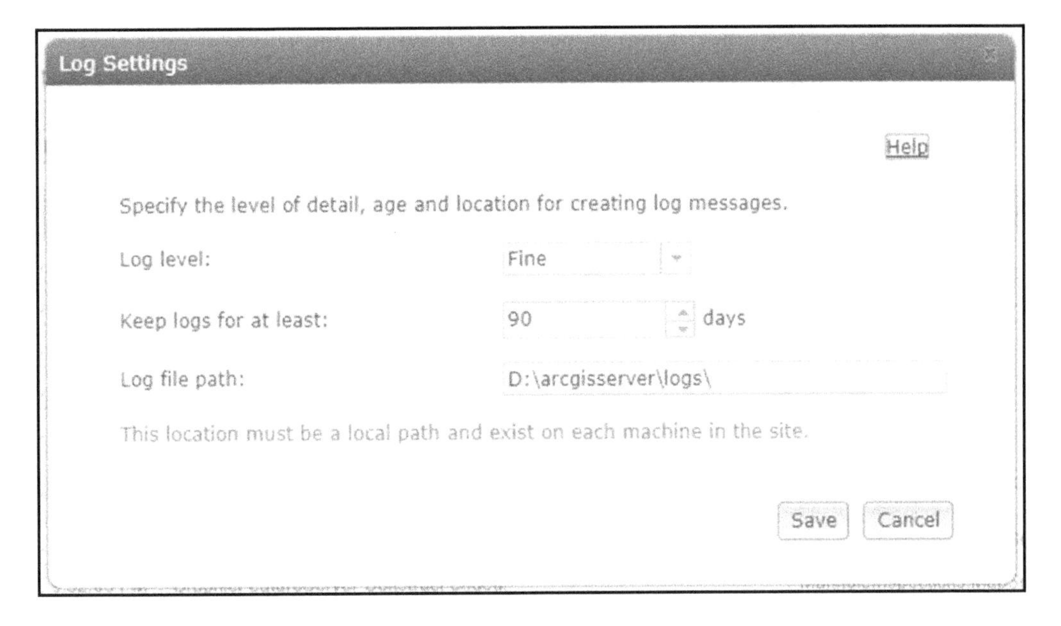

4. **Log level** defaults to **Warning**, which is recommended for most environments. To troubleshoot issues, you may wish to change this level to something such as **Fine** or **Verbose** to capture more information. We discussed log levels in detail in `Chapter 4`, *ArcGIS Server Administration*. **Keep logs for at least** determines how old your oldest log will be; anything older than this value will get purged automatically by ArcGIS Server. Set this parameter wisely, too much retention and you have gigabytes of logs, too little retention and you don't have enough history. **Log file path** sets the path to where the logs are stored in the file system. We discussed best practices for the logs directory in `Chapter 4`, *ArcGIS Server Administration*.
5. **Log Filter**: This dropdown starts the list of dropdowns for query parameters you can use to filter down the log messages you see. When selecting the log level to filter on, know that you will see the level selected and all levels above that in your results, except for **Severe**, as it is the highest of the levels. So, for example, selecting **Info** will return **Info**, **Warning**, and **Severe** messages.

5. **Age**: This determines the time for which you would like to see log messages. The default is **Last Hour**. Selecting **All** will show you all log messages in the current history as set by **Keep logs for at least** in the log **Settings**:

7. **Source**: This is a powerful filter, as it lets you hone in on specific sources or areas of interest. The list can get quite lengthy, but it essentially allows you to only view events related to the server framework, all services, individual services, and the different system and `Utilities` services. This can be especially helpful if you know a certain service is having issues. Select that service from the dropdown and you have eliminated a massive amount of clutter from your search. Search the ArcGIS Enterprise online documentation for **Using log filters to narrow down search results** for more information on the source types.
8. **Machine**: If you have multiple GIS Server machines, you can select to view messages from **All** machines or an individual machine.
9. **Log listing**: This is the area where your log messages are displayed. Each page of results displays 1,000 messages. If more than 1,000 results are returned from your query, the **Newer** and **Older** buttons on the lower right will let you page through the results. The headers are clickable for sorting here as well.

ArcGIS Server logs workflow

Now that we know our way around the ArcGIS Server logs interface, how can we use the logs to help us troubleshoot an issue? If you have a reported an incident and you know the time it occurred, you can first go to the logs in ArcGIS Server and perform a quick query to see if you see anything unusual within that time period. First, look for **Severe** errors. Also, if you know the issue is related to a particular service, select just that service in the source dropdown and see if there are any relevant messages. Still not seeing anything? It might be time to try and reproduce the problem. This is where gathering as much information as possible regarding the issue is so important (see the *Gathering information* section for more details).

If you can reproduce the issue, look again and see if anything was logged of interest. If nothing is being logged still, lower your log level to **Info** or **Fine**, reproduce the problem again, and see if you get any relevant messages in the logs.

Portal for ArcGIS logs

Remember from `Chapter 5`, *Portal for ArcGIS Administration*, that there is no place for viewing logs in Portal itself, but only in Portal Admin. To query and view the Portal logs, follow these steps:

1. Open **Portal Admin** (`https://<webadaptor>/portal/portaladmin`) and log in as an administrator.
2. Go to **Logs** | **Query**.

If you have just read through the preceding *ArcGIS Server logs* section, most of this will sound familiar, with a few exceptions. Let's break down the query parameters available to us in Portal logs and define them in the list below:

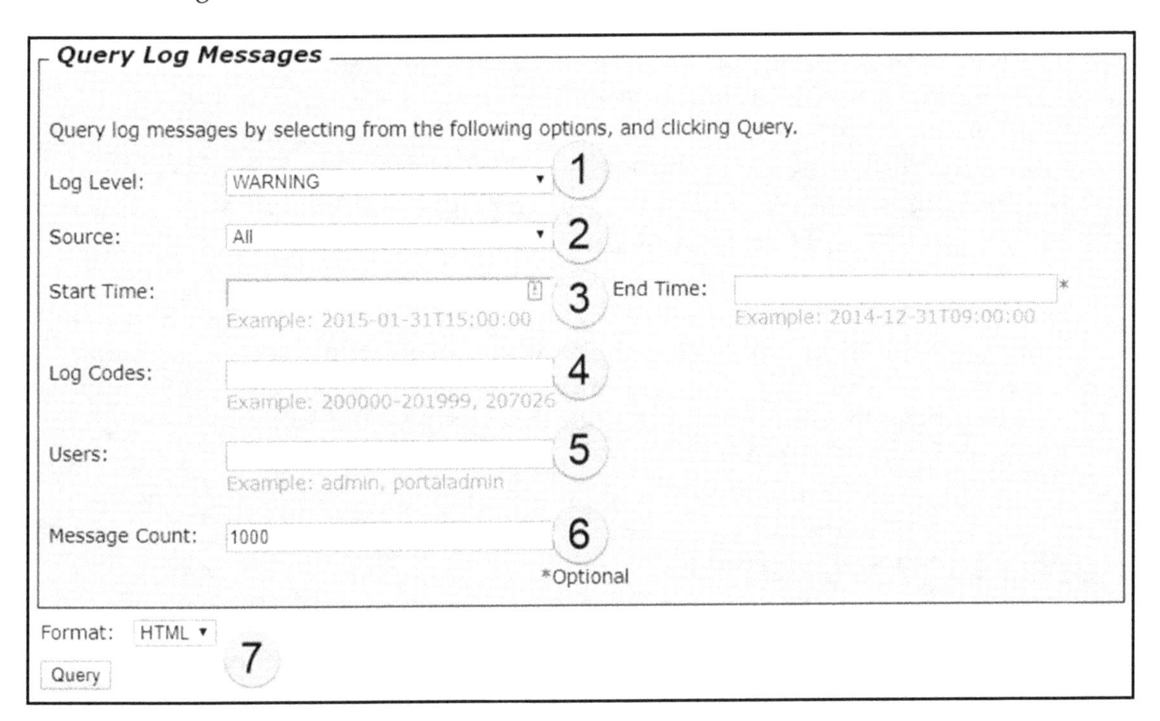

1. **Log Level**: This is required. These are described in detail earlier in `Chapter 5`, *Portal for ArcGIS Administration*, in the *Working with Portal logs* section. Just as with ArcGIS Server, use this to gather additional information when working on an issue.
2. **Source**: This is required. Source informs us which component of Portal the errors were generated within, of which there are four options, as follows:
    - **PORTAL ADMIN**: Security and indexing-related events.
    - **SHARING**: Publishing and user-related events.
    - **PORTAL**: Installation events.
    - **ALL**: All components are queried. The default.
3. **Start Time** and **End Time**: This is optional. It is used to filter events based on a time frame, where **Start Time** is the most recent time and **End Time** is the oldest time. The format here is `yyyy-mm-ddThh:mm:ss`, such as **2015-01-31T15:00:00,** where `hh` is hours in the 24-hour format.
4. **Log Codes**: As with ArcGIS Server, log codes are organized into categories based on ranges. For example, security-related log event code falls into the range of `204000-205999`. You can filter based on a single code, a comma-separated list of code, a range of code, or a range of code and individual code separated by commas. Search the ArcGIS Enterprise online documentation for **Work with Portal logs** for more information on the available log code within Portal.
5. **Users**: This allows you to search for messages on requests submitted by a user.
6. **Message Count**: The number of messages to display in the result set. The nice thing with Portal is that all messages display on a single page, allowing you to use *Ctrl+F* to search for terms in all returned messages.
7. **Format**: Choose the format which your results will be displayed in. Choices are **HTML**, **JSON**, or **XML**. In most cases, you will want to use **HTML** when simply viewing log messages in a web browser. Be aware that you could get the results back as **JSON** or **XML** and then bring those results into a software page such as Excel.

Portal logs workflow

Using the Portal logs for troubleshooting issues will be an experience very similar to that employed earlier with ArcGIS Server. First, do a cursory query and examination of the logs to see if anything jumps out. Next, try and reproduce the issue to see if anything of use gets logged.

One advantage here is the **Users** field; enter the Portal member's username here, the user who is experiencing the issue, or your username if you are the one now trying to reproduce the issue.

If still nothing shows up, lower your logging level, reproduce the error again, and check the logs. Don't forget to rerun your query after you reproduce the error yet again.

Tracking issues

Keeping track of issues that occur within your system can be a lifesaver. The software for doing this is commonly referred to as an **issue tracker**. There are plenty of issue trackers out there to choose from; some are open source and free, some are fee-based commercial but are products. At GISinc, all of our teams use an enterprise issue tracker, and it is downright indispensable for our team. Some reasons to consider using an issue tracker include the following:

- **Accountability**: With issues logged over time, you have a record of what you have been dealing with and spending your time on.
- **Historical reference**: Logging issues consistently and with details gives you a reference to look back through. This could, quite possibly, be the most important reason to have an issue tracker. I can't tell you how many times I've had an issue come up, only to ask myself *Haven't we seen this before?*. I search through the issue tracker for the error code or message, and sure enough, there's a past issue with the same error, often with a resolution.
- **Transparency**: Other individuals within your organization who can see what is going on can grasp the gravity of what you are dealing with. It goes hand in hand with accountability, previously listed.
- **Communication**: Some issue trackers have commenting capabilities built in, often with notifications. This allows communications regarding issues to stay with the issue in the tracker and not in emails. If your issue tracker has this capability, use it, as it is fantastic for historical reference. Emails get deleted, while the conversation thread in the tracker lives on.

Installation and configuration issues

Sometimes, unfortunately, it seems like you cannot even get your system up and running initially without running into issues. The ArcGIS Enterprise ecosystem is much more complex and involves many more components than it did just a few years ago.

To get your system initially set up, all these components need to communicate with one another to work together. Here are some items to be aware of when doing your initial configuration of ArcGIS Enterprise.

Web Adaptor issues

Issues can arise from incorrect configuration of your Web Adaptors for both ArcGIS Server and Portal. If you can access a resource bypassing the Web Adaptor (getting to the ArcGIS Server REST endpoint over port 6443 or Portal over port 7443, for example), chances are your **Web Adaptor** needs to be reconfigured or perhaps even uninstalled/reinstalled. Let's consider a Portal Web Adaptor that perhaps isn't functioning as expected. To get to your Portal **Web Adaptor**, follow these steps:

1. Log on to your **Portal Admin** as an administrator.
2. Go to **System** | **Web Adaptors** | **<web adaptor ID>**, and you will see something like the following. Note the **URL** parameter. If yours is not an FQDN (as shown in the following screenshot), but a machine name instead, that is a problem. So, in the screen capture below, if my **URL** parameter was `https://win-25fpfgemua9/portal`, that would, of course, not be resolvable on the public internet and the **URL** would need to be changed to `https://www.masteringageadmin.com/portal`:

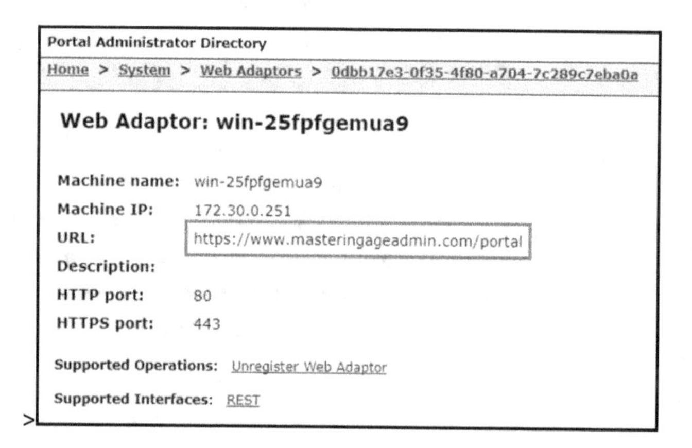

3. To change the URL parameter in a Portal **Web Adaptor**, when at the **System** | **Web Adaptors** | **<web adaptor ID>** page, add **/edit** to the URL in the browser address bar. This takes you into edit mode, where you can change the URL parameter (it is called the **Web adaptor URL** on the **Edit Web Adaptor** page).

Uninstalling/reinstalling a **Web Adaptor** is sometimes just as easy to do and ensures a fresh slate for your **Web Adaptor** settings.

Federation issues

Federation can be tricky, and an incorrectly entered URL can keep servers from properly communicating. To view federated ArcGIS Server information in **Portal Admin**, go to **Federation** | **Servers** | **<serverID>**. You will see something like the following screenshot:

Portal Administrator Directory

Home > Federation > Servers > wIHwsbSfT9H8dY2k

Server: win-25fpfgemua9:6443

Name:	win-25fpfgemua9:6443
Id:	wIHwsbSfT9H8dY2k
Url:	https://www.masteringageadmin.com/arcgis
Role:	HOSTING_SERVER
Admin Url:	https://win-25fpfgemua9:6443/arcgis
Server function:	

Supported Operations: Validate Update Unfederate

Supported Interfaces: REST

The two parameters of interest here are **Url** and **Admin Url**. **Url** should be the FQDN URL to **/arcgis** and **Admin Url** should be the internal machine name over port **6443** to **/arcgis**. Ensure that these are correct; if they are not, you will need to unfederate (via the **Unfederate Supported Operation**) and federate again (see `Chapter 6`, *Security*, for more information on federation). Note that re-federating will require changing ownership back to the publisher account, as re-federated services will be owned by the administrator that performed the federation.

Port issues

In `Chapter 1`, *ArcGIS Enterprise Introduction and Installation*, we discussed the ports that are required to be open for the different components of ArcGIS Enterprise communications. On some networks, all machines on the network can communicate over all ports. In other words, all inbound ports are open and there are no firewall rules blocking any ports. A configuration such as this makes installation and configuration simpler, as all ports you will need are open between machines. This can be made even easier if the **Web Adaptor** server is inside the network. All that said, this is rarely the case. In many networks, internal machines, for security reasons, cannot talk on all ports and the **Web Adaptor** server; again, for security reasons, they do reside on the internal network, but live outside the network in a perimeter network or demilitarized zone, or DMZ, as it commonly referred to. When security is tight like this, ports must be explicitly opened inbound between internal servers and the Web Adaptor in the DMZ. They need to be able to communicate with the ArcGIS Server machine inbound and outbound on ports **6080/6443** and the Portal machine inbound and outbound over **7080/7443**.

For more information on ports and network considerations, search the ArcGIS Enterprise online documentation for **Deployment scenarios**.

To resolve port and server communication issues, work closely with your IT department, possibly a systems administrator or network engineer.

Installation logs

Both ArcGIS Server and Portal log levels are set to **Verbose** during installations and upgrades. If an error is encountered during installation, first try to access the logs through normal avenues; ArcGIS Server Manager for ArcGIS Server or Portal Admin for Portal logs. If, since errors were encountered during installation, those are not available, logs can be viewed directly from disk. Those locations are as follows:

- **<ArcGIS Server install directory>/arcgisserver/logs/<machine name>/server/**
- **<Portal for ArcGIS install directory>/arcgisportal/logs/<machine name>/portal/**

Once your installation or upgrade is complete, logging for both ArcGIS Server and Portal are set to the default level of **Warning**.

Permissions issues

Like any other complex system that is constantly accessing resources, ArcGIS Enterprise relies on having proper access to these required resources. Permissions errors are by far some of the most prevalent sources of issues with ArcGIS Enterprise and they also happen to be some of the most difficult to diagnose. What's worse is that permissions issues typically spring up out of nowhere; one day things are working fine, the next day, your entire site is down.

What to look for

With permissions issues, look for anything that might seem like it's related to ArcGIS Server not being able to access something it needs. Permissions issues can manifest in a multitude of ways, but some to look for include the following:

- The ArcGIS Server Windows service will not start (has the password on the ArcGIS Server account expired?)
- ArcGIS Server services will not start (permissions on the configuration store could be out of sorts)
- ArcGIS Server services cannot be stopped nor deleted (configuration store permissions)

What to do to fix permissions issues

Just like with any other error, you want to ask yourself *Did anything change to precipitate this issue?* Some things to consider are:

- Were Windows updates done on any of the GIS servers recently?
    - If so, could a Group Policy be overriding privileges?
- Were ArcGIS Enterprise updates done recently?
- Could passwords have changed or expired?

- To fix permissions issues, do the following:
    1. Stop the **ArcGIS Server Windows** service.
    2. Run the **Configure ArcGIS Server Account** tool available on the **Start** menu. This tool sets a plethora of permissions, granting the ArcGIS Server account access and granting the ArcGIS Server account log on as a service rights. For more information on the permissions granted by the **Configure ArcGIS Server Account** tool, search the ArcGIS Enterprise online documentation for **the ArcGIS Server account**:

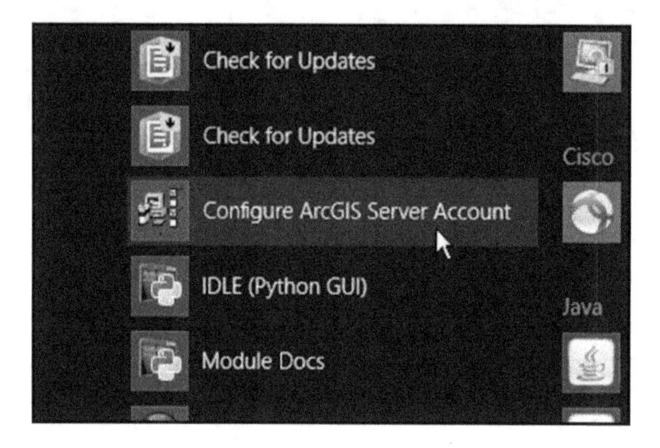

    3. Search the ArcGIS Enterprise online documentation for **What permissions do I need to grant to the ArcGIS Server account** and grant the permissions listed there.
    4. Restart the ArcGIS Server service.

Web browser considerations

Ever since the days of Netscape Navigator and well, any version of Internet Explorer, dealing with web browsers and their different acceptances of standards, bugs, and quirks has been a challenge.

If you've ever done any sort of web development or design work, you know how important it is to test your sites in multiple browsers. Sometimes things look or even behave or function differently, usually not for the better. Well, the same goes in GIS. Got something in an application that you think should look right, but it just doesn't? Try another browser and see what you get.

Passwords

We've talked at great length about passwords throughout this book multiple times, so much in fact that you're probably sick of hearing about them. Here's one last tip:

Make sure you are using the right password.

Not able to access something and you're stumped as to why? Step back and make sure you are using the right password. For that matter, make sure you are using the right username as well. I cannot tell you how many times I have done this over the years, and every time I feel just as ridiculous as the first time. Sometimes all you can do is laugh about it. Seriously, though, the more credentials you have in your collection, the more confusing it gets, and the easier it is to use the wrong ones. Again, this is where a password manager can really save your sanity and up your bus factor as well. By utilizing a password manager, you can use shortcut keys to copy/paste passwords without even having to really know them.

Scripts

As we discussed earlier, scripting is a vital component to an efficient, smooth-running Enterprise GIS system. Scripts, however, typically rely on inputs and outputs that are fixed and well-known; once something goes wrong with those inputs or outputs, errors can start to pop up. Knowing how to best evaluate, diagnose, and resolve script issues quickly is an important skill for any ArcGIS Enterprise administrator.

Troubleshooting in production

Before we go any further here, let's discuss where you should troubleshoot. When something goes wrong with your scripted process in production, you may be tempted to quickly try and troubleshoot the problem there, in your production environment.

Be very careful troubleshooting in production!

Troubleshooting an issue in production can be risky. I'm not going to say not to do it because many GIS shops only have production and don't have the luxury of test and development environments. However, if you must troubleshoot in production, try to isolate your outputs as much as you can.

For example, if your script writes data out to your production enterprise geodatabase, try to have a second enterprise geodatabase stood up for testing and troubleshooting. If you can't do that, you may be able to get by with using a file geodatabase to test against.

Debugging code on a production server can also be tricky, as there isn't often an integrated development environment, or IDE, on the server, but merely a text editor (which is often only Notepad). An IDE can help by allowing you to debug your code and step through the process (see the the *Debugging* section below for more information). Whenever possible, I try to replicate the production environment as closely as possible in my local development environment, that is, my laptop, try to see if I can determine the cause of the issue at hand, and then, once I have found a solution, deploy the fix to production.

In some cases, debugging in production may actually be necessary, as the problem could be within the production environment itself. When this is the case, nothing may be wrong at all with your script, but it could something related to a data connection, the Python environment, or a host of other issues that could arise.

Finding and understanding errors

In `Chapter 9`, *ArcGIS Enterprise Standards and Best Practices*, we talked about logging and how to set up a Python logger with the `daiquiri` module. Using a `daiquiri` logger, it is easy to both log to a file and to standard output at the same time, which is great, but, sometimes, in larger scripts, you need a little bit more than just a logger that prints out the errors. Finding where the error occurred in a larger script can sometimes be your biggest challenge. `arcpy` comes with an array of error messaging tools that work great to provide us with errors related to `arcpy`, but standalone scripts often need a little bit more related to standard Python errors (those not related to `arcpy`). This is where the `traceback`, `inspect`, and `sys` modules from the Python standard library can really help. Together, these two modules can be used to tell us not only what Python error has occurred, but what line it occurred on in our script. Let's look at how to set up a script to do this type of logging and what it can provide us.

First, we will import all of the Python modules our script needs:

```
import arcpy
import daiquiri
import datetime
import inspect
import logging
import os
import sys
import traceback
```

```
from daiquiri import formatter
```

Next, we will set up a function, `trace_error`, that will determine where the latest exception has occurred, what the error is, and what line it occurred on. We will call this function from a `try/except` block in our code momentarily:

```
def trace_error():
    tb = sys.exc_info()[2]
    tbinfo = traceback.format_tb(tb)[0]
    filename = inspect.getfile(inspect.currentframe())
    line = tbinfo.split(", ")[1]
    synerror = traceback.format_exc().splitlines()[-1]
    return line, filename, synerror
```

For logging, we will set up a `daiquiri` logger very similar to the one utilized earlier in `Chapter 9`, *ArcGIS Enterprise Standards and Best Practices*, that will log to both a log file and standard output, giving us an historical record of the run in the log file and providing us with information at runtime. We set up our logger with the following:

```
daiquiri.setup(
    outputs=(
        daiquiri.output.TimedRotatingFile(
            "python_logging.txt",
            formatter=daiquiri.formatter.ColorFormatter(
                fmt="%(asctime)s - %(filename)s - %(lineno)d - " \
                    %(levelname)s - %(message)s"
            ),
            interval=datetime.timedelta(days=7),
            backup_count=10
        ),
        daiquiri.output.Stream(
            sys.stdout,
            formatter=daiquiri.formatter.ColorFormatter(
                fmt="%(asctime)s - %(filename)s - %(lineno)d - " \
                     %(levelname)s - %(message)s"
            )
        ),
    ),
    level=logging.INFO)
logger = daiquiri.getLogger(__name__)
```

Next, we will set up a `try...except` block. `Try...except` blocks work by trying to execute code and throwing exceptions when something goes wrong. We can have multiple `except` statements, allowing us to account for different error types, such as arcpy-specific errors and general Python errors. In this example, we will try to get a listing of files in a directory. If an exception is thrown, it will call our `trace_error` function from earlier, which will examine the currently running code's latest exception and return the filename in which the error occurred, the line on which the error occurred in that file, and the actual error that occurred. Next, our script will look for any errors specific to `arcpy` in the exception and log those via our `daiquiri` logger:

```
try:
    datasets = os.listdir(r"C:\temp\some-dir")
    for dataset in datasets:
        # Do something
        print dataset
except arcpy.ExecuteError:
    line, filename, synerror = trace_error()
    logger.error("\n\n\tError on {0} of '{1}'.\n".format(line,
                                                          filename))
    errs = arcpy.GetMessages(2)
    logger.error("arcpy errors: {0}".format(errs))
except:
    line, filename, synerror = trace_error()
    logger.error("Errors:\n\n\tError on {0} of '{1}'. " \
                 "Message: {2}\n".format(line, filename, synerror))
```

To wrap up our error handling neatly, we call the `finally` statement, which, no matter what happens in our code, will always be executed. The `finally` statement is a great place to complete tasks such as closing database connections that may have been opened in your script or do any sort of clean up that must be done regardless of the script result. Here, we will simply log a message saying that processing has finished. This keeps things uniform in our log file, letting us know for sure where one run finished and where the next will begin:

```
finally:
    logger.info("Finished processing.\n\n")
```

Now that we have our script set up, let's run it and get a listing of the content in `C:\temp\some-dir`. Here, I'll run my code in my IDE, PyCharm Professional 2017.1 (more on using an IDE soon). Note the yellow output in the bottom pane:

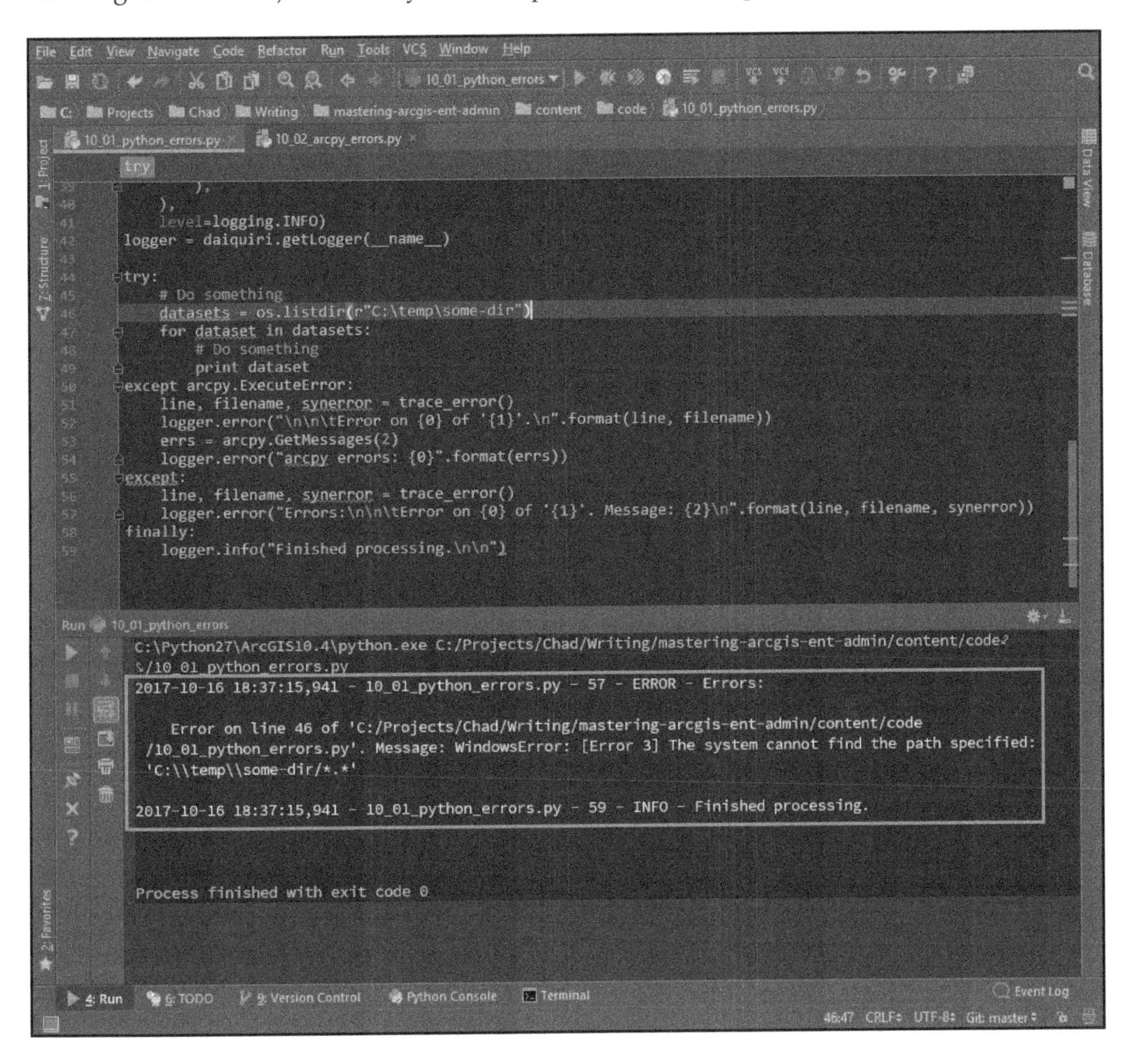

See what happened there? Our full Python error gets logged, including the line in our script on which the error occurred. We received the **WindowsError: [Error 3] The system cannot find the path specified** error as the **C:\temp\some-dir** directory does not exist on my C drive. Oops. But what about our log file? Let's look at that:

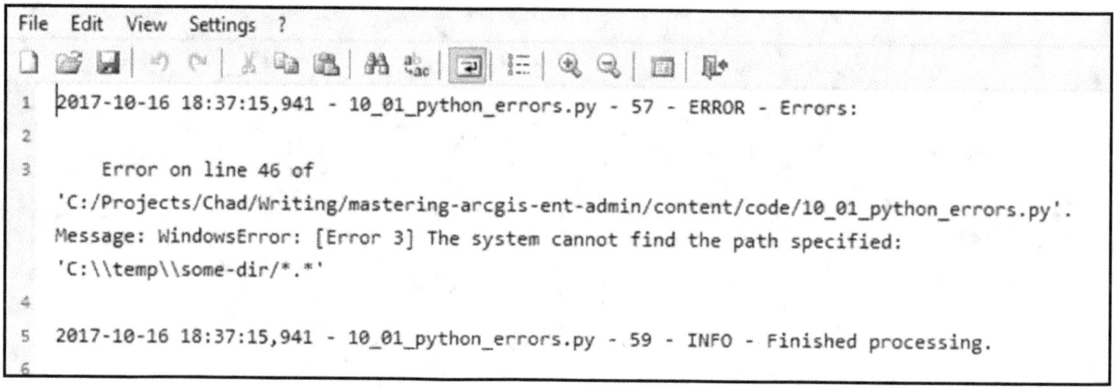

```
File  Edit  View  Settings  ?
1  2017-10-16 18:37:15,941 - 10_01_python_errors.py - 57 - ERROR - Errors:
2
3      Error on line 46 of
   'C:/Projects/Chad/Writing/mastering-arcgis-ent-admin/content/code/10_01_python_errors.py'.
   Message: WindowsError: [Error 3] The system cannot find the path specified:
   'C:\\temp\\some-dir/*.*'
4
5  2017-10-16 18:37:15,941 - 10_01_python_errors.py - 59 - INFO - Finished processing.
6
```

Hey, look at that, it's the exact same message that was printed to standard output in our IDE. That's because we set up our `daiquiri` logger to log the same output to both standard output and our log file. Nice consistency, right?

The previous example covered a standard Python error, but what about errors specific to `arcpy`? What would that error message look like? Remember from earlier that with `arcpy` errors, those also run through `trace_error()`, but we also call `arcpy.GetMessages(2)` to pull in any error messages specific to `arcpy` as well. The example below shows how to log arcpy-specific errors:

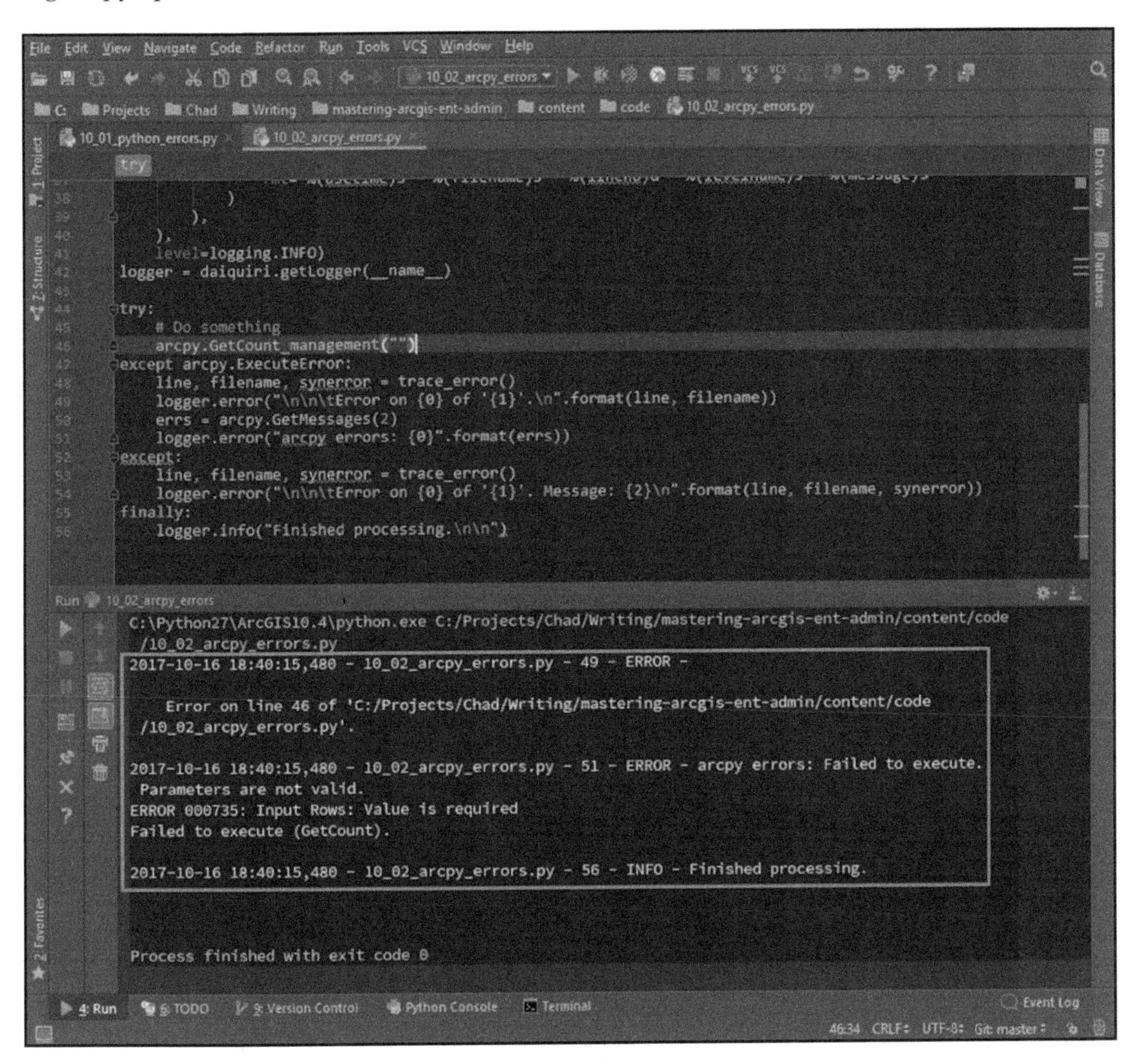

Here, we are informed that the error occurred on **line 46**, but, then, we are also provided the actual `arcpy` error code and message. Imagine that, it looks like I forgot to reference an input dataset in my call to `GetCount()`. And our log? You guessed it, an exact mirror of the text we see in the IDE, as shown in the following screenshot:

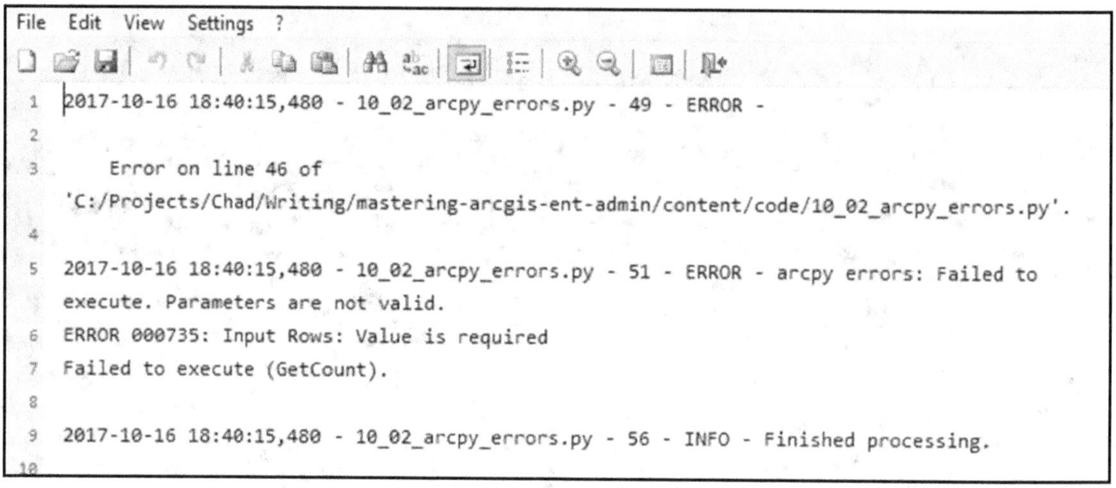

```
File  Edit  View  Settings  ?
1  2017-10-16 18:40:15,480 - 10_02_arcpy_errors.py - 49 - ERROR -
2
3      Error on line 46 of
   'C:/Projects/Chad/Writing/mastering-arcgis-ent-admin/content/code/10_02_arcpy_errors.py'.
4
5  2017-10-16 18:40:15,480 - 10_02_arcpy_errors.py - 51 - ERROR - arcpy errors: Failed to
   execute. Parameters are not valid.
6  ERROR 000735: Input Rows: Value is required
7  Failed to execute (GetCount).
8
9  2017-10-16 18:40:15,480 - 10_02_arcpy_errors.py - 56 - INFO - Finished processing.
10
```

Debugging

Debugging is the process of locating and correcting errors in your code. Debugging can take on many forms, ranging from simple `print` statements, to interactive debugging in an IDE, to unit testing. No matter how you do it, debugging is a critical step, that, when done properly and efficiently, can speed up development and testing of your scripts. For our purposes here, let's focus on the two simplest and easiest ways to learn and implement methods--`print` statements and interactive debugging.

Print statements

Adding `print` statements to your code to print out values or even just to let you know how far the script progresses during execution is an easy way to help you debug your scripts. However, the downside to this method is that it takes time to add all those `print` statements, and then most of the time you aren't going to want to leave them in production code, so you'll have to take them out. Then, when an issue comes up, you must put the `print` statements back in again--essentially, it's a kludgy process.

Continuing with our `GetCount()` mishap earlier, let's look at a basic script that gets a count of the records in a feature class, and if the record count is greater than zero (there are features in the feature class), export the feature class to a new feature class using `arcpy.FeatureClassToFeatureClass()`:

```
import arcpy

result = arcpy.GetCount_management(
    r"C:\Projects\GDBs\Sandbox.gdb\StudyAreas"
)
record_count = int(result.getOutput(0))
if record_count > 0:
    arcpy.FeatureClassToFeatureClass_conversion(
        r"C:\Projects\GDBs\Sandbox.gdb\StudyAreas",
        r"C:\Projects\GDBs\Sandbox.gdb",
        "ActiveStudyAreas"
  )
```

Straightforward, right? The only problem is that when we run the code, it completes successfully, but we don't get our **ActiveStudyAreas** feature class exported, and we are positive there are records in the source **StudyAreas** feature class. What we could do here is insert a `print` statement or two to help us determine what is going on during execution. Let's first `print` out how many records are returned from the `GetCount()` call and then add a `print` statement once (`if`) we have records to export out to a new feature class. Finally, we'll add an `else` block to print out a message if no records were there to export:

```
import arcpy

result = arcpy.GetCount_management(
    r"C:\Projects\GDBs\Sandbox.gdb\StudyAreas"
)
record_count = int(result.getOutput(0))
print "Records: {0}".format(record_count)
if record_count > 0:
    print "Exporting StudyAreas..."
    arcpy.FeatureClassToFeatureClass_conversion(
        r"C:\Projects\GDBs\Sandbox.gdb\StudyAreas",
        r"C:\Projects\GDBs\Sandbox.gdb",
        "ActiveStudyAreas"
    )
else:
    print "No records to export!"
```

Let's look at the results of this script when run in PyCharm:

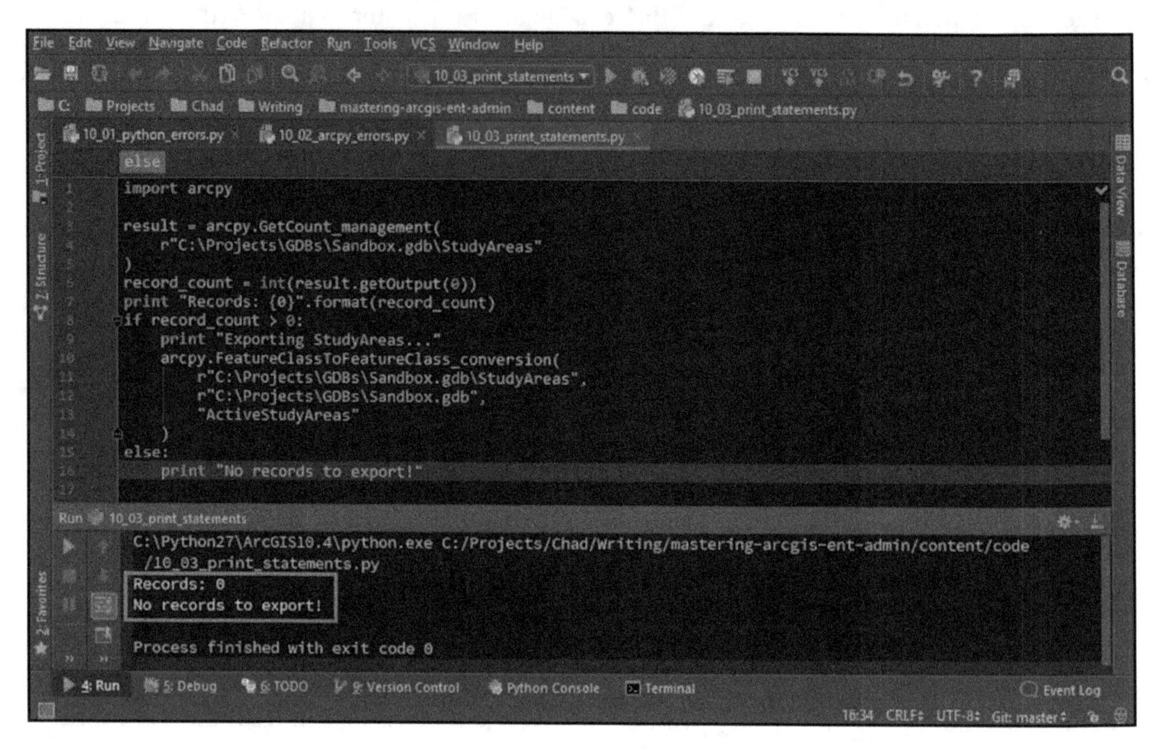

Well, here's why we aren't getting our exported feature class, no records exist in the source feature class. Looks like we should've checked to be certain. Now, although this is a contrived and simple example, it shows how adding `print` statements can indeed help figure out what is going on in your script. It's simple and it works, but let's look at a better way.

Debugging in an IDE

Using an IDE offers many advantages over using a simple text editor to write code. There is an astounding number of Python IDEs available today, ranging from open source solutions (some of which are written in Python) to commercial packages. Of the many advantages IDEs offer, the ability to step through and debug your code is one of the best. We already looked at using `print` statements for messaging and to notify you of progress and execution throughout your script. When debugging, you can set points in your scripts to pause execution, known as breakpoints. In conjunction with breakpoints, you can step through your code either line by line, breakpoint to breakpoint, or any combination of the two, all the while being able to inspect variables and objects within your code.

With debugging, `print` statements scattered throughout your code are no longer necessary as you can interactively inspect your code as it is executing. To illustrate, let's look at our earlier example where we used `print` statements, but, this time, we will debug in PyCharm, a popular Python IDE made by JetBrains (`https://www.jetbrains.com/pycharm/`).

To begin debugging, we will first set a breakpoint in our code (the red dot in the left margin) at the beginning of our `if` statement on line 7. Next, go to the **Run** menu and select **Debug <script name>**. This will start code execution in the debugger at the beginning of our script. Once the debugger hits our breakpoint, execution of the script is paused, and we are presented with the debugger pane in the lower half of the PyCharm window. In the **Variables** pane, we can inspect any currently populated variables and objects. Note that the line that execution has halted on is our breakpoint and the line is selected and highlighted for reference. Also, in the **Variables** pane, note that we can inspect the `record_count` variable and its value of `0`, along with its type of `int`, as shown in the following screenshot:

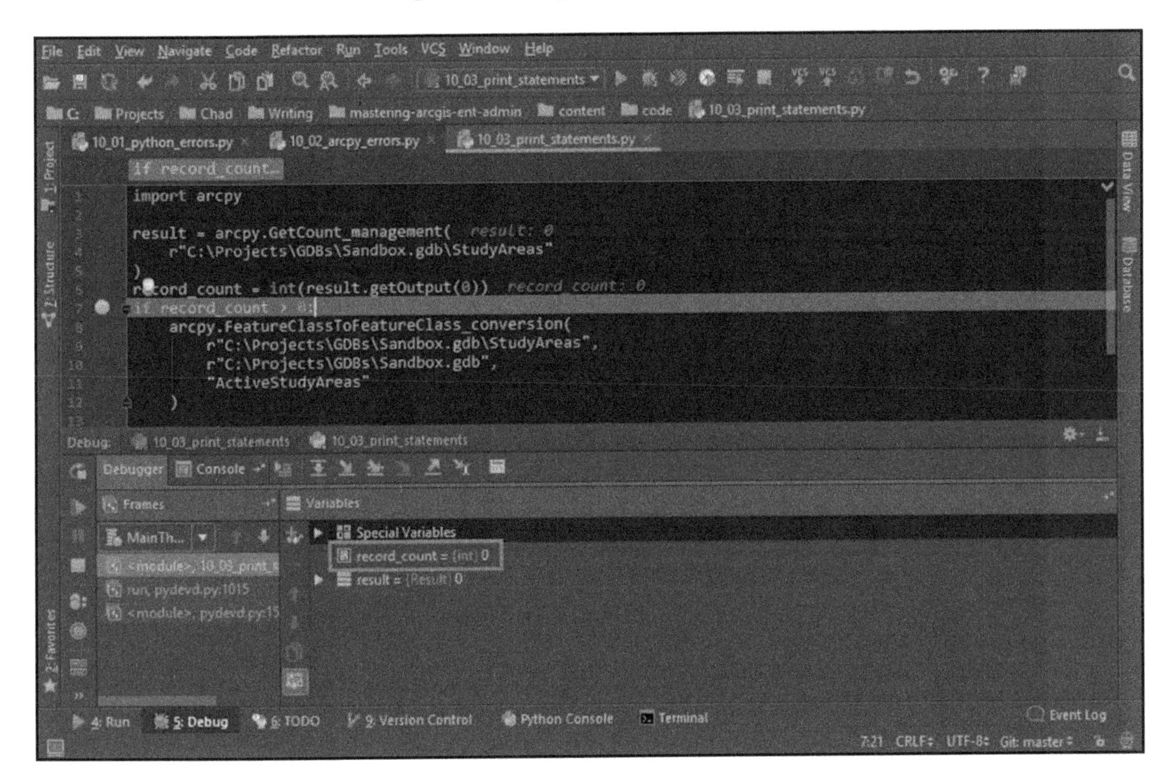

Wow, that was easy, wasn't it? We just found out with a couple of mouse clicks what several `print` statements told us earlier, but we didn't have to add one single thing to our code; all we did was debug in the IDE.

Again, this was a very basic and simple example, but it illustrates the point--debugging your code in an IDE has benefits that far outweigh trying to use `print` statements or even logs for troubleshooting your code. I invite you to try out any of the many Python IDEs out there and learn how to use them; you won't regret it as it will make your code development, testing, and troubleshooting workflows more efficient and streamlined.

Logs

We have discussed logs at length throughout this book; in `Chapter 9`, *ArcGIS Enterprise Standards and Best Practices*, we showed how to set up a `daiquiri` logger, and, earlier in this chapter, we put that same `daiquiri` logger to work. Let's talk about them one more time. If you have a Python script that runs, or is run on any sort of schedule, you need to have logging enabled in it. Having logging in your scripts not only helps you figure out what went wrong last night when it ran and failed at 2 A.M., but logging gives you a history of your runs (so long as you implement your logging to not overwrite the same log file with every run; for example, with a `daiquiri TimedRotatingFile` output) and provides accountability, showing that your processes are in place and running. I really cannot stress this enough. For example, just today, I had a client report an issue with a script on one of their servers. My first inclination was to check the script logs to see what the history looked like and what happened with the last run or the last several runs. When I looked, however, I was disappointed to find out that the logger overwrites the same log file with every run, so there was no history. Bummer, true, but also room for improvement. An hour or two spent improving the logging capabilities of that script could save many more troubleshooting hours over the life of the process down the road, because, as we all know, errors happen.

Tools to help you

Most of the time, unless the issue at hand is immediately resolvable, you will need a tool or utility of some sort to help you determine what is going on with your issue. We've already discussed using logs, `print` statements, and debuggers to help you resolve issues, but what do you do when the issue is with a web application or a call to a web service? There are plenty of ways to tackle those issues as well; let's take a look at a few.

Browser dev tools

All modern browsers now come with some flavor of development tools, or *dev tools* as they are commonly referred to. With dev tools, you can perform a variety of tasks, such as view the source of a page, debug and step through the code of a site, or watch network traffic to see what happens as the code executes, to name a few. Internet Explorer, Firefox, and Chrome, all have dev tools, but here we will be discussing Chrome's dev tools, as they are some of the easiest, most complete, and most intuitive to learn and use.

Don't let the dev tools intimidate you. My son was inspecting pages and checking out the JavaScript behind them in Chrome when he was in the fourth grade. Blew my mind.

Let's say you have a web mapping application that is built on the Esri JavaScript API, such as the Esri Tax Parcel Viewer (`http://solutions.arcgis.com/local-government/help/parcel-viewer/`), part of the many solutions Esri offers for local governments. Whenever any interaction occurs between the JavaScript API and ArcGIS Server, it takes place through the REST endpoint of the ArcGIS Server service. This means that an identify, query, search, or layer draw, to name just a few, are all calls to a REST endpoint that returns a response.

Knowing this, let's see how the ArcGIS Server REST endpoint can help us troubleshoot issues:

1. In Google Chrome, go to `http://solutions.arcgis.com/local-government/help/parcel-viewer/` and click on the **VIEW APPLICATION** button to view the live application.
2. Right-click anywhere in the application and select **Inspect** from the menu. This will open the dev tools, more than likely in a docked window within Chrome. You'll see something like the following screenshot:

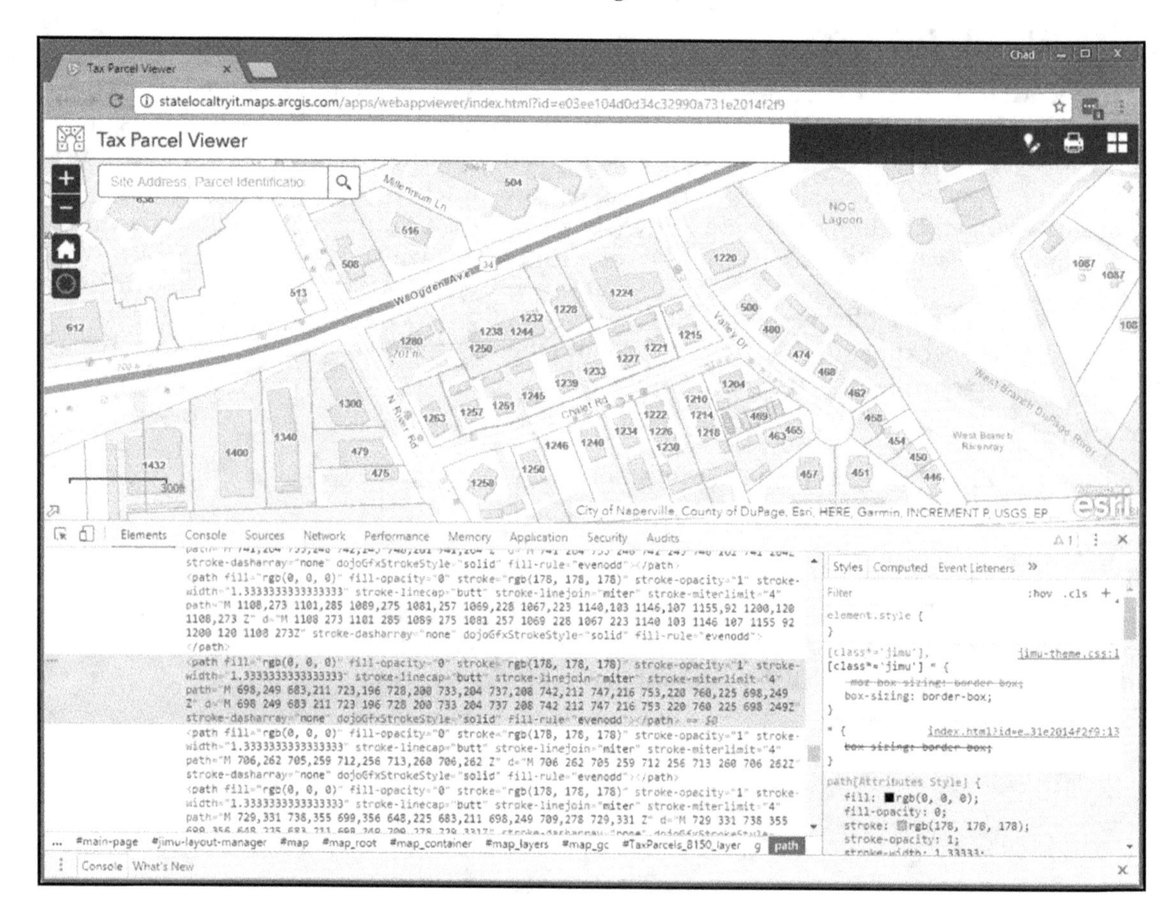

3. In the dev tools, click on the **Network** tab (dev tools more than likely opened in the **Elements** tab):

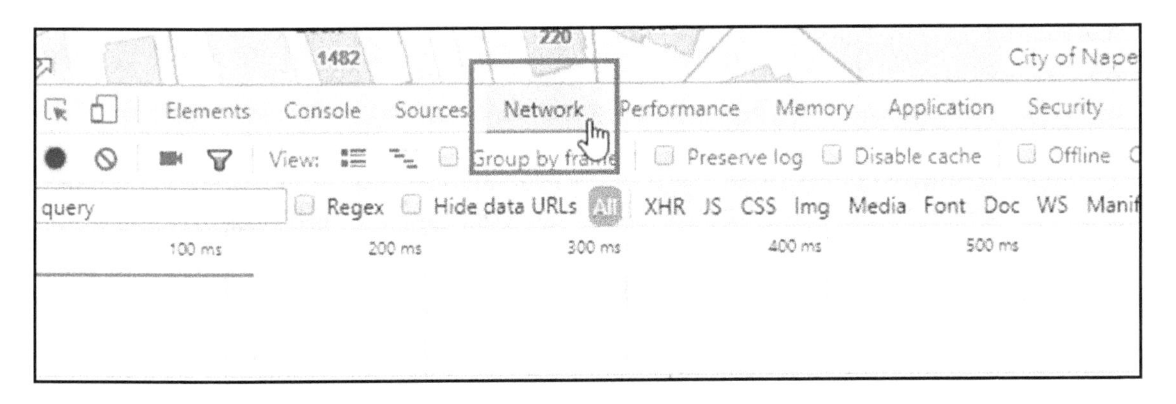

4. In the search bar in the upper right of the application, enter **400 W Jefferson Ave** and hit the *Enter* key.
5. In the **Network** tab, you should see several calls with `query` in them in the **Name** column, but one should have a SQL `where` query in it:

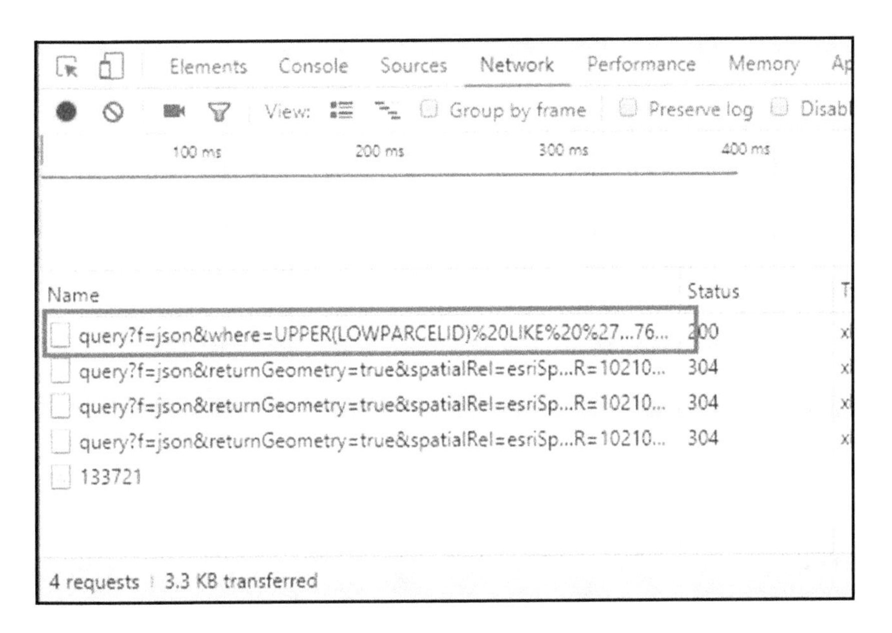

6. Click on the query call with the where clause in it. In the right **Network** pane, you now have four tabs. Poke around these, especially **Headers**, **Preview**, and **Response**, remembering to scroll down to view all the content in each. It's quite a lot, isn't it? This is the request that was sent to the REST endpoint when we queried for **400 W Jefferson Ave** and the response we got back.
7. Go to the **Headers** tab. Scroll down to the **Query String Parameters** section. Here you will see the **where** clause that was sent to the REST endpoint. This can come in handy when trying to determine why a query isn't returning results or might be erroring out. Note **Request URL** under **General**. This is the request that was sent to the REST endpoint.
8. Leave this browser tab open, as we are about to use it again in the next section.

Now that we've talked a (tiny) bit about Chrome's dev tools, let's take this inspection one step further to see how we can utilize information gleaned from the dev tools at an ArcGIS Server REST endpoint.

We are barely scratching the surface of what Chrome's dev tools are capable of. Another tab of interest is the **Console** tab, which will show any errors and warnings that your application code may be throwing as it is executed by the browser on the client side.

Using the REST endpoint

We've discussed the endpoints for ArcGIS Server and Portal administrative tasks in several places throughout this book using the REST Administrator. What we haven't discussed is how to utilize the ArcGIS Server REST endpoint as an invaluable resource for troubleshooting web applications.

In the previous section, we utilized Chrome dev tools to view network traffic, namely the query that goes to the REST endpoint when we search for an address in the **Esri Tax Parcel Viewer**. That's great, and plenty of information can be gleaned from seeing what the application is sending and receiving over the web. However, what about trying to execute that same query? Can we do that? As a matter of fact, we can.

Go back to the **Headers** tab from the query we sent for **400 W Jefferson Ave** in the previous section and look at the **Request URL** in the **General** section. To drill down and begin to look at the query, do the following:

1. Select the **Request URL** and copy it to the clipboard:

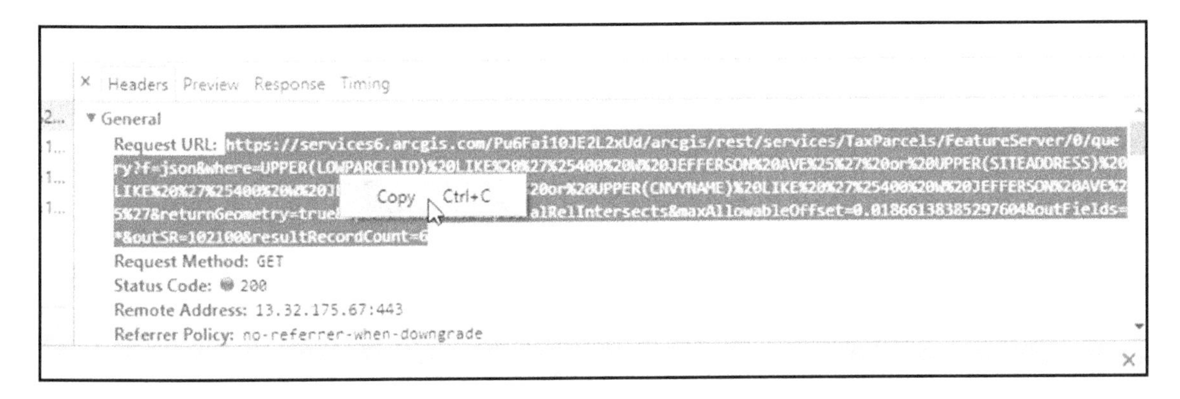

2. Open a new browser tab and paste the URL into the address bar. You should get a screen full of JSON back; this is the query result.
3. In the address bar, look for the **query?f=json** string and replace **json** with **html**. Hit *Enter*.

3. You are now at the `query`operation endpoint for the **Tax Parcels** layer that powers the **Esri Solutions Tax Parcel Viewer**. This is where the queries from the application go for address and parcel searches. The HTML version of the endpoint is for us humans to use. Note that all the query parameters are populated in the forms on this page along with the query result further down near the bottom:

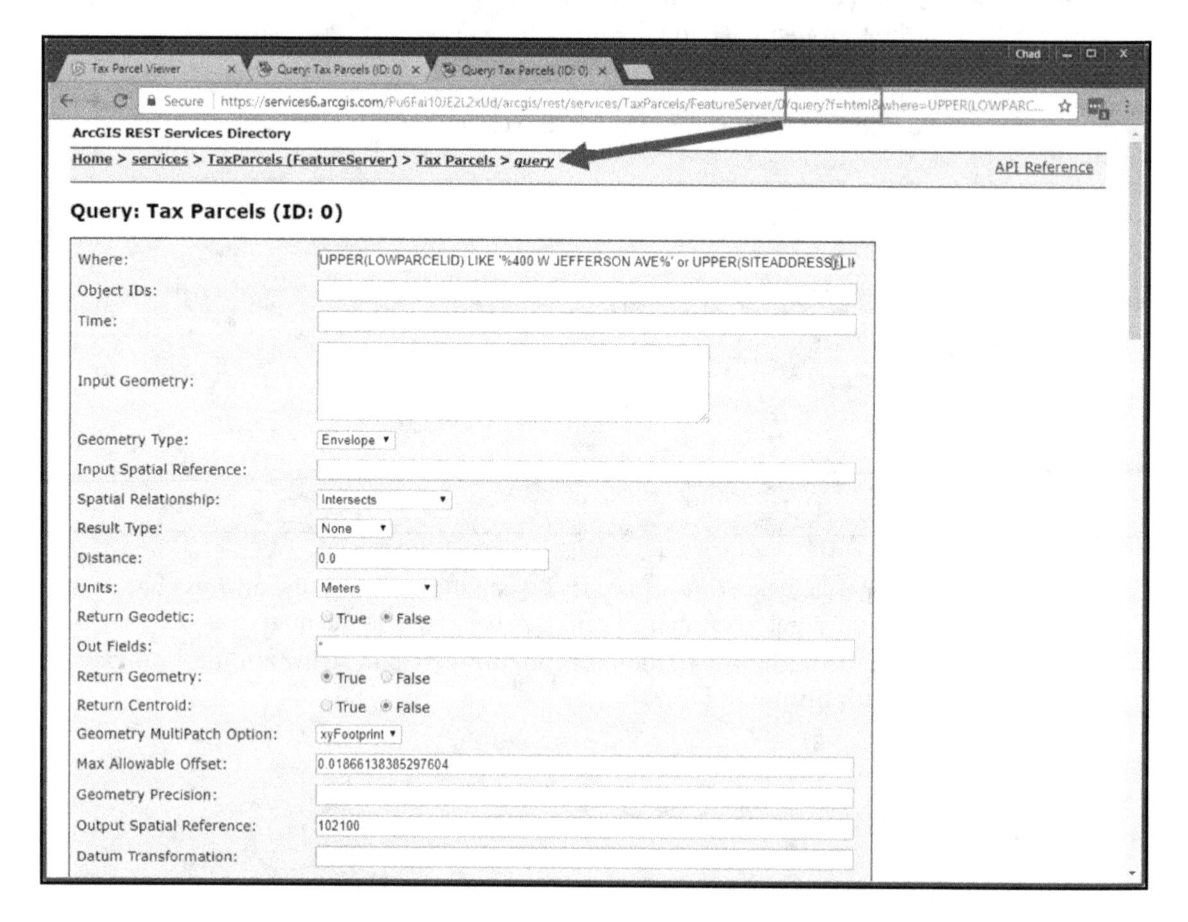

I invite you to spend some time exploring the inner workings of your web applications (or someone else's for that matter) and combine what you find there with the REST endpoints that get called upon. Not only will this help you understand how your application is working on the backend, but it will give a deeper understanding of the REST API and how it is structured and consumed by the JavaScript API.

AGO Assistant

In `Chapter 5`, *Portal for ArcGIS Administration*, we covered ArcGIS Online Assistant, or AGO Assistant for short. There, we discussed how to access AGO Assistant and use it for a variety of administrative tasks. It turns out that AGO Assistant can come in quite handy for troubleshooting issues in Portal and ArcGIS Online items as well.

I have a web map named **MowAreas** in my Portal that is consuming an ArcGIS Server layer from a third party. I made this web map many months ago, but, now, when I try to view it in Portal, I get an error stating that the layer cannot be added to the map:

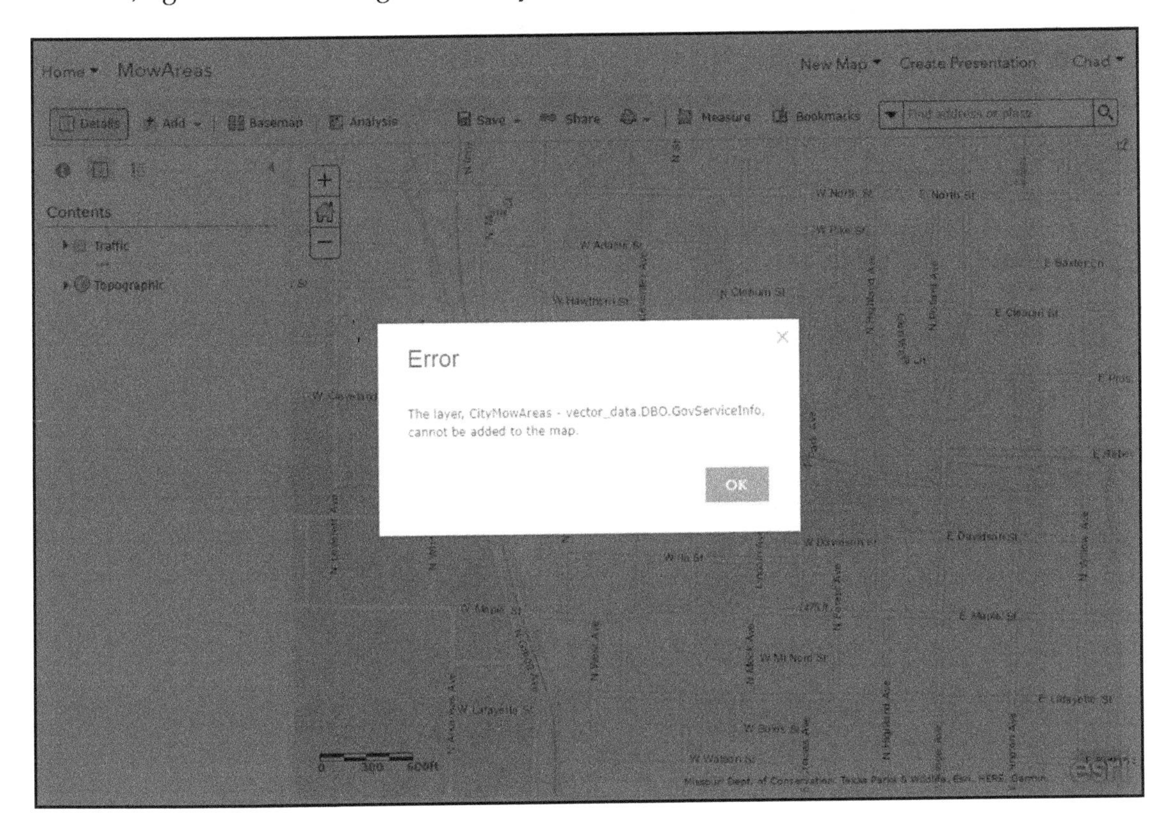

This is a frustrating error in Portal (and ArcGIS Online as well), as the message really doesn't tell us much at all. Remember though that layers in a web map are always referenced with a URL, and, in `Chapter 5`, *Portal for ArcGIS Administration*, we looked at how AGO Assistant can be used to examine and even change URLs in the JSON of a web map. In a situation like this, we can use the AGO Assistant to easily look at the URLs of services in our web map. Maybe that might shed some light on the issue. In logging into the AGO Assistant for my Portal, I select **Update the URLs of Services in a Web Map** from the **I want to...** dropdown, then I select my web map in the left column of items. We can now see the service URLs in this web map:

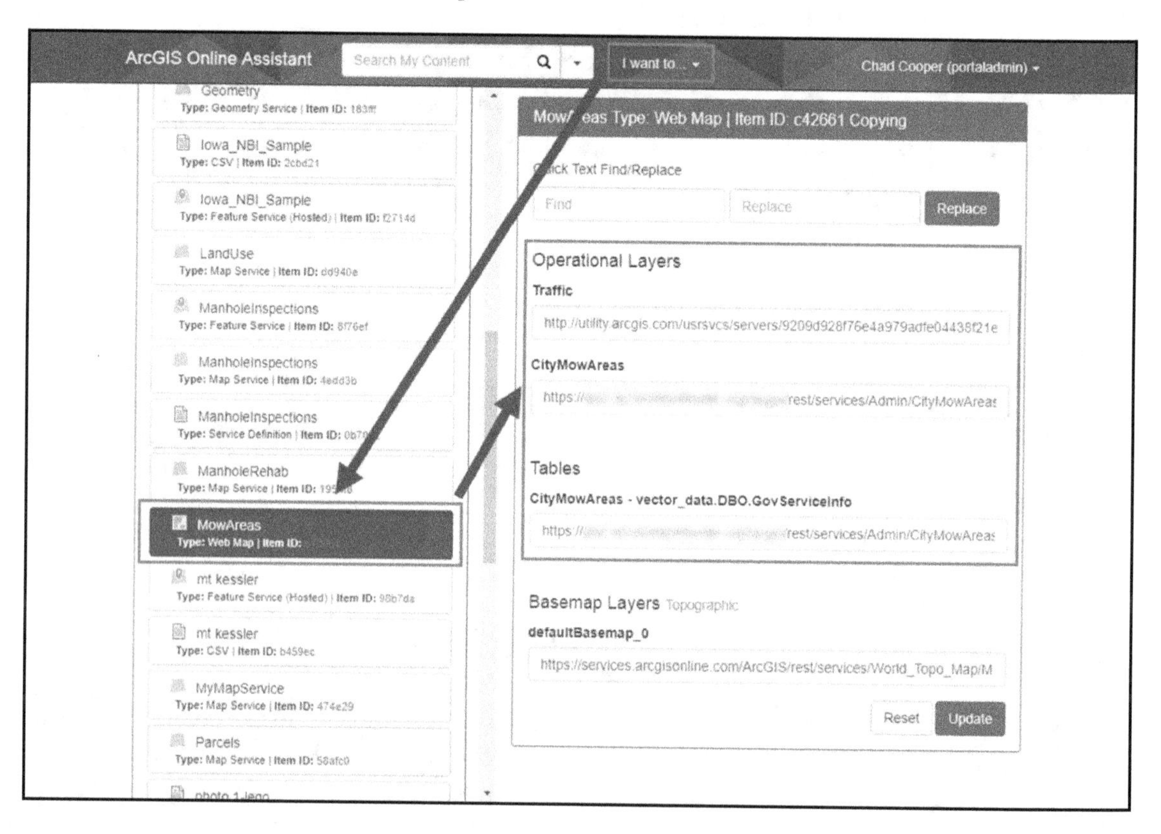

Earlier, Portal informed us that the **CityMowAreas** layer could not be added to the map. Let's copy/paste that URL into the **CityMowAreas** layer into a browser and see if we can access that layer, as shown in the following screenshot:

>

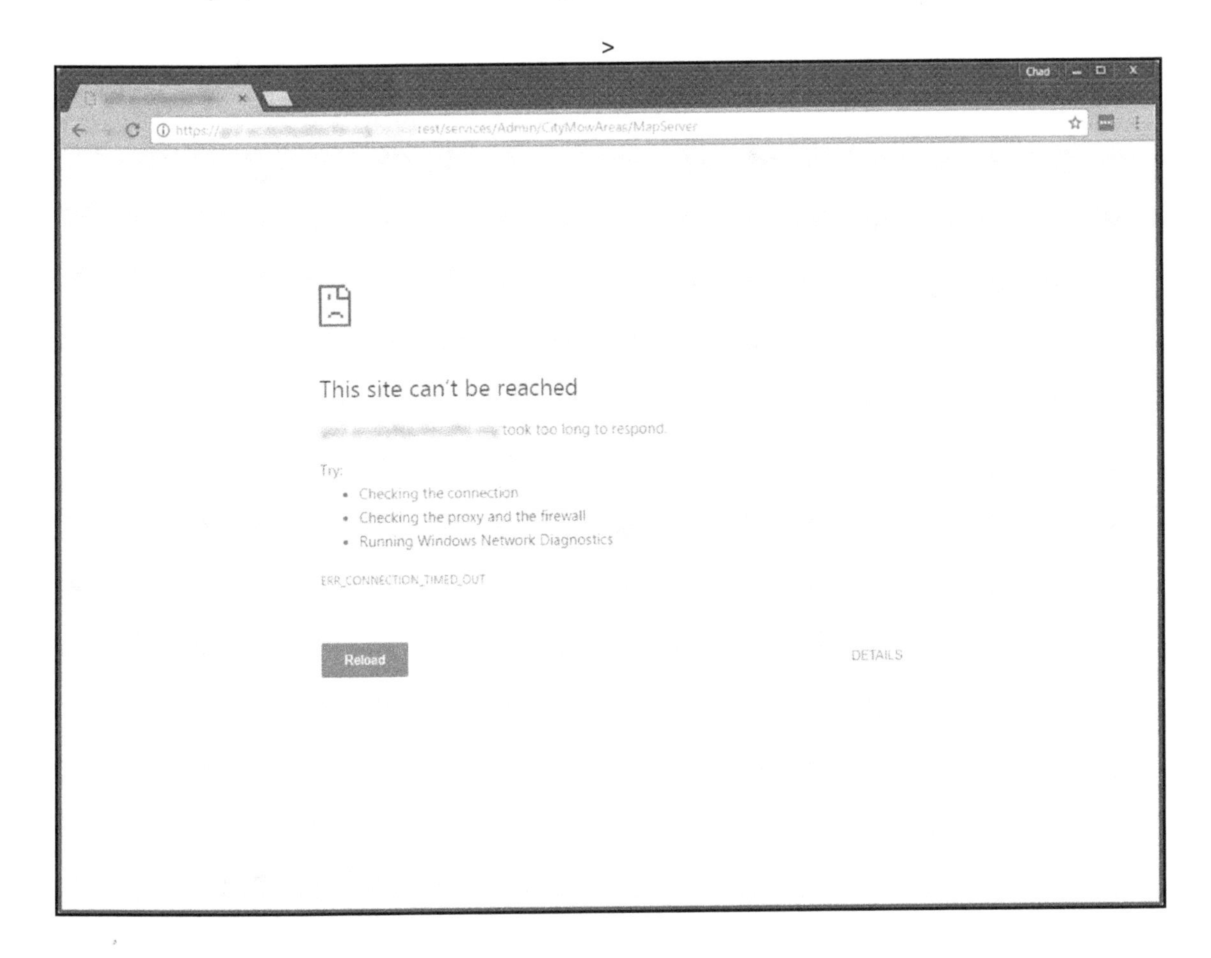

Hmm, the connection is timing out. Also, notice that there is no lock in the address bar to the left of **https**. Is this service even available over **HTTPS**? Trying the same URL over HTTP gets the following response:

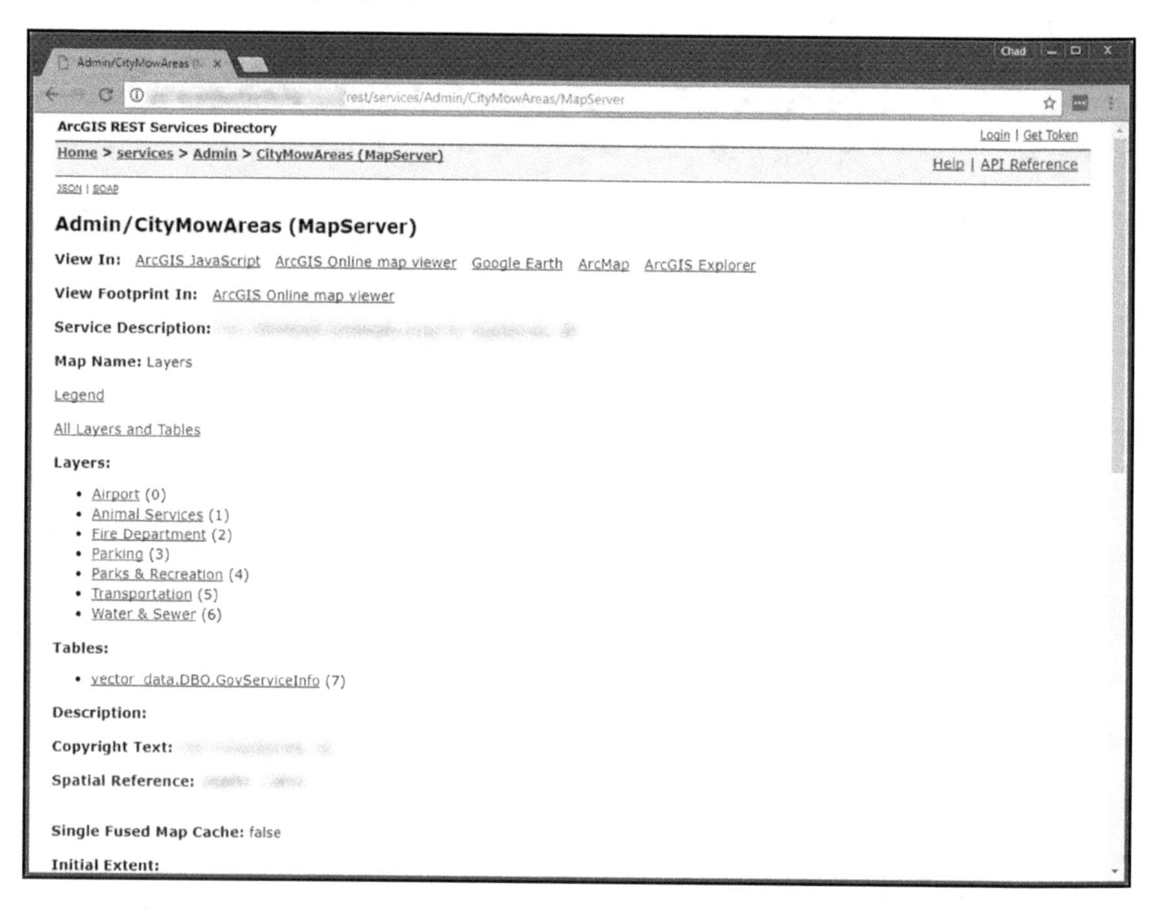

Perhaps this service was once provided over HTTPS, but it no longer is. I can now go back to AGO Assistant, change the URLs from **https** to HTTP, save the edits, and my web map is now working again.

Outage and issue scenarios

Now that we have talked about some of the more common issues you might see and ways to address them, let's play through a scenario that could and does happen in the real world. We will look at a situation where a user tells you that your website is down.

Scenario - the website is down

If this hasn't happened to you already, it will soon enough; a user sends you an email that says nothing more than **The GIS website is down**. What's the first thing to do in this situation? Anyone remember? That's right--don't panic. Now that we've kept our cool, let's work our way through this issue.

What we are about to cover is, by no means the absolute right way to troubleshoot an issue, as there is no silver bullet when it comes to this sort of a task. The following scenario is meant more to provide ideas on the many avenues that can be taken when troubleshooting ArcGIS Enterprise issues.

Vague and ambiguous emails like this from users are often the norm. Remember that your users are using your applications for specific reasons, so no two users view those applications in the same light. That said, you must first determine what exactly they mean by *the site is down*. As a matter of fact, if you have multiple applications that the user could be referring to, you might first have to figure out which site they are even referring to. Maybe you know this user and are familiar with their workflows within your applications, and you know which app to check on. If you're not sure, get back to the user and find out exactly which site it is they are experiencing an issue with.

Once you know which site the user is having issues with, visit the site to see for yourself if the entire site is truly down. Is it throwing a `404` not found, a server error `500`, or any other sort of error? If so, check the IIS logs for those errors. If it's a `500` server error, you might want to check your ArcGIS Server logs as well to see if that is where the error is originating from. Are you getting just a white, seemingly empty page or maybe just an application header? This could be a code issue, perhaps a configuration file issue. Right-click on the page, go to **Inspect**, and then the **Console** tab. Are any errors being thrown here (refresh the page perhaps)?

If the site isn't down but appears to be functioning at first glance, it might be time to reach out to the user who reported the issue and get clarification on just what isn't working right for them. Get them to walk you through what they were doing when they experienced the issue. If necessary and possible, use screen sharing software to view their screen or go to their desk and have them show you what they are seeing. Take notes if necessary so you can reproduce the issue yourself.

Once you have a firm grasp of what the user is experiencing, try to reproduce the issue. If you cannot reproduce the issue, it may be a browser issue; get them to try a different browser. If you can reproduce it, it's time to get to work:

- Make sure services vital to the application are up and running.
- Check the ArcGIS Server logs and/or Portal logs for errors. If none are present, lower the logging level to **FINE** or **VERBOSE**, reproduce the issue, then check the logs again.
- Ensure that any queries that are being executed are successful.
- Is this an access issue? Does the user have access to all the proper resources to complete the task?
- Does anything else in the application seem to be having issues as well? This could show a pattern.
- Bring in another set of eyes. If you keep working on this issue for long enough, you will start missing details. Get a colleague to look and see if they pick up on something that you are not.

Summary

Errors happen, even in the best kept and most well-groomed of environments. Being prepared, remaining calm, assessing the situation, and determining the best path to take to resolution are all key to quickly and effectively solving ArcGIS Enterprise issues when they arise. In this, our final chapter, we pooled a vast array of knowledge together that we discussed throughout this book. Becoming a master troubleshooter takes time; the more issues you see, the better you will become at recognizing, diagnosing, and resolving them. Always remember to stay calm and focused, gather as much information as you can, and check your logs. When those don't quite cut it, remember there is a wide selection of tools out there that can help you save time and get your ArcGIS Enterprise environment back up and running smoothly again.

Index

A

E

F

G

H

I

J

L

M

N

O

P

R

S

T

U

V

W

Printed in the USA
CPSIA information can be obtained
at www.ICGtesting.com
LVHW081209160824
788413LV00004B/25